P9-CMT-652
NEWARK PUBLIC LIBRARY
DISCARDED
5 2125 300895291

BY THOMAS LEVENSON

Money for Nothing:
The Scientists, Fraudsters, and Corrupt Politicians
Who Reinvented Money, Panicked a Nation, and Made the World Rich

The Hunt for Vulcan . . . and How Albert Einstein Destroyed a Planet,
Discovered Relativity, and Deciphered the Universe

Newton and the Counterfeiter:
The Unknown Detective Career of the World's Greatest Scientist

Einstein in Berlin

Measure for Measure: A Musical History of Science

Ice Time: Climate, Science, and Life on Earth

MONEY *for* NOTHING

MONEY *for* NOTHING

THE SCIENTISTS, FRAUDSTERS,
and CORRUPT POLITICIANS
WHO REINVENTED MONEY, PANICKED A NATION,
and MADE THE WORLD RICH

THOMAS LEVENSON

RANDOM HOUSE
NEW YORK

Published in the United States by Random House, an imprint and division of
Penguin Random House LLC, New York.

Random House and the House colophon are registered trademarks of
Penguin Random House LLC.

Image credits appear on page 435.

Library of Congress Cataloging-in-Publication Data
Names: Levenson, Thomas, author.
Title: Money for nothing: the scientists, fraudsters, and corrupt
politicians who reinvented money, panicked a nation,
and made the world rich / Thomas Levenson.
Description: First edition. | New York: Random House, [2020] |
Includes bibliographical references.
Identifiers: LCCN 2020001521 | ISBN 9780812998467 (hardcover) |
ISBN 9780812998474 (ebook)
Subjects: LCSH: England—Economic conditions—18th century. |
Stock exchanges—England—History—18th century. |
Debts, Public—England—History—18th century.
Classification: LCC HC254.5 .L48 2020 | DDC 330.942/07—dc23
LC record available at https://lccn.loc.gov/2020001521

Printed in the United States of America on acid-free paper

randomhousebooks.com

2 4 6 8 9 7 5 3 1

First Edition

Book design by Edwin Vazquez

For

Richard, Irene, and Leo . . .

with love and thanks

Contents

INTRODUCTION

"the great Follies of Life"

LONDON, 1719

The year had begun well enough for London's stock traders, working from their corner of the city, a narrow passage called Exchange Alley. Buying and selling shares—dealing not in things but in numbers—was still new to the city. There was no fixed marketplace for traders in paper. So those who had mastered what was to many still a very dark art concentrated in a few taverns and inns, at Garraway's, a coffeehouse that catered to the gentry, and more of them just around the corner at Jonathan's, the rival coffeehouse that saw the most feverish trade in all the new ways in which it was possible to make—or *be*—money.

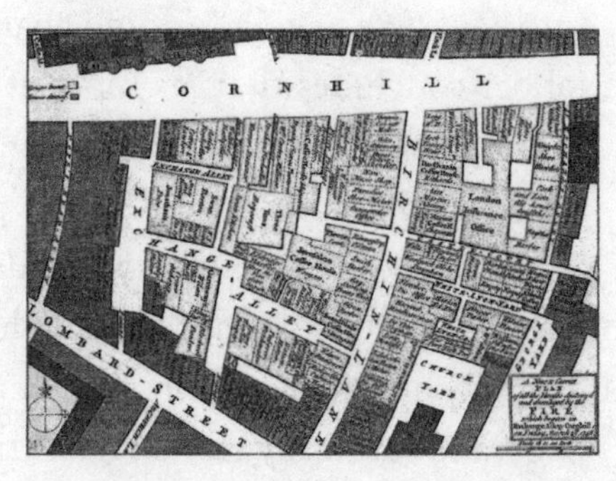

Exchange Alley in its early-eighteenth-century layout

For journalist, propagandist, and gadfly Daniel Defoe, Jonathan's and the rest were familiar—and dangerous: dens of iniquity. Defoe had been warning his fellow Britons about the perils of the Alley for almost three decades. Now, near midsummer, he was ready with his most desperate alarm yet, a pamphlet titled *The Anatomy of Exchange Alley,* written in the voice of "a jobber," or unlicensed dealer in stocks. In part, the pamphlet served as a kind of travel story, leading its readers on a journey to an exotic spot. Exchange Alley was an island in miniature. It could be walked in a minute or two: "Stepping out of Jonathan's [Coffeehouse] into the Alley, you turn your Face full *South,* moving on a few Paces, and then turning Due *East,* you advance to *Garraway*'s; from thence going out at the other Door, you go on still *East* into *Birchin-Lane,* and then halting a little at the Sword-Blade Bank to do much Mischief in fewest Words, you immediately face to the *North,* enter *Cornhill,* visit two or three petty Provinces there in your way *West.*" There, a few hundred paces at most, and the visitor would be almost done: "Having Box'd your Compass, and sail'd round the whole Stock-jobbing Globe, you turn into *Jonathan*'s again." Home again!—but not in safe harbor, for the jobber concludes that "as most of the great Follies of Life oblige us to do, you end just where you began."

And what folly it was! Defoe painted the risk faced by any reckless soul foolish enough to wander into Jonathan's, in the tale of a naïvely avaricious countryman who encounters a couple of con men. They ply him with rumors, urge him to trade on their insider's knowledge, and surgically extract his entire fortune: "his Coach and Horses, his fine Seat and rich Furniture," all sold "to make good the Deficiency."

That was typical of Exchange Alley, Defoe warned his readers. It fostered "a compleat System of Knavery . . . a Trade founded in Fraud." Its tricks were hardly new, of course. In some form, they're as old as human desire, as the book of Proverbs attests: "The getting of treasures by a lying tongue is a vanity tossed to and fro of them that seek death." But this hopeful, nervous year of 1719 held something new, a scheme more ambitious than anything previous attempted by the devilish denizens of the Alley. The South Sea Company had opened for business in 1711. The firm had never really managed to do the work implied by its name, shipping goods and slaves to the

Spanish ports of South America. Instead, it played in what was then just being born, a marketplace for credit, all the notes and bonds and much stranger inventions that the British government was using to build its ever-growing mountain of debt. For several years, the South Sea Company itself had nibbled around the edges of this market, completing a handful of minor deals, but its directors now aimed at a vastly more ambitious project—one that would, if it worked, solve Britain's borrowing problem once and for all. They proposed a heroic attempt at what we would now call financial engineering—taking the whole of the national debt, accumulated over a seemingly endless series of wars, and turning it into shares of a private company—theirs—which could be traded back and forth at will in the nascent stock exchange. In its partisans' view, that would be the saving of the nation. Alternatively, as the skeptical Defoe warned, the clever men of Exchange Alley had figured out a way to get rich off of the public interest: they were "ready, as Occasion offers, and Profit presents, to Stock-jobb [buy and sell] the Nation, couzen [trick] the Parliament, ruffle the Bank, run up and run down Stocks, and put the Dice upon the whole Town."

"Stock-jobb the nation." That was the crux of Defoe's polemic: schemes like this transformed the national debt—a public necessity—into a form that could be manipulated for private profit. That was, he argued, if not treason itself, then treachery's nearest cousin: "Is not all that is taken from the Credit of the Publick, on such an Occasion . . . is not every Step that is taken in Prejudice of the King's Interest . . . a plain constructive Treason in the Consequences of it?"

In the most straightforward account of the events that were to come—known to history as the South Sea Bubble—Defoe would be proved right. In the year 1720, every Briton with two shillings to rub together, it seemed, would hear of the South Sea Company, would buy into its promises, and would be dazzled, for a time, at the prospect of riches beyond imagining. Half of Europe would too, and for a very great many it would end in ruin.

AND YET, SEEN with enough distance (and a comfortable remove from those lost fortunes), it's clear that Daniel Defoe was also wrong. What would happen in Exchange Alley over the next year wasn't simply the work of "a Trade founded in Fraud, born of Deceit, and nourished by Trick, Cheat, Wheedle, Forgeries, Falshoods." The South Sea Bubble—the headlong rise and the sudden collapse of London's nascent stock market—wasn't the original sin of early modern capitalism—or rather, it was never only that.

Instead, as this book argues, if we are to understand the Bubble year, wider histories must come into play, ones that reach both backward and forward in time, from Defoe's troubled days to our own. In this telling, the calamities of 1720 can be read as a watershed moment in the long, tangled process of creating the modern concept of money, and especially of money's most dynamic incarnation, credit—which makes promises expressed in numbers that connect the future to the present. The Bubble is a part of the history of finance but is not confined to it. Rather, it opens a window on the circumstances from which later financial thinking emerged: the grand shift over the preceding century in the way human beings understood their experience of the material world, an intellectual transformation better known as the scientific revolution.

The history of the scientific revolution is usually told as a sequence of discoveries, mostly in mathematics and physics. That picture leaves out a central human fact: those who solved problems of planetary motion or the flight of cannonballs did not confine themselves to natural philosophy. From the beginning, they used the same methods and habits of mind to tackle human questions, to guide the choices made by individuals and societies. In 1719, on the brink of the wild ride to come, the greatest revolutionary of them all worked just a few hundred yards to the southeast of Exchange Alley. There, in rooms built along the outer wall of the Tower of London, Sir Isaac Newton, master of the Royal Mint, produced Britain's supply of the "real" money: gold and silver bullion of precisely defined and authenticated purity, rolled and punched and stamped into disks of legally mandated weight, decorated with the head of the king. He'd been advising the crown on monetary matters since the mid-1690s, and by this time he was an experienced stock market player on his own account—which included, at this moment, a tidy sum in South Sea stock. The year to come would test him as much as anyone else—but his significance to the events of 1720 lies in the way he taught his contemporaries to think, not just about money, but about anything that could be observed, measured, and counted.

THE CATASTROPHE NOW known as the South Sea Bubble is on record as the first and in many ways the archetypal stock market crash and fraud. What happened then and what the British state did in response both have a direct connection to what has occurred (and may soon come again) in the financial life of the twenty-first century. In the moment, it certainly proved Defoe's point about the significant threat that arises when the motives and interests of the financial elite clash with those of national governments and the public. And certainly, pawing through the wreckage left after the Bubble burst uncovered an extraordinary catalog of corruption, a comprehensive guide to all the ways it's possible to subvert the market for private gain.

But to understand how the nation got itself into the Bubble, and to grasp what actually had occurred behind the surface of catastrophe, we must follow the story further back in time—to a garden, an orchard. There waits the young Isaac Newton, a man barely past boyhood. He can see an apple tree, laden at this midsummer moment with fruit—a variety called Flower of Kent.

At any moment, one might fall.

Part One
Counting and Thinking

Natural Philosophy consists in discovering the frame and operations of Nature, and reducing them, as far as may be, to general Rules or Laws,—establishing these rules by observations and experiments, and thence deducing the cause and effects of things.

—Isaac Newton

Chapter One

"The System of the World"

Woolsthorpe, Lincolnshire,
Midsummer, 1665

He had walked for three days to escape a farmer's life just four years before. There'd been no hint he'd ever return. Yet, here he was again, about to open the gate. Little had changed: the main house, stone-built, comfortable; the barn behind; and, just across the track from the front door, the little patch of garden with its stand of apple trees, warming in the early-summer sun. He'd been nineteen when he left, a country boy, awkward and unsociable, much given to scribbling in his notebook. He was twenty-three now and had a place in the world: a scholar of Trinity College, Cambridge, with rooms and a stipend and his own place at table. But now, smack in the middle of term, he was coming up the lane, crossing to the door, and stepping over the threshold, into the house.

Isaac Newton had come home.

Or rather, he had been chased back to Woolsthorpe's quiet corner of the Lincolnshire countryside. Cambridge had been emptying since late spring, as each of its seven thousand or so residents who had anywhere to go desperately fled the approaching danger. That threat had been carried in its turn by those coming up the roads from London, on the run from the terror that had already reached the capital.

No one knows precisely how the plague reached London in the early months of 1665. Attentive observers had been nervous throughout 1664. The diarist and navy bureaucrat Samuel Pepys kept his eyes on Amsterdam, where a full-blown epidemic had begun in the previous fall. Pepys first took note of the danger in October, and then in November recorded in his diary that ships coming from infected ports now faced quarantine along the Thames. But some still arrived, smugglers carrying cloth, or, after the new year, with the start of the second Anglo-Dutch War, vessels bringing prisoners home. There were always rats on board, and rats carry fleas. Fleas host the *Yersinia pestis* bacterium.

Y. pestis causes the plague.

Winter came, the slowest season for infection. For a time, it appeared that England might escape the scourge suffered on the other side of the North Sea. A few cases were reported. The bill of mortality for the crowded parish of St. Giles in the Fields records one plague death on Christmas Eve, 1664—a "Goodwoman Phillips." When she fell ill, Phillips, her husband, and their unnamed and unnumbered children were placed in quarantine, and their home was shuttered, guarded, and bedaubed with the plea "Lord Have Mercy on Us." But the infection apparently stopped there, and the next weeks were quiet. Plague was endemic in England, and every year saw a few come down with the disease. This appeared to be just one more case, and not the first of a new tidal wave of infection.

The winter ended, and spring began kindly. Two more plague deaths show up in London's weekly bills of mortality in the last week of April, again in St. Giles in the Fields. The first week in May saw none . . . but then the numbers turned. Nine in the second week, three in the third. Fourteen for the week of May 23. Seventeen reported on the thirtieth. Forty-three on June 6—and the tally leapt from there. The official record topped a thousand on July 28, then doubled, and doubled again. By September, one thousand Londoners were dying each day. By year's end, the official toll would approach seventy thousand, as many as one in eight of the city's residents.

Everyone who could get out of town did. On June 21, the ubiquitous Pepys, secretary to the Admiralty, had a meeting in Whitehall, trying to unravel King Charles II's perpetual money troubles. Achieving nothing more than usual, he set out across the city to Cripplegate, one of the ancient openings in London's city wall. There, he recorded in his diary, he found "all the towne almost going out of towne, the coaches and waggons being all full of people going into the country." He stopped awhile at the Cross Keys tavern—long enough to enjoy the company of the barman's wife—but by the next day he was ready to decide "whether to send my mother into the country today." She didn't want to go, but "because of the sicknesse in the towne, and my intentions of removing my wife," he finally managed to put her on a coach that would take her east, toward Cambridgeshire.

London's exodus had the predictable result: refugees from the capital carried the contagion into the countryside. Some towns barred their gates to keep the disease at bay. It didn't work. In Cambridge, the blow fell on July 25. John Morley, five years old, was found dead at his home in the parish of the Holy Trinity. There were dark spots on his chest. When the plague inspectors came, they found Morley's younger brother already showing black irruptions on his face. The child was taken to the pesthouse, where he died ten days later.

Ann Fisher, a child from All Saints Parish, died on the same day, confirming that the disease had spread beyond a single neighborhood. More cases followed, more deaths, and from there Cambridge followed London's pattern. Businesses shuttered. Stourbridge Fair, one of the greatest open-air markets in Europe, was canceled. The university scattered. On August 7, the College of the Holy and Undivided Trinity acknowledged the obvious and decided to pay its members an allowance whether or not they remained in residence after that date. No record for Isaac Newton appears in Trinity College's accounts for the extra stipend. The newly passed bachelor of arts had already fled, traveling the sixty miles north and a little west to Woolsthorpe.

THERE HE REMAINED for almost two years, cut off from every other scholar or mathematician. The isolation suited him. "In those days," he would recall half a century later, "I was in the prime of my age for invention & minded Mathematics & Philosophy more than at any time since."

Those twenty months are now known as Newton's *annus mirabilis*—his miracle year. In that brief time, he would solve several problems at the leading edge of contemporary mathematics. The decks thus cleared, he would go on to invent whole new ideas, laying the foundations of what we now call calculus, the mathematical tool used to analyze (among much else) change over time—where a cannonball might be at any instant, or a planet, for example. He then turned to what we now call physics, beginning with mechanics, the study of bodies in motion. Here too, before he could uncover specific results he had to work out fundamental concepts, mastering the first modern understanding of inertia, for example, an idea he first encountered in Descartes's work, and thinking deeply about what it means to be a force—two ideas so essential to the future development of our understanding of the physical world that he in large measure had to construct them himself before he could go any further. Then came the first glimpses of what would become his theory of gravity and all that flowed from that epiphany. And, still not done, he dove further into an investigation of light, color, and optics that would yield his first great public triumphs. The record does not reveal when—or whether—Isaac Newton slept.

This seemingly superhuman accumulation of ideas during his plague-imposed exile has created a mythology of superhuman genius, conjuring worlds of thought out of country air. It's not quite that simple, of course. Newton's definitive biographer, Richard Westfall, points out that the program for the work to come was laid down in 1664, when Newton, just twenty-one, was still enrolled at Trinity College. That's when he first dove into the mathematical inquiries that would dominate his first several months back home, and when he produced an extraordinary series of forty-five queries in which he grappled with fundamental issues of time, matter, motion, and much more. He didn't need the plague, that is, to launch him into his comprehensive assault on the whole of natural philosophy. But it is true that when he reached the farm he was ready to move beyond anything Cambridge could teach him. He was "consistently concerned," as Westfall wrote, "to develop general procedures" that would, in the end, produce not just new mathematics but a new way of thinking about how math insinuates itself throughout the material world.

One of the first problems to catch his eye was how to calculate the area marked out by a curve. The study of curves was a central fascination of seventeenth-century mathematics, and Newton had plunged into the field when he read a translation of René Descartes's *Geometry* a year before the plague hit. Descartes's approach helped link together the approach of classical geometry, which explored shapes and their properties, and the ideas of algebra, with equations whose solutions could be mapped onto a particular curve.*

One of the key advances Newton encountered in his copy of *Geometry*—the Latin translation that spread Descartes's ideas throughout learned Europe—was the coordinate system now known as Cartesian coordinates. It provided a way to map any point in two dimensions with just two numbers, corresponding to its horizontal and vertical positions. Using two lines perpendicular to each other—the familiar cross of every graph in grade school math classes—and a standard unit length applied to both axes, Descartes created a systematic way to measure and map any shape a geometer wanted to study—including the classical curves, circles, ellipses, and the rest.

* For example: the geometrical definition of a circle is a curve on which every point is the same distance from a single central point, a distance called the radius. Algebra produces that same circle as the solution(s) to an equation: $(x - a)^2 + (y - b)^2 = r^2$, in which x and y are the coordinates for any point on the rim of the circle, a and b are the coordinates of the central point, and r is the radius of the figure.

When Newton came to study this work, and then more contemporary mathematics, he soon turned one of its approaches on its head. In classical geometry, the starting point for most European mathematicians, the curve or the shape is the object of inquiry. Even though Cartesian coordinates offered a new and powerful way of representing equations as shapes on his coordinate system, many of Newton's contemporaries saw such equations as a property of a given figure, a line or a circle or some more complicated form. But it took Newton just a few months after encountering Descartes to realize, as his biographer, Richard Westfall reports, "The equation is more basic than the curve; the equation defines, or as Newton put it, expresses the nature of the curve."

That sounds like a technical point, or even one of taste: some folks think in pictures and, if they are mathematical, dive into the relationships between shapes and volumes, while others play the game of manipulating those abstractions. But Newton's insight—starting first with the equation, rather than the shape—was foundational because it would, first hesitantly and then through centuries of development, yield a new way of seeing the world through mathematics. For his predecessors, the classical geometers, the curve was there, complete, a synoptic view of the object. But in Newton's work, in the early years of what is now known as analytic geometry, a curve is built up as a calculation reveals the solutions to the equations that generate any given geometrical object. The accumulation of specific answers to these calculations—points on a curve, plotted on the page to produce a geometric object—can be interpreted in various ways. The interpretation that Newton would develop focused on arguably the most important implication: equations describe the evolution of a system—how its solutions build a picture on a page. That picture is a map of the relationship of variables—things that can change. If one of those variables is the passage of one moment into the next, then the abstract play of symbols and shapes becomes a portrait of change in action.

Ultimately, this mathematical insight is at the heart of modern physics, the science that Newton, more than any other single thinker, would create. In its simplest form, the idea is this: the full picture, the complete geometrical representation of all the available solutions to a system of equations, can be understood as all the possible outcomes for a given phenomenon described by that mathematics. Each specific calculation, fed with observations of the current state of whatever you're interested in, the flight of a cannonball, the motion of a planet, how a curveball swerves, how rapidly an outbreak of the plague might spread, makes a prediction for what will happen next. In his twenties, working on his own, with almost no systematic experience of the study of the real world, Newton did not yet grasp the full power of the ideas implied by the way he had begun to think about math. That would come in time. But what made his *annus mirabilis* so miraculous was the speed and depth with which Newton forged the foundations of his ultimately revolutionary way of comprehending the world.

THE NEXT STEP in that revolutionary path came as Newton worked on new ways to analyze and solve mathematical problems. Only the simplest algebraic equations can be solved just by plugging in numbers and doing the arithmetic. Seventeenth-century attempts to analyze more complicated expressions often employed a particular mathematical tool, the infinite series—endless sequences of terms (for example, 1, ½, ¼, ⅛ . . . and so on). Some of the works Newton had already read used infinite series to attack a variety of questions—for one, chasing down ever more precise values for that bane of middle school geometers, π (the number used to calculate the circumference of circles). Newton, with his exceptional ability to speak both algebra and geometry, began to use infinite series to work out how curves behave. One of his favorite tricks was to think about the area under a curve—all that space on the graph between the curve and the x axis—and then build out a series that would add together smaller and smaller patches of that area until the sum of all those terms approached the entire territory and number being sought. Newton applied that idea to a wide range of different curves. He wrote out sequences. He plugged in numbers. He exhausted himself in calculation, cranking his exercises out to fifty decimal places and more.

He totaled his sums—and then discovered what modern mathematics calls the generalized binomial theorem. This result allowed Newton to solve a wide range of specific algebraic equations, including, most significantly, the problem of the area beneath a curve (called quadrature), not just for one shape at a time, but for whole classes of curves. It was a discovery that became one of the pillars of modern mathematics.

As he played with his series, he noticed that in some of them each step in the calculation added a smaller and smaller amount to the total. Extending the operation by hand—row after row of numbers, a strangely beautiful triangle, growing across the page—produced a better and better fit to the ultimate answer. The endpoint, well beyond the stamina of even so heroic a numbers-cruncher as Newton, was obvious: the last terms in such series must dwindle toward nothing. Toward, but never all the way there, an infinitely small approach to zero.

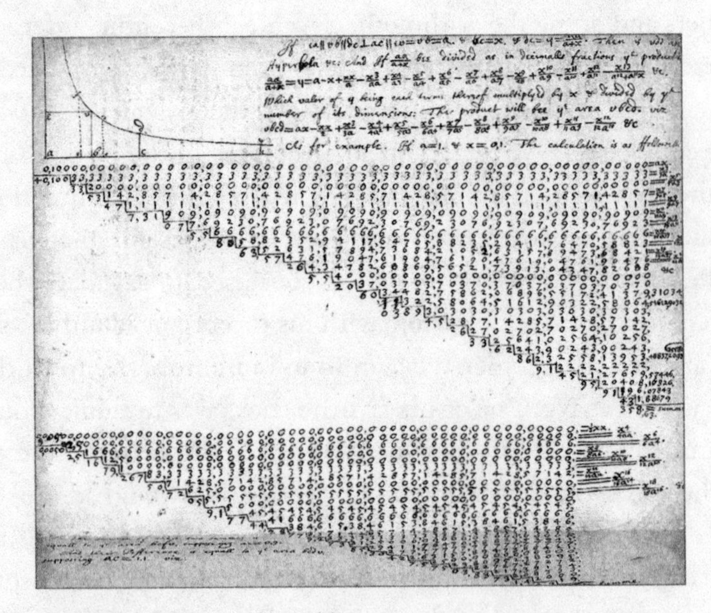

Newton's calculation of logarithms

Newton was not the first to ponder such infinitesimals. The Greek philosopher Zeno had played with the idea in his famous paradox: the race between the hero Achilles and a tortoise. With a fine sense of fair play, fleet-footed Achilles gave his opponent a head start. According to Zeno, that meant that no matter how much faster Achilles ran, he'd never overtake the tortoise. His reasoning was that in the time it took him to reach where the tortoise had just been, the reptile would move a little farther. When Achilles moved to that point, the tortoise would have moved again, and so on, forever. That increment of distance could get as small as you like, Zeno said, but it would never quite disappear. Hence, the tortoise would beat Achilles every time.

That's obviously absurd: in real life, an Achilles would charge past a tortoise, no matter how generous the head start. As early as Aristotle, logicians offered formal arguments to refute Zeno. But neither philosophical rigor nor common sense could erase the uneasiness produced by the idea of ever-smaller quantities. Many, like Descartes, simply didn't want to wrestle with an increment so tiny that it was effectively but not quite zero. Galileo knew that there was something vital about that infinitude but quailed before its mystery, "incomprehensible to our understanding." That is: there was, as yet, no established mathematical procedure that could use quantities indistinguishable from nothing to demonstrate that yes, in fact, Achilles smokes the tortoise every time. Newton himself, in his first months at Woolsthorpe, was often perplexed, straining to interpret the difference between almost zero and zero itself. But he didn't linger in the tangled metaphysics of a nothing that wasn't quite nothing. Instead, he put it to use.

In one case, Newton wanted to be able to identify how much a curve was curving at any point: how steep it might be, and how that steepness—Newton called it "the crookedness in lines"—changed at each point along the figure. Here, he used infinitesimals to produce a straight line whose slope could be calculated and that touched the curve at just that one point and no other—what's called a tangent.

Such problems took Newton unequivocally beyond classical approaches, in which the curves he was trying to understand had been examined as whole, finished phenomena, parabolas or ellipses or anything else of interest as the object of study. But Newton's thinking in the last months of 1665 employed his genuinely new way of seeing, in which the mathematical objects he analyzed emerged in the solutions to equations, point after point accumulating along the figure. He wasn't completely free of the older view as he pondered the tricks that schoolmasters used to construct the canonical curves without bothering with any algebra: a string attached to a peg that could be used to generate a circle; the same string fitted to two pegs to trace out an ellipse; and so on. He thought about more elaborate "mechanical" ways complex curves can emerge—the cycloid, for example, a form traced by a point on the rim of a wheel that rolls in a straight line—and others, still more complicated.

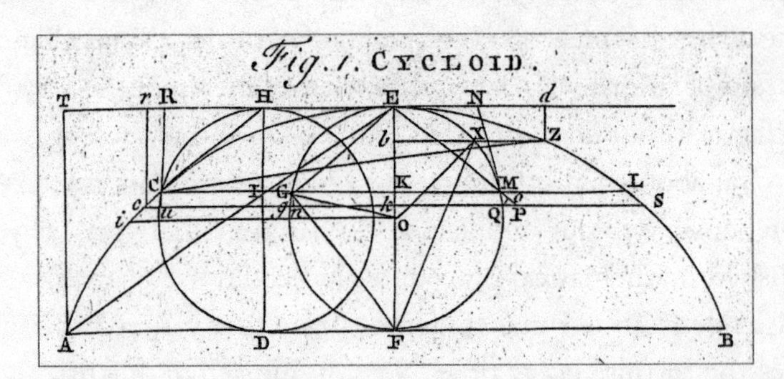

The making of a cycloid, from an eighteenth-century encyclopedia

In all those ways of ending up with a curving line on a page, there was one common theme: every curve was a map of motion, a mathematician's travelogue. A point travels through space, and its trail, its trace, creates the stuff of geometry. Crucially, sometime during these months, Newton realized that this approach, the "generation of figures by motion," could apply not just to abstract travel, the path of points in Cartesian spaces, but to the real stuff of the real world. In other words, motion in the universe, and not just in the mind's eye of the geometer, could be expressed in the mathematics he was inventing.

Newton did not grasp the full implication of this work all at once. He understood at least the mathematical side of his breakthrough by as early as November 13, 1665. In the paper he wrote then, he described the "infinitely little lines" that accumulated at each infinitely brief instant of time as his figures evolved. His breakthrough came when he realized that his two seemingly separate questions—how a curve bends and how much of the Cartesian plane it encloses—are actually twin faces of the same problem. Every change in the slope of a curve affects how much that shape encloses beneath it, and the same is true in reverse: the accumulation of territory under a curve reflects the shifting trace of that geometrical figure.

With that insight Newton arrived at a discovery that, on its own, would have made him one of the most famous thinkers who ever lived. Figuring out how to characterize how the shape of a curve is changing at any point in time is the core of what is now called differential calculus, which he then extended to integral calculus, which addresses the questions relating to the areas bounded by such curves. Taken together, those two interrelated ideas, as developed and extended, remain the foundational mathematics of material experience.

Newton never underestimated his own powers. He had to have grasped the importance of his accomplishment in those few months of enforced seclusion on his farm. Yet for most of the next two decades, he kept this new mathematical insight almost entirely to himself.

Still, this was the inflection point, after which the way humankind understood its circumstances would be irreversibly altered from what had been known before. What is motion but change over time? And what is the world but matter in motion, an ever-transforming flux, continuously transforming as the instants pass into seconds, hours, years?

THE INVENTION OF the calculus, the mathematics of change, was one of the keys to what we now call the Newtonian revolution—and Newton in his miracle year put his breakthrough into almost immediate use. As 1666 began and the plague continued to rage, Newton turned from pure math to questions of material experience. At the heart of his inquiry lay the problem of gravity. As he told the story sixty years later, the essential clue to his ultimate theory came to him during the summer of 1666. One day he found himself in his garden "in a contemplative mood." The tree in front of him was heavy with fruit. Suddenly, an apple fell—an utterly ordinary occurrence. And yet, it nagged at him. "Why should that apple always descend perpendicularly to the ground?" he recalled asking himself. "Why should it not go sideways or upwards? but constantly to the earth's center?"

Why indeed? The myth of genius has asserted that this was all it took: in the not-quite-infinitesimal slice of time it took for the apple to drop to the ground, Newton grasped the ultimate prize, his theory of gravity. In the moment, so the story goes, he knew that matter attracts matter in proportion to the mass contained in each body divided by the square of the distance between them; that the tug is between the center of each mass; and—the ultimate prize—that the power "like that we here call gravity . . . extends its self thro' the universe."

This much is true: the tree itself was real. After his death, the original at Woolsthorpe was still known in the neighborhood as Sir Isaac's tree. Every effort was made to preserve it as long as possible, until it finally collapsed in a windstorm in 1816. It rerooted itself and can still be seen at Woolsthorpe, while grafts from the tree have been used to propagate clones of Newton's apples since the 1820s.

But even if Newton watched the apple fall (and thought about gravity as it plummeted) it still took him decades to work out his ultimate theory. He used his newly gained mastery of the mathematics of circular motion to discover why things—including us—don't simply fly off the surface of the earth, given the Copernican realization that the earth doesn't sit still at the center of the cosmos but rather travels at an impressive speed, spinning on its axis as it tracks around its central sun. He calculated the strength of the so-called centrifugal force that should be hurling us into space. He put together that number with a rough approximation for the earth's size—a number refined over the previous two centuries of European exploration by sea. Taken together, that was enough information to estimate the outward acceleration experienced at the surface of a revolving earth—how strongly any of us are being pushed out into space.

Then he performed the other half of the analysis, considering the downward tug at the surface of the earth of what he called gravity in something like the modern sense of the term. Galileo had already observed the acceleration of falling bodies, but Newton trusted no measurement so well as one he made himself, so he reworked that earlier result by studying the motion of a pendulum—an experiment that brought him close to the modern value for the earth's pull. He knew that his data were still imperfect, but, he wrote, he "found them answer pretty nearly"—by which he meant that he was able to calculate a result that made sense of the evident reality. The gravitational force holding our feet to the ground is (clearly) more than strong enough to do the job—in his calculation, approximately three hundred times stronger than any centrifugal urge to launch us upwards.

That result, at once imprecise and spectacular, would also have placed Newton in the vanguard of European natural philosophy if only anyone had heard about it. He was not yet fixed in his habit of silence, a determination reached a few years later, after a few bruising exchanges with other learned men. But isolated on his farm, he remained focused on the work at hand, applying his almost daily expanding mathematical skill to physical questions. Applying numbers to a concrete question—why stuff sticks to our planet's surface—transformed the pure mathematical reasoning within his calculus into a literally down-to-earth experience. Newton's work now became one of the early examples of what we would call a mathematical model, a representation of some aspect of nature abstracted into a form that could be manipulated, extended, and solved. Today we are utterly immersed in the Newtonian worldview, in which these models, systems of equations, are understood to be properties of the universe. During Newton's miracle year, there was no such recognition, not yet. His next move, though, would push him ever closer to the ultimate triumph, his demonstration that the book of nature is written in the language of mathematics.

The fall of the apple had produced a breakthrough, but not a fully realized theory of gravity itself. In a leap of imagination that is still astounding, Newton realized that whatever pulled that piece of fruit toward the ground must have been the same phenomenon that held the moon in its orbit—tracing a path around the earth as our planet's inward pull counteracted the moon's urge to shoot off into space. At some distance from the earth, those two impulses must balance. Sitting there, an object would fall forever, tracing a (nearly) circular path around the center of the earth. Our moon is held to its course by the same phenomenon, the earth's gravity, that drew Newton's apple to the ground.

The insight was there, but his first attempt to write down the mathematics of gravity wasn't quite right; he would arrive at his famous "universal law of gravitation" only in the mid-1680s. But the apple (if Newton's late-in-life tale is to be believed) did give him the critical piece of the puzzle: laws of nature are universal. Abstracting experience into equations, such laws penetrate beneath all the surface confusion of experience to reveal common patterns and deep truths that govern the cosmos. Most important, this new quantitative approach to nature offered the gift of prophecy: calculate now, and you can find where the moon or Jupiter or whatever will be, days, years, centuries from now. Though Newton called what he did in that Lincolnshire farmhouse "natural philosophy," this was science in its (early) modern form.

THERE WAS ONE more move Newton made to complete this new approach to nature. A system of equations could yield a portrait of some aspect of the wide world, but any such model still needed reliable knowledge extracted from the natural world to make the connection between math and experience. Observation, measurement, and especially a commitment to rigor were keys to Newton's new approach to natural philosophy: it was essential to test the world precisely and reliably enough so that any mathematical analysis of what was going on could yield a real insight into reality. One of the reasons Newton is remembered as perhaps the greatest genius in history is that from the beginning he did just that, plunging deeply not just into mathematics but into measurement of the world—at times to his own peril.

For example, in the early 1660s he wanted to know how the shape of the human eye might affect the perception of color. To find out, he turned to the nearest experimental subject, himself, and stuck a bodkin—a blunt needle—into the bottom of his eye socket and levered up. He meticulously recorded his results, including defining the curve he induced in his eyeball ("ye curvature a b c d e f") and noting that the colored circles grew brighter "when I continued to rub my eye with ye point of ye bodkin." His sketch of what he did to himself is at once meticulous and stomach roiling, a measure of both the urgency of his hunger for data and his utter recklessness.

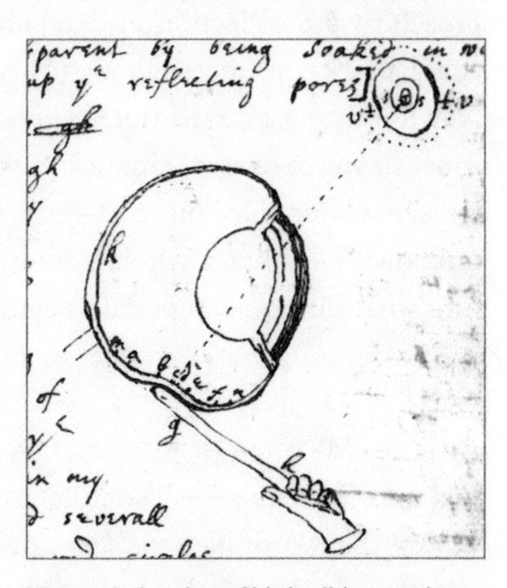

Newton's drawing of his bodkin experiment

Such mad hunger would seize Newton over and over again—he would later pursue his alchemical experiments so relentlessly that he drove himself to the point of physical exhaustion and, at least once, in 1693, tipped all the way over the edge into true mental collapse, months of silence and paranoid misery. But such excess shouldn't obscure his deeper and lifelong commitment to a cooler empiricism:

the need to base any scientific speculation on the solid ground of rigorous and systematic observation.

That's what occupied him in the last months of his plague exile: optical investigations that demanded hands-on effort in a series of organized, rigorous experiments with prisms and other apparatuses to tease apart the properties of color and light. The work ultimately produced what he called the *experimentum crucis.* That was his crucial demonstration—that sunlight, so-called white light, is actually a blend of distinct individual colors, the rainbow spectrum from red to blue.

The creation of new knowledge of the world required theories, ultimately mathematical accounts of the relationships between phenomena and events. It also needed a careful, logically coherent approach to the collection of data, observations and experiments that could probe the material world. As his miraculous year unfolded, Newton found himself tracing this arc: through each particular advance, from pure mathematics to what he was able to deduce as he twisted a triangular piece of glass, he uncovered not just previously unknown facts but a whole new way of organizing that knowledge into what he would later call, accurately, a "System of the World."

BEYOND WOOLSTHORPE AND its scenes of revolutionary intellectual transformation—all completely unknown outside the farmhouse walls—the rest of Newton's world, haltingly, began to reassemble itself. By mid-March 1666 the city of Cambridge marked six weeks without a single plague death. The university reopened, and Newton returned to his rooms on or around March 20. Then, on Wednesday, June 6, Jane Ellingworth, a seamstress living on Penny Farthing Lane, felt poorly. Her father brought her a cup of ale and urged her to bed. She died the next day. The infection spread, with deaths reaching double digits in the city within two weeks. The colleges closed again, and Newton retreated to the countryside once more. The plague burned itself out over the summer, and by the end of the year Newton felt it safe enough to travel back to the university. He would remain there for the next thirty years.

By the time Newton resumed his almost-cloistered life in Trinity College, the onetime center of the epidemic had been utterly transformed. London's parish records reported a total of 68,596 plague deaths for 1665. The true figure was certainly higher. But the toll dropped during the winter, and stayed low, with only 2,000 more deaths from the disease tallied through all of 1666. Then, just as it appeared that the epidemic was truly done, came the four days in September when the Great Fire of London destroyed most of what lay within the old city walls, along with some newer neighborhoods to the west. Four hundred and thirty-six acres burned; at least thirteen thousand homes were destroyed. So were 87 out of the city's 109 churches, including old St. Paul's Cathedral. When that giant building caught fire, the tons of lead in its roof melted, creating a river of liquid metal flowing into the Thames.

A new London began to rise almost immediately. The fire seemed to obliterate the plague, though that could well have been a coincidence in timing rather than the result of any lasting impact on the city's rat population. Christopher Wren—often with the aid of his colleague in the nascent Royal Society, Robert Hooke—took the lead in restoring sacred London, building fifty-one parish churches along with his crowning monument: the new St. Paul's, with its glorious and technically sophisticated dome.

Life in the capital soon returned to an approximation of its preplague normal. The savants, Hooke and Wren among them, resumed weekly meetings at the Royal Society. Their conversations overflowed into the invisible university housed in the still-exotic coffeehouses and inns of the rebuilt city. Some of that talk was more enthusiastic than rigorous: at early meetings, the Society heard reports on "a Very Odd Monstrous Calf" and "Of the Way of Killing Rattle-Snakes," presented alongside "A Spot in one of the Belts of Jupiter" and "General Heads for a Natural History of a Country." Newton himself took no part in that eager, hungry, small "c" catholic pursuit of new knowledge. It would take him twenty years and more to organize the results of the plague years into a fully realized body of work. He did so mostly in silence. He had some contact with members of the nascent Royal Society in the early 1670s. But he soon disappeared from the view of learned Europe. That was partly because he resented challenges to his results, partly because of his determination never to share a discovery before he was certain, and partly because for many of those "missing" years he pursued lines of inquiry that he actively wanted to keep secret: inquiry into heterodox religious beliefs and into the ancient pursuit of alchemy, which he saw as one more way to investigate change in nature. He wrote of his alchemical experiments to a handful of fellow searchers, but he rarely communicated in any public way with his fellow natural philosophers in London, and he visited the capital even less.

But such silence did not mean that he was unmoved by the same impulses driving the early Royal Society men, with their public commitment to knowledge for its own sake—and to the application of whatever could be discovered to practical uses. From the beginning, he too recognized that natural philosophy could comprehend daily life as well as the broad sweep of nature. As early as 1664, for example, before he plunged into the question of gravitation, he laid out a geometrical approach for calculating compound interest—his first contribution to the mathematics of money. A decade later, he turned his quantitative virtuosity to the service of his home institution, helping Trinity College's bursar analyze how much rent he should charge for land—farms that the college owned. Newton could count and Newton could think, and his work here—pricing an asset that offered payments over time—already hinted at the possibility that those two skills could make a man rich.

If that thought crossed his mind in his Cambridge years, he didn't act on it. His full immersion into the world of money—on his own account as well as in service to the crown—would come a full three decades after his miracle year, when he took up new duties as an officer of the Royal Mint. Others, though, were beginning to recognize that there might be a connection between quantitative reasoning and wealth. They would pursue both riches and yet another new discipline: a science of change over time that took not the planets but people as its object. Perhaps the most complete representative of these new men was someone much less remembered than he should be: a polymath and voraciously money-hungry parvenu named William Petty.

Chapter Two

"to make a *Par* and *Equation* between Lands and Labour"

William Petty lived the cliché: he was a genuine Renaissance man. By the time he reached his thirties he had been a music professor, an anatomist—a physician eager to dissect more or less any mammal that came his way—and a more than competent chemist. He was a constant inventor, turning his hand to everything from farm implements to ship design. Mathematics was an early love, and he had a gift for practical calculations. That would prove the skill that would make him rich, through his work as Ireland's first comprehensive geographer.

In his middle age, just as the young Isaac Newton began to build his system of the world, Petty's ambition narrowed, increasingly focused on a single goal, shifting his "Trade of Experiments from Bodies to Mindes," he wrote—bodies here being inanimate matter, celestial objects and the like. This was recognizably kin to what would become the Newtonian synthesis, with one key difference between their two aims. Where Newton focused on the analysis of matter in motion, Petty aimed "to have understood passions as well as fermentations"—human feeling as much as any natural process. But, he emphasized, his program was to be as rigorous as any purely physical inquiry. He had little interest in mere opinion or vague verbal descriptions. Rather, he declared, he would "express myself in terms of *Number, Weight, or Measure.*" This was a credo: the only way to make sense of the human world, Petty argued, was to quantify, and to reason only from that solid, empirical starting point, so as "to consider only such Causes," in his words, "as have visible Foundations in Nature."

THAT WAS THE grand vision of his later years. Little in Petty's origins suggested he would rise to such grand ambition. He was born on May 26, 1623, in the town of Romsey, to a father who was an unsuccessful clothier and dyer. He left his local school at thirteen, already the master of "a competent smattering of Latin, and . . . entered into the Greek." Before his fifteenth birthday he made his escape from Romsey and its meager possibilities, signing on with a merchant ship working the cross-Channel trade with France. Here, fortunate ill-fortune intervened. No natural sailor, he broke his leg aboard and was turned ashore at the Norman port city of Caen. The Jesuits of the University of Caen took him in and exposed him to some of the new, humanistic learning then spreading across Europe.

Petty left Caen in the late 1630s, still in his teens, and returned to England, already showing the magpie-like mental agility of the polymath he was becoming. Before his twenty-first birthday he managed to write up astronomical ideas, publish poetry, complete "Severall paintings and drawings," and perform some jobs for the navy. He paused to train as a physician—a curriculum that exposed him to yet more new ideas, about mechanical cause and effect, anatomy, and chemistry. When he was done, he was both a doctor and an example of a new kind of man, a "virtuoso," as the growing network of London's eager investigators styled themselves. He and they were committed to what was already being called natural philosophy, with its program of discovery through experiment, observation, and the unequivocal testimony of nature. There was only one problem. He was still poor. His wit kept him fed and clothed, but he wanted more.

His fortunes would change when he became a winner in one of the most vicious episodes in two brutal decades of misery on the wretched island of Ireland. The Irish Rising of 1641 had been a violent mess that led to the creation of the Irish Catholic–dominated Confederation, a short-lived, contested, but more or less autonomous government with at least nominal authority over much of the island. The Confederation existed in a state of near-continuous conflict with English royalists in Ireland and Protestant power centers aligned with the parliamentary side of the ongoing English Civil War. Ultimately, when King Charles I offered official toleration of Catholic worship in Ireland, along with other provisions that amounted to recognizing Irish autonomy, the Confederation agreed to place its armed forces under English officers. The deal didn't hold. Charles tried and failed more than once to import Irish soldiers to aid him against Oliver Cromwell's New Model Army, Parliament's military force. With the king's final defeat and execution on January 30, 1649, the victors saw in the Confederation a rebellious province that had, more or less, lined up with the losers. Ireland to the new English Commonwealth was both threat and prize.

Cromwell invaded seven months after the king's death, landing at Dublin on August 15, 1649. It was a bloody, brutal campaign. After his first major victory, the capture of Drogheda, two thousand Irish Royalist soldiers were executed, an atrocity followed by a general massacre of both combatants and civilians at Wexford. By the spring of 1650, the invaders had managed to retake most of Confederate Ireland, and Cromwell himself returned to London. There was just one problem: he left behind an army that hadn't been paid, as almost a decade of civil war had left Parliament as much as £3 million in the hole.

The new government couldn't rely on the usual habits of overstretched monarchs and simply borrow the money. In the turmoil and uncertainty after Charles's beheading, traditional lenders had become skittish. That forced the English back to another old trick, the brigand's solution: trade the chance of loot to come for cash in hand. It worked. The soldiers who did the actual fighting went without pay, and the same went for the army's suppliers—to whom were added the so-called adventurers, those bold souls who were willing to lend money to the new English government. In return they received the promise of . . . Ireland, almost all of it, to be taken from Catholic landowners along with anyone else who had chosen the wrong side during the Civil War.

After the victory, though, there was an obstacle between the English and their spoils: figuring out what Ireland was actually worth, piece by piece. By 1653, with the countryside at last pacified, the need to do so became acute. A first attempt to catalog the property to be forfeited was merely a summary accounting of the total amount of available land. The second was primarily a document review, existing records of boundaries and the rents and other measures of value of each plot of land. This "Civil Survey" was widely believed to be inaccurate at best, and, by the end of 1654, thousands of trained men with guns remained without their promised reward. Almost two years into the process, they were becoming just a bit impatient.

ENTER WILLIAM PETTY. He had wangled the post of physician-general to the army in Ireland, and moved to Dublin in 1652. Initially, he alternated practicing medicine with more of the natural philosophizing he'd enjoyed in London, anatomizing local fauna with the Anglo-Irish aristocrat Robert Boyle, for example, or diving into studies of the Irish harp. But as the early surveys faltered, Petty saw his chance. In what amounted to an intellectual coup, he pitched the army leadership a new approach to the mapping problem. The incumbent surveyor-general had opted to measure only land marked for confiscation. In place of the "absurd and insignificant way of surveying" of the Civil Survey, with its primary focus on land ownership records and not on a direct account of the assets to be seized—

improvements to the property, livestock, tools and the like—Petty offered to measure everything: the whole of Ireland, its ownership, its geography, all of its "rivers, mountains, ridges, rocks, sloughs and bogs," along with the houses, barns, fences, and all the rest that made up the island's real property. And he would do all this for a fee to be paid mostly in thousands of acres of Irish land.

William Petty in 1651

THE AUTHORITIES ACCEPTED Petty's proposal in December 1654. He began immediately, swiftly surpassing (as he'd promised) the incomplete and disputed work of his predecessors. He committed to direct empirical observation. He sent to London for new surveying tools, to be produced by modern methods—"The said Petty, consideringe the vastnesse of the worke, thought of dividinge both the art of makinge instruments, as also using them into many parts." That is: he set it up so that one craftsman would make measuring chains, another compass needles; woodworkers would concentrate on building cases; and so on. This isn't quite Adam Smith on the manufacture of pins, but Petty's idea is recognizably kin to that vision of rationalizing human behavior to some greater (and more interchangeable) end. Armed with these standardized kits, teams of surveyors and assistants set out, guarded by soldiers, and working with designated "bounders"—men who could confirm local property lines set out across Ireland.

It wasn't all smooth sailing. Some surveyors quit in the face of (understandably) hostile locals. The work itself was trickier than anticipated, as the patchwork of tiny landholdings found in many districts demanded more fine-grained and hence more time-consuming cartography than Petty had expected. He had to fight off enemies within the Dublin government, especially some who accused him of cheating, to the point that he had to appear before military commissions to defend himself against charges that he relied on "drunken surveyors" whose work defrauded deserving soldiers by classing "unprofitable land . . . put upon the army as profitable."

Even so, Petty pushed forward at a tremendous pace. He included an unprecedented amount of information in his maps, not just numbers and symbols, but drawings as well: green hills, church steeples, the buildings and streets of villages and towns, and a constant tally of significant details—"unprofitable mountain and Turfe bog," Capenaheny's "Timber Trees with some under Wood," along with such vital distinctions as "Edward Dungan Jr. Pap(ist)," who worshipped in a different (and disqualifying) form than did his near neighbor to the east, one "Nenabbey Protestant."

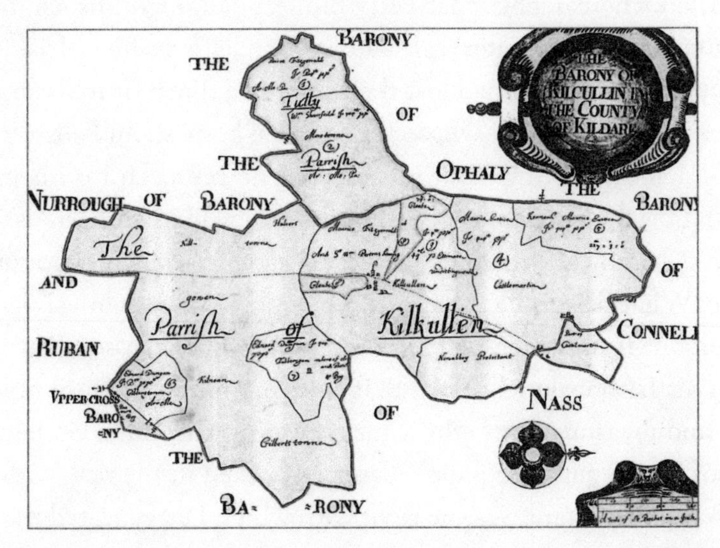

The Down Survey map of the Barony of Kilcullin in the County of Kildare

It would take until 1659 to fill in the last corners in the far west, but the Dublin authorities began handing land to soldiers by the middle of 1655, less than half a year after what came to be called the Down Survey had begun. Its maps do not, of course, measure the human misery that flowed from their information. The Cromwellian settlement enabled by the Down Survey wrecked the social world of Catholic Ireland. In 1641, Catholics had owned 60 percent of Ireland. By the end of the century, Irish Catholic ownership of Irish land had fallen to just over 20 percent, and in another fifty years less than 5 percent remained in Catholic hands.

That outcome was, of course, the point of Petty's masterwork. He delivered what he'd promised: a blueprint for what was in essence a form of ethnic cleansing. It defined who was to be expropriated and who would take their place. At least ten thousand Catholic landowners were driven from their land in the first wave of the survey, driven into exile with their laborers and Catholic town dwellers, transplanted to the far west of the island, or abandoned to die. The human cost of the Down Survey capped a decade-long demographic catastrophe that killed as many as one of every three Irish men, women, and children.

There's no evidence that Petty himself delighted in such immiseration. He would later calculate that half a million Irish men, women, and children had lost their lives as a direct or indirect result of the years of war, "for whose Blood some body should answer both to God and the King." But he accepted no reproach for his part in separating the survivors from their property. He had been asked to solve a technical problem: measure England's newly reconquered prize. What was to be done with his report was a political decision (albeit one from which Petty himself profited). The lesson Petty drew from his Irish experience was that applying formal rigor to observation and measurement—the same concept his London circle of virtuosos had begun to explore—actually worked in the real world.

Most important was the rigor with which Petty created the tools of data collection for his survey, standardizing them to complete one of the first modern works of political geography. This was the new natural philosophy brought (literally) down to earth, Ireland anatomized on paper. The final result rendered all the prolific variety of a landscape into a form—maps—that could be read by anyone. Most important, the Down Survey caught the same kind of insight Newton would later capture to transform natural science. It combined observation with numbers. In this, Petty's first great project, it was still a very simple relationship, just assessments of land that could be expressed as the pounds, shillings, and pence owed to each creditor. Soon, though, Petty would come up with a much more ambitious vision, one that imagined politics itself, the art of statecraft, as a form of mathematical argument.

PETTY HAD COME to Dublin a poor, clever, cocky twenty-nine-year-old. Seven years later he returned home a rich man—master of the thousands of acres of Irish land that formed his surveyor's fee—and a famous one, acknowledged as the author of one of the first great practical applications of natural philosophy. Back in London in the early 1660s, his scope of action necessarily shrank: he no longer commanded an army of surveyors across an entire nation. But in their place he found friends, just as absorbed as he was in the application of mathematics to human society.

To casual observers, London's community of virtuosos in the first years of the restored Stuart monarchy were heroic talkers but seemingly little more than that. Scornful witnesses to such gatherings reported that such endless colloquy was merely "a confused way of gabbling" in which "the Noblest Speculations, the Divinest Truths, become as Common . . . as stones." But among the learned there were those who understood the power of organizing even something so mundane as a conversation. A few of them began to gather after lectures at Gresham College. Frequent encounters became regular ones until, on November 28, 1660, the group "did, according to the usual manner withdraw for mutual converse."

There were about a dozen men present—Petty among them—and that afternoon they decided to turn their informal encounters into something more, an institution "for the promoting of Physico-Mathematicall Experimentall Learning." They drew up a list of about thirty other learned men to invite and agreed to a fixed schedule of meetings: every Wednesday at three o'clock. Sir Robert Moray, the best connected among them, took the news to court and reported back the next week that Cromwell's replacement, the recently restored King Charles, had been told of the plan and "did well approve of it." Charles II granted the group a charter on July 15, 1662, which thus became the birthday of what could now be called the Royal Society—dedicated, as its founding members had declared, to the creation of new knowledge, tested against the hard check of direct investigation.

The stated aims of the new institution suited Petty perfectly. The Down Survey had focused on one particular problem: how to value Ireland at a single moment, to find the momentary exchange rate between land and cash. In these first years at the Royal Society, Petty began to look to the future in order to find a way of analyzing social life much more broadly that would enable him to predict—and shape—political outcomes over time. As his thinking advanced, he gave his program a name, "Political Arithmetick," for what had become a comprehensive attempt to use his old touchstones—"Number, Weight or Measure"—to construct a quantitative science of society.

At its simplest, Petty's political arithmetic was an exercise in applied demographics. He began with what we would now recognize as a census: a count of how many people lived in any given territory; who they were, identified by religion or national origin or custom and tradition; and what resources they possessed and how productively these were exploited. Given such knowledge, Petty argued, the state would know—to a certainty never before realized in statecraft—what it should do to achieve a given political end.* His most exten-

* There is another obvious example of spectacular political surveying in British history: the Domesday Book, compiled in the 1080s at the order of William the Conqueror to assess the value of the kingdom he'd just acquired. There are similarities in both attempts to assess the value of their domains. But what distinguishes the Down Survey is its focus on reducing Ireland's contents to a clear, abstract, quantitative account, one that would allow the direct comparison of the value of one grant or award with another. The Domesday records may have aspired to precision but didn't come close to achieving it.

sive attempts to apply political arithmetic in the real world landed—unsurprisingly—on the luckless Irish. Over the last two decades of his life he would produce a variety of plans intended to secure English control of Ireland (and his own interests as an Irish landowner), torquing his analysis in response to the tangled politics after the restoration of the Stuart dynasty. Through each variation, though, his essential procedure did not change. First, he would make an approximate population estimate: in 1672, he determined that there were eight hundred thousand "Papists" versus three hundred thousand Protestants, themselves made up of two hundred thousand English and one hundred thousand Scots. Then, he would tease apart the first-order categories within that undifferentiated data. In the 1672 proposal, for example, he noted that there were almost eight hundred thousand people able-bodied enough to work, of whom only half were doing anything productive. The rest he split between those who "do follow the Trade of Drink" or were simply malingerers, "Casherers and Fait-neats" (Do-nothings). And from there he would begin to perform simple calculations that he believed would reveal the profitable ways to engineer particular political or social results.

In this early version of his arithmetic applied to people, that meant proposing a change to who lived in Ireland as a way to make the territory more productive and its people less restive. He wrote, "If an exchange was made of but about 200,000 *Irish* and the like number of *English* brought over in their rooms, then the natural strength of the *British* would be equal to that of the *Irish,*" which would mean, Petty added, that "the *Irish* would never stir upon a National or Religious Account," while the like number of Irish who were shifted across the Irish Sea would be far outnumbered by the English at home.

Petty went on to suggest a refinement to that initial plan of forced relocation. Among the displaced, he said, there should be twenty thousand marriageable Irish women to be distributed among England's parishes each year. To keep gender ratios stable (and with them, the number of marriageable couples), the same number of

English women should be sent to Ireland. These would marry locals—and here we come to the point of Petty's exercise in social engineering—and thus civilize Irish bachelors, creating Anglo-Irish households conforming to English culture, domestic behavior, and allegiance.

Petty didn't stop with the simple manipulation of populations. He argued that the government should gather statistics to enable the leaders of a mostly agricultural economy to manage their affairs. "To make nearer approaches to the perfection of this Work," he wrote, "'twould be expedient to know the Content of Acres of every Parish, and withal, what quantity of Butter, Cheese, Corn, and Wool, was raised out of it for three years consequence"—to count in each plot of land how much could be produced by how many workers. Repeat the exercise over the entire realm, and the result would be a whole new insight into the wealth of the nation: a level of quantitative detail that would enable the arithmetician "to make a *Par* and *Equation* between Lands and Labour." Petty's approach wouldn't offer what we would recognize as a measure of gross domestic product in anything like a modern calculation, but it is recognizably an ancestor to such national accounting. This was a modern idea, born of the conviction Petty shared with his Royal Society peers: everything was or could be made to be expressed mathematically, thus revealing relationships between everything that empirical experience could count, weigh, or measure.

Fifteen years later, Isaac Newton would publish his great account of celestial motion. It's easy to exaggerate the connection with Petty's attempt to quantify economics and politics, certainly—but the resonance between them is no mere coincidence either. Newton and his colleagues advanced mathematical models of motion to the point where they could test them against nature—and generate useful, accurate, *true* predictions. Petty never approached that kind of accuracy, nor were there a handful of axioms about human experience that could support an account of the human cosmos analogous to the compact set of physical laws Newton would use to build his universe.

Still, he argued that even simple quantitative connections—acres to butter—exposed authentic patterns of human experience.

PETTY WOULD TURN his political arithmetic to other challenges over the next fifteen years: weighing the relative wealth and power of England, France, and the Dutch through various statistical measures, for example, and attempting to calculate the prospects of the young English colonies in North America. Near the end of his life, he came up with a last, terrifying proposal for Ireland. He would cut the island's population down to a tiny fraction, leaving just a handful behind to act as drovers on what would become a giant cattle ranch. Everyone else would be pressed into the service of another attempt to engineer nations via demographics: almost all the Catholics in Ireland would be brought to England in support of King James II's doomed hope of re-Catholicizing his realm. (Petty wasn't the only Englishman to prosper through the tumult of England's bloody seventeenth-century politics, but it's a mark of his flexibility that he was able to serve both Cromwell's ends at the beginning of his career and James's passions at its end.)

Nothing came of these later Irish schemes, either Petty's dream of taming unruly papists in their homes or his terrifying last vision of an emptied land. Ultimately, he was a better intellectual evangelist than he was a social engineer. His success lay in convincing colleagues and successors of his core idea: that the systematic, reliable measurement of human populations, studied with increasingly powerful mathematical tools, could transform the way people and states did business. His close friend John Graunt advanced Petty's attempts at demographic analysis with his landmark text *Natural and Political Observations . . . upon the Bills of Mortality* in 1662. In that work Graunt reduced life and death to data and, among much else, arrived at the first rigorous estimate of child mortality in England. Thirty-six percent of the nation's children, he found, more than one out of three, would die before their sixth birthday.

After Petty's death in 1687, new thinkers continued to act as disciples, advancing what they too called political arithmetic. The prolific prototype-economist Charles Davenant defined the term as the "Art of Reasoning, by Figures, upon things relating to Government." Davenant's friend Gregory King used that framework to argue for the importance of a seemingly disinterested gathering of statistics on just about any aspect of life. King wrote on births and deaths, estimates of the fractions of national income earned or held by each occupation, public tax receipts—and even, as historian Ted McCormick uncovered, an accounting of the totals of large and small flowers embroidered in "Mrs Kings fine Calico Gown."

These later writers were careful to avoid some of Petty's most grandiose impulses, proposals to move whole populations back and forth across the Irish Sea. King offered careful advice rather than proposing wholesale national transformations. For example, he used his analysis of the wealth of England and France to suggest, cautiously, that England under its new King William might not be able to afford the Nine Years' War against Louis's France. (He was right.) Such seeming modesty concealed the broader ambition Petty's successors pursued in the use of quantitative social knowledge—as when, for example, King produced a rough model of national resources and the capacity to fight and then used it to project the fate of English arms years ahead.

This was the critical step: to take the approach that the nascent Royal Society championed—the "Physico-Mathematicall-Experimentall" approach that could track the motion of a falling stone or the flux of the tides—and apply it to the realm of human experience. As it matured, the branch of natural philosophy that sought to anticipate social outcomes faced a question that pure physics did not: not just to anticipate the future but to put a price on it.

Here again, Petty had shown the way in his Down Survey when he converted Ireland into a uniform measure counted in pounds sterling. But that was just a first step. The fact that human societies alter over time made the problem of putting a price on the interplay of human action and resources as complex as any other problem tackled by scientific revolutionaries. Solving it would evoke both new math, especially ideas about probability and risk, and new ways of thinking about experience. Creating those tools occupied some of the finest minds in England throughout the last decades of the seventeenth century—including one still-young man, a leader within the Royal Society's second generation, poring over lists of the dead.

Chapter Three
"Very probable Conjectures"

In 1693, Edmond Halley, age thirty-seven, was, for all his relative youth, recognized as one of the Royal Society's most formidable talents.

Like many of his contemporaries, Halley didn't confine himself to a single discipline. He was an astronomer, of course, among the best observers in Europe and the author of that great triumph of theory, the prediction that the comet of 1682 would return in seventy-six years. He was an inventor, numbering among his creations the diving bell. He was a courageous explorer, taking a fifty-two-foot Royal Navy vessel across the Atlantic to produce the first large-scale ocean survey of variations in Earth's magnetic field. He mastered Arabic to the point where he could translate mathematical texts into Latin. He generated a "Solution of a Problem of Great Use in Gunnery" and even offered a hypothesis about the cause of Noah's Flood: a glancing collision with a comet sufficient to knock the earth on its side.

To that catalog, add one more triumph: he was the first to create a rigorous, reliable way to put a price on a human life. "Comet" Halley can be seen, with only slight exaggeration, as the inventor of the mathematics of what we now call life insurance. Halley's interest in demographics and the quantifying of human births and deaths came suddenly, lasted just a short while, and turned (appropriately enough) on chance—the discovery of a letter from an obscure German clergyman in the correspondence of a dead man.

CASPAR NEUMANN IS known to history for one great skill: he could count.

Between 1687 and 1691, Neumann, a philosophically inclined pastor in the Silesian city of Breslau (now Poland's Wrocław), meticulously tallied births, deaths, the ages of those who died, and the genders of those who entered and left this world on his watch. His was a simple record of events, lives reduced to numbers, carefully tabulated.

In time, as his data accumulated, he looked to share what he had found. He sent word to the preeminent natural philosopher in the German-speaking world, Gottfried Leibniz, then a member of the Hanoverian Court. In 1673 Leibniz had been welcomed into the Royal Society as a foreign member, elected after he demonstrated his "step reckoner," the first mechanical calculator able to perform all four of the arithmetical operations: addition, subtraction, multiplication, and division. His advice to Neumann? Send his figures to London, where they would reach the critical mass of natural philosophers gathering at the Royal Society's weekly meetings.

Neumann did, his data traveling a tortuous road. It went first to Henri Justel, a French Protestant who had come to London to escape France's increasingly harsh persecution of non-Catholics. Justel was no deep thinker, but he eagerly maintained connections across Europe, a node in the web of learned men already being called the Republic of Letters. Justel may have presented Neumann's work to the Society himself, but if so, no one seems to have paid any attention. To most of the fellows, a list of deaths in a provincial German town would have been just one more of the hundreds of earnest observations that earnest amateurs had sent off to the Royal Society since its founding—everything from that account of "a Very Odd Monstrous Calf," published in the first issue by the great Robert Boyle himself, to an investigation of water pressure at depth, performed by "a Person of Honour" in 1680, to Robert Boyle's questions for a Doctor Lower about blood transfusion, wondering, among other hypotheses, "whether a *fierce* Dog, by being often quite new stocked with the blood of a *cowardly* Dog, may not become more tame," or, from Paris, Monsieur de la Quintinie's "*some further directions and observations about* Melons," which concludes with the still-sound advice to "trouble not your self to have *big* Melons, but *good* ones."

Every week, letters reached London, to be mentioned at a weekly meeting if they tickled someone's fancy, and then to be lost to memory. Neumann's accounts could easily have fallen prey to that common fate but for chance. When Justel died in 1693, Edmond Halley had a rummage through his papers. What he encountered in Neumann's "curious Tables of the Bills of Mortality at the City of Breslaw" inspired him to apply his full powers of mathematical argument to fix the expectations for a human life, anywhere, at each moment between cradle and grave.

AS HALLEY DOVE into Neumann's numbers he discovered within them this fact: the town was a sleepy place, quiet and rather insular. By the numbers, Breslau had very little interaction with the outside world: not many people left town, and few immigrated. Neumann's accounts of deaths and births balance neatly, and the pattern of deaths by age remained basically constant across the five years of data. That made its vital statistics ideal for Halley's attempt to investigate the mathematics of mortality—how many would die and at what ages in every year—in an environment far less complex than his own London, with its ceaseless flow of men and women rolling into town.

Observing that stable population, Halley first extracted the simplest facts from Neumann's raw data: over the five years surveyed,

6,193 babies were born while 5,869 men, women, and children were buried. Diving into those summary figures, Halley discovered a slaughter of the innocents: "348 do die *yearly* in the *first Year* of their *Age*," he wrote, "and likewise . . . 198 do die in the *Five Years* between 1 and 6 compleat." The arithmetic was implacable: just 692 of the 1,238 children born each year—56 percent—survived to celebrate their seventh birthday. After the holocaust of infancy, Breslau's children could persist in the reasonable hope that they would live to have babies of their own, as the death rate fell to roughly 6 percent through youth and young adulthood. That proportion rose slowly until the citizens of the city reached fifty, after which death would capture higher and higher fractions as each age's residue of survivors continued to age.

From there, Halley reexamined those mortality data at a still finer level of detail, calculating the percentage of each year's age cohort who died before their next year—how many thirty-year-olds remained alive at thirty-one. He placed those results in another table, a grid that displayed Breslau's age profile in one-year steps from age one to one hundred.

Those numbers led to a nifty piece of algebra by which Halley, using only birth and death records, calculated the total population of the town in the period covered by Neumann's data: thirty-four thousand souls, a number that remained pretty much constant over the years as human beings entered and left this earth with predictable regularity.

From this baseline, Halley came to the questions that allowed him to dissect the lives of strangers hundreds of miles distant. In a conventional exercise in Petty-style political arithmetic, he used Neumann's figures to assess Breslau's military potential. He added up the number of residents of the town between eighteen and fifty-six years old, cut that number in half to extract the number of men in the sample, and thus arrived at the proportion of men able to "bear the *Fatigues* of *War* and the Weight of *Arms*" that Breslau could call upon for its defense: just over one-quarter of all who lived there.

Halley then moved into the heart of the data, examining "the differing degrees of *Mortality,* or rather *Vitality* in all *Ages,*" which he defined as "the *odds* . . . that a Person of [a given] age does not die in a *Year.*" That was a simple calculation. Of the 567 alive at twenty-five, 560 were still living at twenty-six, producing odds of 560:7, or 80:1, that anyone starting their twenty-sixth year would see its end. Next, Halley showed how the same reasoning allowed for the calculation of the odds for each person reaching any age—for example, our chances of making fifty, from a current age of twenty-five.

This far, Halley remained within the bounds of what Petty and his heirs had been doing for more than thirty years: applying simple formulas to one set of measurements or another. What came next, though, took him into new territory: instead of simply identifying patterns within Neumann's numbers, Halley added the dimension of chance, of risk in the modern sense. What, he asked, should anyone bet on the odds of his or her survival for any given length of time? As Halley put it, at what point is it "an even Lay that a Person of any Age shall die"? How much of life can each of us aspire to before the odds even up on whether we will live or die next year?

Edmond Halley in the early 1720s

The answer (for seventeenth-century Europe) emerges directly out of the table Halley made with Neumann's numbers: all he had to do was start at any point and run down the years until he found the median, the age at which only half of those present at the starting line of any given cohort—one-year-olds, twenty-somethings, gray-haired oldsters—remained alive. That's an almost trivial finding, barely an extension of the work of his predecessors. But for all the mechanical simplicity of the calculation, the idea behind it was at once subversive, powerful, and central to its time and place. Having established that it was possible to set the odds on when a life might end, Halley immediately asked how to apply mathematics to a measurement—Neumann's observations—and answer the problem of setting not just the odds at any moment but how much a life was worth in time and in cash. He sought a method to work out the monetary value to be placed on someone's existence at any moment along the journey: no random cast of the dice but a true assessment, a fair price for years one might still see.

If that question sounds familiar, that's because it is. It is a bit of a leap to call Halley the father of life insurance. But it's not far wrong—and it illustrates a basic truth about how the scientific revolution worked when it was at home. Halley's dive into the seemingly mundane problem of insurance may seem surprising now, but he, like his Royal Society colleagues, pulled on his clothes each morning and walked out into the daylit world of people, places, and things—and turned his prepared mind to what he found there.

In this case, Halley had stumbled on questions that were just beginning to percolate through an industry in its infancy. In London, the earliest organized insurance schemes appeared after and in response to the Great Fire; after some false starts, the first company to successfully sell fire insurance to the public—"the Insurance Office for Houses"—began to write policies in 1681.* Marine insurance—covering the risks of trading by sea—has a very long history, with roots in classical antiquity, and an evolution in Italy and then the northern European maritime centers that extends from the late Middle Ages forward. Attempts to insure England's expanding sea-

* These early companies faced more problems than the quantitative one of figuring out what to charge for protection. Individual insurance companies created their own firefighting teams, which led to times when a brigade would lay down their tools and let a house burn if it was discovered the property hadn't bought protection. The conversations that followed such actions can be imagined, though probably not re-created in full voice and vocabulary.

borne wealth took shape over the seventeenth century, culminating in the founding of Edward Lloyd's coffee shop in 1686. Lloyd catered to his seafaring clientele by offering shipping news updates regularly—which catalyzed the emergence of a marketplace in which shipowners and insurance underwriters could strike their bargains.* Those deals were worked out by experience, or hunches—the best guesses available to knowledgeable people.

Offering coverage for individual humans against the risk of death—or, at a minimum, its financial consequences—was a relatively late addition to London's financial landscape. The first life insurance schemes appeared in London in the 1690s, just as Halley was working up his findings. Those experiments shared a common problem: they had to match their revenues to the payouts and thus had to work out how much they might have to pay out in any given year. They needed, that is, to work out the odds that any given human of one age or another would die in a particular period of time, and then to find a model that would tell them how much to charge each person who sought their coverage.

That's what Halley set out to solve: what risk each of us—and hence our insurers—faces as each year passes, and, once that's understood, how to take that knowledge to find out how much it should cost now to purchase a given payout when we do leave this earth. He did so by examining the sequence of probabilities that enabled him to compare the risk that one person might die over one term of years versus another. This analysis, Halley wrote, exposed "the difference . . . between the price of ensuring the *Life* of a *Man of* 20 and 50." He laid out the mathematics of the obvious next question: how much one should pay for such insurance (or, for an annuity, payments over the years of life remaining to the investor).

* One of the earliest specialized lines of insurance bought and sold at Lloyd's was coverage of the slave trade, a reminder of the extent to which the early British Empire formed a transatlantic market in humans.

With that, Halley made the final leap. Up till then, every analysis, every example, had begun in the present and looked forward to the future: if I, sixty as I write this, want to know my chances of dancing at my son's wedding, I can look at a modern version of Halley's table and transform hope into expectation, present circumstances into some range of possible later outcomes. But when Halley calculated the price someone should pay now for a promise of money in the future, he was actually asking what mathematical relationship could connect an expectation that could come true years or decades down the road with a decision to be made in the here and now. This was like a present-value calculation—but with probabilities determining the end result. In other words: Halley's calculations were among the first to examine risk in a financial transaction.

The balance of Halley's report to the Royal Society on the Breslau data lays out in detail his mathematical approach to putting a price on insurance and annuities, taking into account the odds of having to pay out in any given year. He had managed to connect the first result he had extracted from Neumann's work, the odds or risk of dying over a given term, to a different equation, a compound interest calculation. His Breslau paper wasn't the first to investigate this concept, how interest, the return on a given investment—in land, or a joint-stock share, for example—builds on itself ("compounds") over time.* Newton himself had performed a similar calculation in his early mathematical explorations, and the concept had at least a century-long history within England alone, with European studies of compounding dating back at least to the fourteenth century.

* In an investment where the interest compounds, each new payment calculates the interest owed on the starting investment—£100, say—plus all the previous amounts of interest due. So if the initial investment had commanded a 5 percent return, then £5 would have been due at the first interest date. Add that £5 to the principal, and by the next due date the total investment would be £105 and the interest due would be 5£ 5s. . . . and so on.

But—as Halley himself remarked—it was only with the mathematical discoveries of the preceding couple of generations that it had become possible to compute compounding quantities easily. Compounded investments (or debts) update at regular intervals. Take an annual rate of 10 percent that compounds daily. That would mean a smidgeon of interest—1⁄365 of that 10 percent—would be added to the total each day and would in turn begin earning that same 10 percent rate. Repeat that every day as the account updates, and very soon it becomes clear that the effect of such compounding is profound: a 10 percent return, compounded daily, doubles an investment or a debt in a little over seven years. (That's what inspired the line, unreliably attributed to Albert Einstein, that the greatest invention of all time was compound interest.) Translated into a different mathematical form, Halley's manipulation of just five years of demographic records from a drowsy, distant town created the basic framework that could describe how much a payment now would be worth as an investment for the purchase of a policy for any number of years, based on an assumption of a given rate of return. It works like this: you, an insurance company, would want to know what you would have to charge today to make a profit on a promise of a £100 payout in ten years. You'd then invest that payment in anything you liked for those ten years, and the rate of return on that investment would create the pool of money to make the payout required when the policyholder died. So, assuming a (generous) 10 percent return, compounding quarterly, an initial payment of £38 would yield £102 at the end of a decade—leaving £2 over for you. If your customer lived longer, that meager profit would grow—quickly. (The year 11 return would bring in £10, the next year would add £12 more, and so on.) Of course, if the policyholder died too soon, the loss would bite. But if you spread out your risk over enough people so that your assumptions about how many of the demographic group—people of a certain age, gender, or what have you—would die in any interval of time would be more likely to be confirmed, then the risk of catastrophic losses would drop. And with that you've completed the simplest form of an analysis that modern insurers make for every policy they write.

HALLEY PRESENTED HIS results to the Society in March 1693. He published them a few months later in the Society's *Philosophical Transactions*. For his part, Halley seemed to draw moral lessons as much as practical ones from his investigation of Breslau's lives and deaths: "How unjustly we repine at the shortness of our Lives," he wrote, "and think our selves wronged if we attain not Old Age." We should, he wrote, think philosophically, that is, mathematically, and consider that, "instead of murmuring at what we call an untimely Death, we ought with Patience and unconcern to submit," adding that we ought to rejoice to reach "that Period of Life, where the one half of the whole Race of Mankind does not arrive."

For all its insight, though, Halley's actuarial work had very little impact on those in London's new insurance industry. The early experiments in life insurance were mutual aid societies: men would sign up to pay a monthly subscription that would earn a payout for their widows on their death. The first such enterprise, the Amicable Society for a Perpetual Assurance Office, started operations in 1706, with two thousand members paying in to launch the business. Anyone aged twelve to fifty-five (later lowered) could subscribe, and if they died, their heirs would get a payout at the end of the year as a share of a sum allocated to the survivors of every other member who had died, regardless of age or how long they'd belonged to the society.

Other schemes followed, but the idea of using math to set a rational price for an insurance policy remained almost exclusively the province of natural philosophers. It wasn't until 1762 that the first London-based company to adjust premium payments by age opened its doors, thus just beginning to make use of Halley's insights. Similarly, until near the end of the eighteenth century the British government did not use compounding calculations to help set the prices it charged for annuities—investments on which the government committed to making payments until the initial lender died. Instead, when it offered annuities—which can be understood (and modeled) as a kind of insurance—the Treasury made mistakes like charging the same price to a twenty-year-old buyer as to one who was fifty. That was clearly a better deal for the younger buyer—and a much worse one for the Treasury, which would want to minimize the total amount it would have to pay out.

But even if the nascent business of gambling on lives was unready for the rigorous methods that Halley and others had developed, the thinking behind that work was diffusing through more and more minds over the last decades of the seventeenth century. The passion for transforming experience into figures spread from places like the Royal Society, appearing wherever it seemed that a facility with numbers might put cash in one's purse at the expense of those slow to catch on to the mathematical measure of daily life.

This shift in how people thought about money can be seen in the person of Isaac Newton himself. By the 1690s, no one in Europe was more thoroughly identified with mathematical mastery. That was a recent revelation: after the sudden explosion of ideas in the plague years, and a brief series of exchanges with members of the Royal Society for a few years after that, Newton remained in Cambridge, working as he pleased on a range of questions with very little contact with the virtuosos in London. That near isolation lasted until that day in the summer of 1684 when Edmond Halley paid his fellow comet enthusiast a visit.

Halley had come to Trinity College to settle a question of the sort more commonly associated with the work of the Royal Society, a problem of cosmic motion: Did Newton know what orbit a planet would trace if it were bound to its sun by a gravitational pull whose tug depended on the square of the distance between the two bodies? This was, of course, the same phenomenon Newton had attacked two decades earlier, how bodies move under the influence of gravity. By now he was able to answer immediately: a planetary orbit traces an ellipse.

When Halley asked for a mathematical proof, Newton promised him one soon—and delivered. But he didn't let Halley publish that first brief sketch of what he realized could become a whole new science of motion. Instead, he spent much of the next three years developing the physics that governed the movement of matter, culminating in the series of demonstrations in which Newton showed how mathematics could analyze and predict the behavior of the entire known cosmos, from its farthest reaches to the ebb and flow of tides lapping against Dover's white cliffs.

That labor became the *Philosophiae Naturalis Principia Mathematica—Mathematical Principles of Natural Philosophy,* better known simply as the *Principia.* It is—correctly—recognized as the culminating work of the seventeenth-century scientific revolution, both an explanation for a host of physical phenomena and a demonstration of a method and a worldview that drives the investigation of the material cosmos to this day. But as vital as its demonstration that problems of motion could be resolved by mathematics, so was the persuasive rhetorical power of Newton's enormous achievement. Once the *Principia* decisively demonstrated that nature obeys number, claims by people like Petty—and soon, Halley—that *human* nature must do so as well took on much greater force.

The *Principia* was an instant sensation. Its first printing in 1687 of roughly 350 copies sold out immediately, and its author, having spent decades hiding in Trinity College, was equally swiftly recognized as *the* thinker of his generation. (Samuel Pepys was then president of the Royal Society; he authorized the book's publication in 1686—which landed his name on the title page.) In 1689 Newton came to London to reap the rewards of fame, being elected as one of Cambridge University's representatives to the Convention Parliament that would transfer power from the deposed Stuart king, James II, to the dual monarchy of William and Mary. Newton enjoyed his time in the big city; his encounters with men like John

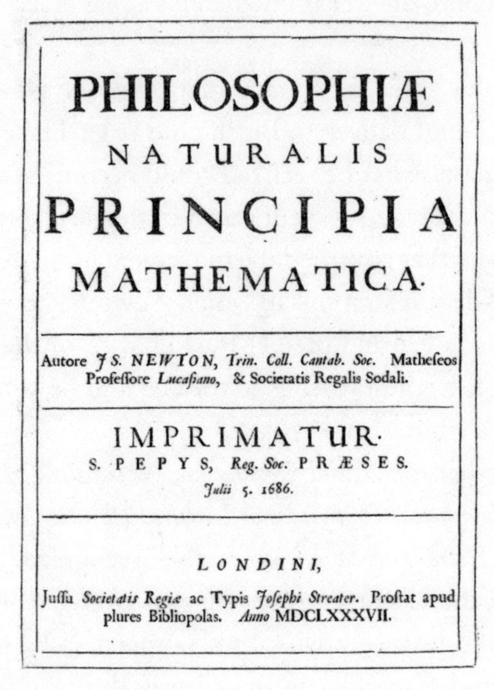

PHILOSOPHIÆ
NATURALIS
PRINCIPIA
MATHEMATICA.

Autore *J S.* NEWTON, *Trin. Coll. Cantab. Soc.* Matheſeos Profeſſore *Lucaſiano*, & Societatis Regalis Sodali.

IMPRIMATUR.
S. PEPYS, *Reg. Soc.* PRÆSES.
Julii 5. 1686.

LONDINI,

Juſſu *Societatis Regiæ* ac Typis *Joſephi Streater*. Proſtat apud plures Bibliopolas. *Anno* MDCLXXXVII.

The title page to the first edition of the *Principia*

Locke, Robert Boyle, Christopher Wren, and many others were vastly more interesting than anything the confined and often dull company in Cambridge could offer. It's a matter of record that on his return to Trinity College after his parliamentary term ended he implored his new friends to help him find a job that would get him back to London.

It would take five years for them to find him a suitable post, but in the meantime he stayed in touch with the contacts he'd made during his year at the Convention Parliament. One of those correspondents was the ubiquitous Samuel Pepys, diarist, voracious conversationalist, and gambler. In November of 1693—not long after Halley published his Breslau findings—Pepys wrote a brief series of letters to his new friend that illustrate just how swiftly the realization that math could make sense of daily life had penetrated even fashionable London. In those messages, Pepys sought Newton's help to win a game of dice.

Newton took the question very seriously, responding in a series of three letters over the next month that calculated specific answers for each of the three different scenarios Pepys proposed. To no one's surprise, he got his sums right and showed that Pepys was a lousy gambler: he had laid his bet at the worst odds and was likely to lose his stake. Armed with Newton's analysis, he announced he would renege on his wager and refuse to pay.

Samuel Pepys around 1690

One lesson to draw from Pepys's cheerful willingness to dodge his debt was that you never know who's going to pull a fast one. Another, certainly, was that Newton really could find his way around an equation. The deeper insight, though, flows from the fact that Pepys knew to ask the question in the first place. Gamblers in his day and many since would speak of "hazarding" one's fortune on a throw or having a fling, or perhaps would invoke God's kindness ("Baby needs a new pair of shoes!"). Such punters abandon themselves to chance, to forces beyond their control or ken. Pepys, by contrast, expressed a radically different and in 1693 a still new notion of risk, gain, and loss. He refused to rely on his flawed intuition, not when he knew someone who could work out the probabilities that could guide his bet with exquisite precision.

This wasn't Newton's first excursion into questions of probability, risk, and chance. In 1670, in the midst of a series of subtle and difficult observations of optical phenomena, he faced the problem of experimental error: how to reduce the probability that his measurements missed the true value by more than a certain, acceptable amount of imprecision. His solution then was to perform experiments multiple times and to average the results to produce the strongest possible claim that what he had observed was real. That attempt to quantify and constrain error was a radical move, a critical step toward the modern scientific method, and he appears to have been the first to address such uncertainty—the chance that any given measurement might be wrong—in this way.

Newton sent Pepys the last of his three letters on odds in December 1693. Halley's work on mortality and the price of lives appeared a few weeks later. As they both wrote, others applied the same kinds of arguments to very different questions. In the mid-1690s, the political arithmetician Charles Davenant made a probabilistic argument for believing in the results of political arithmetic, writing, "*Very probable Conjectures* may be form'd, where any certain Footing can be found, to fix our Reasonings upon," even or especially when one is analyzing such seemingly unphilosophical questions as the value of the East India trade. As the historian Carl Wennerlind points out in his masterful *Casualties of Credit,* this association of probability and knowledge was a qualitative argument. For example, Wennerlind describes the philosopher John Locke's probability-centered theory of knowledge as a four-tier hierarchy of judgment, an approach to using reason to analyze the evidence supporting any given claim. There was an awareness of risk, that is, in Locke's arguments, or Davenant's, but not yet a necessarily mathematical account of it. Except when, like Pepys, one could draw upon a true expert to help work out gambler's odds, most of Newton and Halley's contemporaries drew on the *idea* of systematic analysis of uncertainty but not on the evolving mathematics of such analysis.

But by the 1690s the notions of risk, moment by moment (as in Newton's response to Pepys's need to analyze each roll of the dice), and over time, as in Halley's quantitative analysis of what a life is worth over an uncertain future, helped create a formal framework for thinking about money. For all that the insurance industry itself would not take advantage of these new analytical tools for almost a century, the central idea seeped into public consciousness much more quickly: those in London who made their living buying and selling grasped that it was now possible to think of money not just as a fixed total but as a quantity that could evolve over time, gathering returns, encountering risk—and that could hence be studied, modeled, as a mathematical problem in units of pounds, shillings, and pence.

THE HISTORICAL MOMENT demanded such radical ideas. By 1695, England had been at war with France for six years. Paying for an army in the field in Flanders was proving ever more difficult, to the point where the English faced a plausible risk of financial defeat, a surrender forced not by any triumph by Louis XIV's soldiers but by simple lack of cash. Several expedients had been tried over the preceding few years. Finally, in September 1695, the Treasury sought outside help to deal with one aspect of the crisis: the disappearance of England's hard currency due to its stock of silver coin flowing across the Channel to Paris and Amsterdam. Among those consulted were four fellows of the Royal Society: Davenant, Sir Christopher Wren, the philosopher John Locke—and Isaac Newton. Here the scientific revolution, in the persons of some of its greatest minds, collided directly with a financial transformation, at a moment when the fate of a nation hung in the balance.

Chapter Four

"mere opinion"

Isaac Newton had first made his way down Cambridge's main avenue on June 4, 1661. It was late, and most of the gates along the road were already shut against the night, so that the nineteen-year-old refugee from a Lincolnshire farm had to wait until the next day to present himself at the College of the Holy and Undivided Trinity. Having been entered in the college register, he would then have passed into the Great Court on his way to what would remain his home—barring the plague season—for the next thirty-five years.

On March 19, 1696, that impoverished undergraduate, long since transformed into the Lucasian Professor of Mathematics, received a letter from Charles Montagu, Earl of Halifax and chancellor of the Exchequer, offering "to make Mr. Newton Warden of the Mint"—a post that came with a house and a salary four times what he earned as a professor.

The letter liberated Newton. University life had become intolerable. As the story goes, Newton was walking outside Trinity College after the *Principia* was published, when a passing student said, "There goes the man that writ a book that neither he nor anybody else understands." Even if the incident gained in the retelling, it captures what Newton felt after his sudden rise to celebrity in the capital. He was bored; he was lonely; he was, he felt, profoundly undervalued—and as long as he remained mired in Cambridge, he would remain so.

So when Montagu's offer arrived, Newton could not move fast enough. He left for London on March 23, met with Montagu no later than March 25, and, that same day, received his official appointment as an officer of the Royal Mint. He returned to Cambridge almost immediately and settled his affairs in less than a month. He left again for a job seemingly as far removed from the life of a professor as it was possible to imagine. No letter has been found written to anyone he had so swiftly left behind.

Initially Newton's new life was largely confined to the Tower of London. The inner walls, court, and towers of the old fortress remained the province of the army, but the outer wall housed the Mint's narrow warren of offices, workshops, and stables for the horses that powered the largest machines required to transform molten metal into coins. When he arrived, Newton lived in the warden's official residence, backed against the northeast corner of the complex, until its narrow rooms looking out to the tall, blank, sun-blocking face of the Tower's inner wall drove him to a more pleasant situation in Westminster, near St. James's Church.

As a professor, Newton had been required to deliver just a single series of mathematical lectures once each year. At the Mint he became, in effect, the operations manager for what was almost certainly the largest metalworks in England. The warden wasn't the most important Mint official. By long-standing practice, the master ran the actual production of coins, earning a percentage of every coin struck on his watch. The warden governed the mint's physical plant, which is to say that officially, Professor Newton had to make sure the holes in the roof were patched, the horses were fed, and the machinery in the workshops was maintained in good running order. That formal list of duties wasn't supposed to trouble the (presumptively) unworldly new warden. For the better part of a century, almost every incumbent had treated the post as a sinecure, leaving the actual work to deputies. His patron, Montagu, had told Newton as much: he would, he promised, face "not too much bus'nesse to require more attendance than you may spare."

Newton ignored the precedent. He took his oath of office on May 2, and from that day forward his name appears constantly in Mint records. By midsummer, more or less by force of personality, he seized control of the Mint's coining operation from the current master, the gambler, spendthrift, and speculator Thomas Neale. It wasn't really a fight, and there's no evidence that Newton directly challenged Neale. Rather, he just started doing the work that Neale avoided, and in very rapid order the Mint's officials simply looked to him, and not their nominal chief, for orders.

This bureaucratic coup was perfectly timed: he had taken charge just as the English currency system was at the point of collapse.

KING WILLIAM HAD adverted to what was about to happen six months earlier, in his speech at the opening of a new session of Parliament on November 26, 1695. William had led England into war with France in 1689. Now in its seventh year, that war showed no signs of coming to an end anytime soon. It was hugely expensive—so much so that the king reported that Parliament's prior grants of "so many, and such large, Aids" had run out, and that "the Funds which have been given, have proved very deficient." It was hardly news, even then, that rulers might lead their realms into fights they couldn't afford. But it wasn't just the price tag that bedeviled the English military effort. There was, William reminded Parliament, a more fundamental problem. The nation's money was no good. The silver coins that formed the basis of England's currency system were either hopelessly debased—containing much less precious metal than the law promised—or simply missing, not to be found in circulation. That—the king implied—meant the nation might go down to a *monetary* defeat in a winnable war, all because of the "great Difficulty we lie under at this time, by reason of the ill State of the Coin."

William hadn't exaggerated. Shortly after Newton came to the

Mint, he found that more than one in ten silver coins were counterfeit, while many of those that were genuine had been clipped down to far less than their legal silver content. Combine the fakery and the extent to which legal coins had been debased, and the depth of crisis becomes clear: the money used for all daily exchange, the day-to-day business of the nation, contained less than half of the silver required to make up its face value. The danger was obvious and was being played out in markets and shops all over England: Who would willingly accept a shilling coin that contained less than a sixpence worth of precious metal?

At peak output the Royal Mint needed fifty horses on its production line.

The edging machine applied a key anti-counterfeiting measure.

Two fundamental causes had combined to rob England of its metal money. One emerged in 1662, when the Royal Mint installed the first machine-driven coin production line. The process began with giant, horse-powered rolling mills to flatten sheets of silver and gold—the raw stock for the lever-driven punching machines next up the line. These produced coining blanks to be passed to the most carefully guarded of the Mint's apparatuses, edging machines that etched difficult-to-copy deco-

The coining capstan

rations around the rim as an early security system for the coinage. Finally, a capstan-driven press, powered by four men, finished the coins, ramming top and bottom dies into the blank with enough force to drive a deep design into both faces of the finished coin.

The newly mechanized mint produced beautiful money. As intended, the machine-inscribed edging made it extremely difficult to debase the currency by clipping or trimming coins—the process by which, by the 1690s, currency criminals had stolen roughly half of the metal originally minted into England's coinage. Similarly, while skilled metalworkers could make passable fakes, it was, at least, much harder than before to do so, given the depth of the designs struck into each face of a coin. There was only one problem: older, handmade coins dating back to the reign of Queen Elizabeth and before remained legal tender. Those hand-hammered pieces suffered all the ills the machine-made currency was intended to cure. They were relatively simple to fake, and, critically, these older coins could be easily clipped, with the trimmings melted down into ingots to be sold as raw metal. And yet, by law, a misshapen, thin, underweight silver shilling with a barely legible portrait of James or Elizabeth was worth exactly the same as a shiny, bright, full-weight one struck on the new assembly line. From the start, then, it made obvious sense for all those who laid hands on one of the newer coins to hoard it as a reserve of silver bullion and to spend only their stock of increasingly debased pre-1662 money.

That would have been bad enough: every year some fraction of the new, machine-edged, full-weight coins minted at the Tower would drop out of circulation. But then it got worse. As the decades passed, England grew ever more enmeshed in what was increasingly a global network of trade, and by the 1680s the exchange rate between silver and gold differed from place to place in Europe. Critically, if you melted down the silver in a full-weight, post-1662 English coin, it would buy more gold as bullion in Paris and Amsterdam than the face value of that coin would say it was, measured by the gold in English guineas. In other words, treated simply as metal, every new shilling was worth more in Paris than it was in London.

The inevitable followed. A complaint by the goldsmiths' guild in 1690 claimed that in just six months traders had sent 282,120 ounces of silver across the Channel, enough for about 10 percent of the total Mint production of the previous five years. Some of that weight might have come from candlesticks and silverware, but a parliamentary investigation confirmed that most of it had started out as the king's coins, melted down and smuggled out. Destroying legal currency was a crime, but that didn't deter those who realized they now possessed the financial equivalent of a perpetual-motion machine, one that could turn English silver into foreign gold that could buy more silver coins that would yield yet more gold, as long as England kept on producing machine-made coins that could be ferried across the narrow sea.

THIS WAS GRESHAM'S Law—bad money drives out good—with a vengeance: clipped, underweight currency and fakes were literally driving the realm's full-weight modern coinage out of the country. As the Victorian historian Lord Macaulay would write, the Exchequer received as revenue no more than ten good shillings in a hundred pounds of tax payments—one in two thousand by value: "Great masses were melted down; great masses exported; great masses hoarded; but scarcely one new piece was to be found in the till of a shop or in the leathern bag which the farmer carried home after the cattle fair."

It was a perfect example of what economic historians have dubbed a crisis of small change. The silver coinage was both the legal tender of the realm—you could pay your taxes in silver—and a basic, daily currency with which people paid for their beef and beer and everything else that makes up the financial lives of most people on most days. There were twelve pence (pennies) in a shilling, twenty shillings in a pound—and then there were the gold coins, guineas and half guineas, not yet fixed in their value in silver money. A laborer earned about just over a shilling a day, while a skilled worker might command as much as a pound a week. At the height of the currency crisis, a golden guinea traded for about thirty silver shillings—one and a half pounds, and close to a month's wages for those at the bottom of the working world. Meanwhile a pound of steak at Spitalfields market cost about three pennies, and a gallon of ale could be bought for a shilling or less. For a very rough analogue, imagine trying to get through a day with nothing smaller than thousand-dollar bills or five-hundred-pound notes (and no credit cards). It wouldn't—it didn't—work. Without an adequate supply of small denominations, the silver coinage that was the engine of daily life, trade suffered and then almost stopped: "Nothing could be purchased without a dispute," Macaulay wrote, and "the simple and the careless were pillaged without mercy." Worst of all, there was seemingly no prospect that the problem would be solved. The Mint made nearly half a million pounds' worth of silver currency between 1686 and 1690. But so much silver poured out of England in the next five years and so little bullion was available that between 1691 and 1695 the Mint made just a total of £17,000 in new silver coins.

That shortage of good coin was enough to threaten national finances. English subjects paid their taxes in silver money—and anyone who could would offer the debased older coins to make up whatever sum was due. At its height, the war with France consumed 80 percent of the government's income, and the debased state of English currency meant no foreign banker would accept the bad old shillings at face value. In the summer of 1695, King William's army was poised to take the Belgian fortress city of Namur. Absent a ready supply of money, though, it wasn't clear that the English could keep their army in the field. The army's paymaster, Richard Hill, tried to raise a loan in Brussels, but the sorry state of English money and credit forced him into months of work to secure the relatively modest amount of 300,000 florins. In that delay, French artillery nearly caught up with him, as he wrote, "I stayed so long about it that the house I lay in was beaten down about an hour or so after I got out of it."

Hill survived, the money made it to William and the army, and

Namur finally fell on the fifth of September. But it was clear to every informed observer in London that the nation was on the ragged edge of monetary disaster. That was when, in September, the Treasury secretary, William Lowndes, reached out to his handpicked group of wise men—among them, Isaac Newton—for advice. Newton's response would put him at odds with other, more experienced financial thinkers and, read in hindsight, offers a glimpse of the comprehensive, radical shift in the meaning of money taking place behind the rage of war and the murmur of the market stall.

LOWNDES'S QUESTION WAS brutally simple: How could England's silver coinage be saved? Newton's answer was equally straightforward. The nation's stock of coins needed to be rebuilt from scratch, with the two parallel forms of currency—the modern, machine-made version and the old, hand-hammered, debased one—replaced by a single, consistent standard. On that point, more or less everyone Lowndes queried agreed: England needed to undertake a complete recoinage, calling in all the old silver money, melting it down, and reminting it with the anticlipping and counterfeiting tricks deployed by the post-1662 coins.

But once the overhaul of the money supply was agreed to, at least in principle, the next decision was far less clear. A single, properly manufactured coinage would address the clipping problem that had cost the old coins half their silver. But as long as the new coins were made to the then-current specification, they would still contain more silver than their face values were worth in gold—and would thus be just as likely to wind up across the English Channel as existing machine-made money. That's why Newton (and Lowndes himself) argued that any recoinage would have to reduce the amount of silver in each denomination of coin. Newton wrote back to Lowndes that it was essential "to make Milled money constantly of the same Intrinsick & Extrinsic value, as it ought to be and thereby to prevent the Melting or Exporting it."

That is, instead of leaving England's money with two different possible values—the "intrinsic" market price of its metal and its "extrinsic" denomination as some number of shillings and pence—those separate measures must be brought into agreement. To make this happen, Newton proposed raising the face value of the coinage by about one-quarter, so that, for example, the crown, a five-shilling piece in the old system, would now be worth six shillings three pence. Critically, Newton argued, "In Stating the Value of Gold and Silver, care should be taken that they bear nearly the same proportion to one another at home and abroad"—thus eliminating the price differences exploited so successfully by the coin-and-bullion smugglers.

To modern ears, Newton's proposal sounds utterly conventional. If something that can be traded is worth more in one setting than another—what would now be called an arbitrage opportunity—simply adjust the price until the disparity disappears. In London in 1695, though, this was genuinely disruptive reasoning. By Newton's logic, units of currency—shillings, half crowns, guineas—were not immutable statements about how much any given amount of precious metal contained in a coin was worth as a shilling or a pound. In his response to Lowndes, Newton didn't state this as clearly he would a few years later, but the implication of his argument was plain: "Tis mere opinion that sets a value upon money," he would write. "We value it because with it we can purchase all sorts of commodities, and the same opinion sets a like value upon paper security."

Mere opinion? Fighting words! John Locke, Newton's friend and—in this instance—his intellectual antagonist, concurred with Lowndes and Newton and others that England needed to remake its silver coinage. But, he argued, the number on a coin—a shilling, five shillings (a crown), or whatever—was a promise: that the coin contained the lawful weight of silver of a given purity. Changing the number associated with a coin, calling a crown-weight piece of silver seventy-five pence instead of sixty, Locke wrote, would not make that coin more valuable—contain more bullion—than it had before.

"I am afraid no body will think Change of Denomination has such a Power."

So far, Locke merely restated the obvious: a shilling minted to Newton's new scale would contain and buy less silver metal than the older standard would. That was true enough, but beside the point. Silver flowed into European cities because of the arbitrage opportunity: English silver coins bought more gold in Paris than they could command in the amount of gold contained within a guinea of the same face value. That meant that English silver could buy gold in France that, when brought back to London, would buy more silver shillings than it had taken to get that gold in the first place—a pure arbitrage play. Locke didn't care. He offered instead a subtle argument. Where Newton sought to force the silver coinage to adjust to shifts in the market for silver and gold bullion, Locke argued that silver, and silver alone, had a special property that made it the sole standard, the foundation on which England's money rested. He wrote, "Some are of the opinion that this measure of commerce [the coinage], like all other measures is arbitrary, and may at pleasure be varied, by putting more or fewer grains of silver in pieces of a known denomination." Not so, he claimed: "But they will be of another mind, when they consider, that silver is a measure of a nature quite different from all other." It was, he said, "the thing bargained for as well as the measure of the bargain"—the one fixed point against which the value of everything else, including gold, should be measured. Silver money, according to Locke, derived value not simply from the amount of metal within each coin but from its official, almost sacred nature as a unique measure of value.

Locke and others raised a number of other arguments in defense of the old monetary standards, concerns about the effect on prices, whether a devaluation might set a dangerous precedent, and more. But the essence of the battle came down to Locke's view of what money was against Newton's. For England's silver-based coinage, that meant that the amount of silver, rounded and flattened and edged and struck with the monarch's likeness, was *real,* immutable.

As such a concrete fact, it could no more shift in its essential properties than a cat could birth a cow.

Newton's version of money was an altogether different species from Locke's. "Opinion" set money's value, he said—which is to say that the worth of a shilling should be understood in relation to the things it could be exchanged for: how much beef it bought, or gold, today, tomorrow, whenever. In modern terminology: a coin with its statutory amount of silver was worth what it could buy. Money made of precious metal, that is, was never simply a unit of value; it was also always a medium of exchange, with what could be thought of as its price also determining the prices of anything else one might buy or sell.

TO THE HARD men of monetary politics, though, for all that Newton's logic may have made sense, Locke's touched the heart of the matter. Devaluing English coins would cost a very particular group of people a great deal of money. If new shillings were suddenly to contain about 20 percent less silver than the coins that preceded them, landowners would lose that much out of their rents. Taxes, customs payments, and excise payments to the crown would face a similar cut, measured in the bullion contained in money. As Locke put it, devaluation would "only serve to defraud the King, and a great number of his subjects, and to perplex all." (The fact that as long as clipped money circulated, government takings were about half their legally mandated weight of silver was ignored in the political maneuvering.)

Unsurprisingly, Locke's view prevailed. Parliament approved the recoinage on January 17, 1696, mandating that the new coins conform to the standard weights of the prior issue—meaning that once again the new coins would hold more silver, measured in the amount of gold it could buy in Amsterdam and beyond, than they were worth when exchanged for gold coins in London. The process was supposed to begin almost immediately, which put the feckless Neale in charge. It was a disaster. The Mint produced only a pittance under his direction until, beginning in May, barely any small change was circulating in England. "No trade is managed but by trust," Edmund Bohun, formerly the licensor of the press—the official censor—wrote to a friend. "Our tenants can pay no rent. Our corn factors can pay nothing for what they have had and will trade no more, so that all is at a stand." This wasn't mere inconvenience, Bohun went on; real misery followed the dearth of coin. "The people are discontented to the utmost; many self murders happen in small families for want." In Plymouth, the government tried to pay the army in old, worn coin but backed down at the threat of mutiny, satisfying soldiers with provisions instead of cash. The diarist and scholar John Evelyn feared the worst: "Tumults are every day feared, nobody paying or receiving money."

In the nick of time, enter Isaac Newton. Neale had no chance against his intelligence, preparation, and simple presence. The new warden showed up at the Mint more or less every day; Neale did not. By early summer, the coup was complete. The Treasury had set Neale a production target of £30,000 to £40,000 a week, which, as one clerk wrote, "was looked on as a thing impossible." With Neale out of the way, Newton ramped up the production line, ordering eight new rolling mills and five coining presses. At the height of the recoinage, in late 1696 and through 1697, Newton commanded about five hundred men and fifty horses driving the giant rolling mills. To ensure that his army of laborers wasted none of their efforts, he conducted perhaps the first time-and-motion study on record. As he observed, it took "two [rolling] Mills with 4 Millers, 12 horses two Horskeepers, 3 Cutters, 2 Flatters, 8 Sizers one Nealer, three Blanchers, [and] two Markers" to move enough silver from the melting rooms all the way down the line to feed two coining presses. Each coining press consumed seven men more—six to turn the capstan bars that drove the die onto a blank while one brave man fed those disks into the striking chamber itself.

As he deployed his forces, Newton imposed the same empirical rigor on his new job as he had with his pendulums and prisms. The Mint could not operate any faster than his men could spin their capstans, and every other step had to be timed to match the work of his presses. So Newton watched to "judge of the workmen's diligence." He saw how quickly the brutal effort needed to turn the press wore out its team. He observed just how nimble the man loading blanks and pulling finished coins from the press had to be to keep his fingers. Eventually, he identified the perfect pace: if the press thumped just slightly slower than the human heart, striking fifty to fifty-five times a minute, men and machines could stamp out coins for hours at a time. By autumn, Newton had the Mint's output up to £100,000 every working week—a century ahead of Adam Smith, and more than double again before Henry Ford showed the world just how powerful time-and-motion rigor could be.

Newton continued to drive his horses and men for the next two and a half years until the nation's entire silver money supply had been remade. In all, under his command, the Mint recoined over £6 million—£6,722,970 0s. 2d., to be exact. As that last tuppence indicates, Newton, having spent the whole of his prior life as an essentially solitary thinker, proved to be a truly extraordinary administrator, bringing the effort home with accounts accurate to the penny and stunningly free of corruption.

That accomplishment, brilliant by any technical measure, was in a larger sense a failure. Newton the counselor had anticipated the fate of Newton the warden's best efforts: making coins that were worth more as metal than their face value could buy was a fool's errand. The illegal money trade across the Channel resumed, ultimately forcing Britain—after Newton was able to revalue the golden guinea coin in 1717—to switch from silver to gold as its currency standard. The operation had been successful, but the patient died.

But even if the recoinage provided only a short-lived reprieve for the silver coinage, the debate that it had sparked reveals how swiftly financial ideas were changing in London. Most important: a range of thinkers and gamblers were discovering that there was more than one way to increase the stock of what could pass as money in a nation. As Newton was managing the manufacture of almost £7 million of new coins, the English government matched that achievement by manipulating not metal but paper in a new species of loans that brought almost exactly the same sum onto the Treasury's books.

THIS BORROWING WAS something genuinely new, even revolutionary. It may have resembled the old habit of kings borrowing to pay for wars they couldn't otherwise afford. But what the English Treasury set in motion in this period shifted ideas about money—away from a simple connection to the material world, to lumps of valuable metal or to the sheep and grain that monarchs could raise on their own lands or tax away from their subjects. In place of such mundane sources of value, English financial authorities created newly abstract forms of money to cover the cost of William's wars, sums increasingly divorced from a direct connection to a coin jangling in a purse. Each pound the Treasury borrowed became for a lender a stream of income—the interest to be paid every year on that borrowed money, payable to anyone willing to trade hard cash for a piece of paper.

A notion of money defined as a stream of income could be interpreted—manipulated—by mathematical models like those that men like Halley and Newton himself had already used to investigate the behavior of any system that evolves over time. In the beginning, in the 1690s, such a science of money was still very much an aspiration. But the urgency of the moment drove what was becoming a financial revolution forward, as England's monetary officials turned to more abstract—and more complicated—ways to extract funds to spend right now by the use of new forms of credit built on "mere opinion."

The driver behind such inventions was the fact that the Nine Years' War was not an isolated event. It was, rather, the first episode in an almost constant century of war. From William and Mary's coronation in 1689 to Napoleon's final defeat in 1815, soldiers and sailors from Britain and France fought each other for 76 of those 126 years. England had never been a stranger to bloody business, but the conflicts William and his successors waged were very different from those that had come before, in both the space they covered—the whole globe—and the numbers involved. In the last decades of the preceding dynasty, the Stuarts had kept only about fourteen thousand men under arms. Following the Glorious Revolution, the average muster of the army William sent to fight the Nine Years' War topped seventy-six thousand.

That was the challenge. England had to pay, feed, clothe, and arm its burgeoning military complex. The process required a second army of clerks, bookkeepers, men armed with pens and dispatches, all needed to maintain the apparatus of large-scale conflict. They managed everything, from the brigades of ovens required to produce a loaf of bread per man per day, to the baggage trains that could haul the half ton of stuff it took to maintain each soldier in the field. Every last ounce of it cost money, which had to come from somewhere. The question was where? And even with a reformed currency after 1698, the answer was . . . not obvious.

Bluntly, William had taken on more than he could handle. Under his predecessor, James II, England had spent about £2 million a year. In the Nine Years' War, annual spending ranged between £5 and £6 million—and even though the new monarchy managed to raise more revenue than the Stuarts had, official income averaged just £3.64 million per year. Over the entire war, the Treasury paid out over £49 million and took in just under £33 million, barely two-thirds of what was needed. As the political arithmetician Charles Davenant wrote, "The whole Art of War is in a matter reduced to Money" and "the Prince who can best find Money to feed, cloath, and pay his Army, not he that has the most valiant troops, is surest of Success and Conquest." If England's armies—unpaid, unfed, and unequipped—were not simply to melt away before the contest could be settled in the field, the king's ministers, aided by his cleverest subjects, would have to raise unprecedented amounts of money, much, much more than any prior English government had commanded.

They did just that, and in doing so bought themselves time—and then years of trouble.

ENGLISH MONARCHS HAD always borrowed. Even for the most prudent of rulers, governing costs money every day, while revenue from taxes, fees, harvests on crown lands—or forfeited property seized from unruly lords and too-worldly abbots—arrives only in its season, and not necessarily when it's needed. In extraordinary moments, times of war, the need to come up with ready cash—to borrow—only intensifies. Though, as it evolved, England's Parliament imposed restrictions on the crown's prerogatives, including its ability to extract money from the realm by fiat, still, before the Glorious Revolution, those debts ultimately fell to the English throne and its occupant. The layers of officials and expedients between a ruler and his or her finances grew more elaborate over the years, but the underlying principle remained: if a king or queen's government spent more than it possessed, it would have to borrow, and, in principle, the ruler him- or herself was ultimately responsible for making the creditors whole.

Some failed. The resumption of the Dutch wars left King Charles II's treasury unable to cover its bills in 1672—leading to the infamous Stop of the Exchequer, in which government suppliers and workers were told they could not be paid for the foreseeable future. It took decades to settle the claims produced by the Stop—and even then, the luckless lenders to the crown had to settle for just ten shillings to the pound, a cut of 50 percent of what they were due. Even though candlemakers and coopers and all the rest would never have gotten close to their nominal customer, ultimately they were owed—and stiffed—by the monarch himself.

In this context, the Glorious Revolution was much more than a mere dynastic change, a shift away from an unacceptably Catholic and autocratic monarch to his Protestant daughter and her husband.

In 1689 the Convention Parliament offered a bargain to James II's would-be successors, the Dutch stadtholder, William, and his wife, Mary Stuart, James's elder daughter. England's new rulers—and all their successors—had to surrender any claim to the kind of absolute power asserted by the Stuarts. Instead, English sovereigns were required to "Governe . . . according to the Statutes in Parlyment Agreed on." Most important: they were explicitly forbidden to raise money without permission from the legislature. Henceforward, Parliament would be the sole source of public funds. While this would seem (and was) a clear loss for any ruler who, like the Stuarts, asserted unrestricted freedom of action, it put more power into the hands of the state as a whole: the entire apparatus of ministers, legislators, and their emerging bureaucracies. If and when Parliament allowed the government to spend money, the authority behind it was now national, in the form of an at least somewhat representative legislature, and not just of an executive personified in the single figure of the monarch.

Kings still mattered, of course, and William could and did use his considerable royal power and influence to lead his domains into European conflicts. But the new arrangement meant that in early 1692, three years into the Nine Years' War, it fell to the House of Commons to figure out how to pay the swiftly ballooning bill. Ten members were directed to solicit proposals, and they reported back on two, both of which they deemed unacceptable. One tried to use the current crisis to settle an old score, seeking a full payment for the remaining money due after the Stop. That was an attempt to get a new ruler to pay a prior king's debts, which was unpalatable to many members of the House of Commons. The other scheme, submitted by William Paterson, a serial financial projector (as seventeenth-century entrepreneurs were called), was not, at first, entirely clear to the members. The committee asked Paterson to resubmit his idea, telling him to come up with something that would behave like what we would now recognize as a government bond, offering a "Fund of Interest . . . where they [the creditor] might assign their Interest, as they please, to any who consented thereto." In other words: create a kind of loan that people would be able to buy from or sell to others. Paterson accepted the challenge and came back with an unusual form of lending called a *tontine*.

A tontine, named for its inventor, the Italian banker Lorenzo de Tonti, is the bastard child of a bond and a kind of insurance, its payouts triggered by the deaths of individual bondholders. Investors paid in their money in return for interest payments; Paterson proposed the rich figure of 10 percent per year, later lowered to 7 percent. Those payments would be guaranteed by a presumably secure stream of government revenue, new levies on beer and other drinks. That was the bond or loan side of the new financial instrument. The insurance came with what would happen over time: as each bondholder died, his or her interest payments would be added to the returns paid to surviving tontine investors, increasing the payout with every death until the last investor was buried. The principal, the loan itself, would never be repaid, but those who bought in and lived long enough could do very well indeed. Best of all, each share of the tontine could be sold to someone else, allowing original investors a way to cash out at any time they chose.

Tontines had been attempted in the Netherlands and France but never in England. There, the scheme had to be modified so that investors intimidated by the complexity of its evolving payout could choose a much more bond-like experience, in which they received annual payments at a fixed rate of 14 percent that would expire when they died. (The higher rate was the alternative to the gambler's chance at the larger payoffs to the longer-surviving tontine adventurers.) With that amendment, Parliament swiftly passed the measure with a target to be raised of £1 million, which received its royal assent on January 26, 1693.

As it turned out, the tontine structure was not terribly attractive to the English public. Almost 90 percent of the investors opted to receive the guarantee of the annuity, with its set, high rate of return, and tontines were rarely attempted again in Britain. Even then, placing the loan was slow going; it took until February 1694 to raise the full million pounds. The next version of this same experiment went better. The ubiquitous Thomas Neale was a better pitchman than the more sober Paterson. He was a gambler to his core, from his days of chasing sunken treasure through his role as the king's groom porter, in effect the court's gaming floor boss. The fiscal crisis of the early '90s played to his intimate knowledge of his fellow Englishman's love of a flutter. To feed the Treasury's need for more cash, he proposed a scheme he called the Million Adventure. The Adventure was a lottery, offering one hundred thousand £10 tickets in a drawing for a top prize of £1,000 a year for sixteen years. Neale's twist was that after the prizes were distributed the tickets would turn into bonds, earning 10 percent annually for that same sixteen-year term. From the government's point of view, this was a very expensive proposition—but given the urgency of the war, Treasury officials had almost no power to bargain for lower rates.

Neale had a fine sense of his countrymen's appetites. All one hundred thousand of the Million Adventure's tickets sold swiftly, the lottery was drawn and winners were named, and the Treasury got its million pounds, all in ample time for the summer campaigning season. But there were troubling hints for anyone paying close attention. The rates of return—double-digit interest on both the first two loans—reflect the sense of crisis evoked by the unprecedented cost of the war. But in the moment, what mattered was that England had discovered something genuinely new that could pay for its ambitions: in place of the personal obligations of untrustworthy kings, these Parliament-authorized and -backed loans amounted to the first true and permanent national debt.

The success in raising those first two loans produced the inevitable sequel. A couple million pounds was great—but the fighting was still going on, and the escalating costs of the army deployed in the Low Countries soon exhausted those funds. One response was to create the Bank of England, which opened its doors as a joint-stock company on July 27, 1694. It existed for a reason: the price of its charter was a £1.2 million loan to the state. That requirement turned the fledgling institution into something new on Europe's financial landscape. Because those who made deposits with the bank could use the banknotes they received in return as cash in any transaction where the other party would accept it, the Bank performed what we now call fractional reserve banking. That is: previously, nationally chartered banks like the Bank of Amsterdam had functioned as pure deposit institutions. You could hand in gold and get paper receipts for your deposit, which could circulate as money—an early banknote. But that was as far as Amsterdam bankers were willing to go.

This early Bank of England note features a beehive in its logo (upper left)—a symbol of industry.

The Bank of England, by contrast, both gave its depositors banknotes in return for deposits and could turn around and lend that same money to the government (following a practice already used by some private goldsmiths' banks). That meant that the amount of hard currency actually held in the Bank's control was only a fraction of the amount of outstanding paper—notes and loans—that it issued; it held only a *fractional* reserve of gold and silver to cover all of its obligations. The assumption was (and is) that everyone won't want to pull their money out of the bank at the same time, and it is thus possible to put the same pound or dollar of deposits to more than one use.* Such a fractional reserve approach puts more money into circulation, and it was one of the key innovations that helped William pay for his war.

On its own, though, the task of keeping the British Army in the field kept running ahead of the Bank's ability to raise new loans. So the Treasury continued to follow Neale's lead and use any expedient to extract funds from the public. In all, using slightly different forms between 1693 and 1698, England was able to borrow just over £6,900,000. It wasn't always easy. The Treasury missed some payments on the Million Adventure tickets and on some of the other new debts. That made potential creditors wary, and in 1697, when the Treasury went to the public with another lottery offer, a £1.4 million draw that was to be funded by taxes on malt (and hence on beer), that Malt Lottery failed. Just 1,763 of the 140 thousand tickets went to paying customers. The Treasury held the rest as cash, forcing the tickets on captive creditors, those who supplied stores to the military, or Royal Navy sailors, who were in no position to reject them. Compounding the fiscal crisis in early 1697, the recently chartered Bank of England couldn't meet a new request for another long-term loan. The danger was clear. Experiments in money were just that. They could fail. And if they did, consequences would follow: a government unable to raise the cash it needed also wouldn't be able to keep up with its existing debts. That would mean it couldn't fight and might fall altogether.

IT DIDN'T. THE war ended in 1698; the government resumed paying the interest due on both the national debt and its short-term bor-

* This assumption turns out to be true, except on the rare occasions when it has proved to be very false indeed: episodes like bank runs during the Great Depression that inspired both new regulatory demands and institutional changes, like the creation of deposit insurance, to reduce the risk to the public.

rowings. It had been a near-run race, but the idea of a permanent debt had worked—for two reasons. The first was obvious on its face: making the king's debts the nation's, guaranteed by a legislature chosen by the same people who could lend money to the state, gave lenders power that they had never had before: they could vote out England's leaders if they dallied with the idea of reneging. A nationalized debt was one the nation was more likely to honor than any beleaguered monarch would.

The second was more subtle but at least as vital. The various forms of borrowing invented over the last years of the seventeenth century shared one critical feature. They were all forms of property that, though vastly more abstract than a piece of land or a bolt of cloth, still existed as a claim on the future. Anyone who provided capital to the new Bank of England, anyone who bought a Million Adventure ticket or placed a tontine bet, received in return certain rights: shares in banking profits, or interest payments to which the Treasury committed itself with each of its borrowings. By design, those rights could be transferred. The huge sums involved were divided up into manageable shares—a baker or a London seamstress could and did buy a ten-pound lottery ticket. Those shares could be traded. Anyone who owned such paper promises and wanted cash in hand could seek out someone else who wanted the income involved—in sharp contrast to the older forms of loans to the state, which paid interest (most of the time) but could not readily be bought or sold.

Contemporary observers already recognized that this mattered. The fact that the government's creditors could get their money out whenever they chose (rather than merely collect a long and on occasion interrupted series of interest payments) meant that more people would be willing to lend in the first place, and England could thus more swiftly and easily raise funds as and when it needed. That is: when the Treasury first began to borrow in this new way, it created not just a new, mostly secure form of state borrowing. The novelty, that fractions of a government loan could be bought and sold in a market in money, attracted clients who would not previously have been able to engage in national finance—effectively expanding the pool of those who would lend to the government, which in turn enabled further state borrowing and the expansion of the national debt.

The desperate experiments of 1693–94 had set out to solve a narrow problem: how to shovel lots of cash to the army. They succeeded in that. But it was the secondary market in financial paper that made it possible to turn emergency measures into ordinary statecraft. Such an exchange had begun to emerge before the national debt burst onto the scene, mostly trading in the shares of a handful of private businesses. The national debt supercharged that nascent financial market, distributed along Exchange Alley, a narrow, twisting passage just east of the Bank's original headquarters. To most Englishmen and -women, the Alley was an exotic, perilous place—and nowhere more so than at Jonathan's Coffee House, inside those rooms in which, as the ever-skeptical Daniel Defoe would warn, the unwary and the ignorant could fall prey to "some of the great Follies of Life."

Chapter Five

"more paper credit"

Nothing remains of Jonathan's today. Exchange Alley is itself a ghost: a narrow, deserted pedestrian corridor flanked by faceless office buildings from the High Banal epoch in British architecture. Even its name has shrunk, trimmed over the years to mere 'Change Alley. Where Jonathan's stood there's now a black, blank, windowless doorway. One of London's blue historical markers is placed, oddly, about knee high. Few visitors pass this way most days, but anyone who bends to read it would discover that "the principal meeting place of the city's stockbrokers" was once found there.

That's true, in the same way that it's fair to say that somebody once flew a plane off the beach at Kitty Hawk: a completely valid statement that misses most of the point. The story of how London's financial markets emerged in Jonathan's and a handful of similar establishments has a much more elaborate origin, a tale that begins with an Englishman abroad, addicted to an alien drug.

Daniel Edwards has little outward claim on history. The most exotic period in his otherwise unexceptional biography came in the 1640s, when he moved to the port city of Smyrna, now Izmir, as the agent for his family's trading business. There, Edwards employed a manservant, a young man named Pasqua Rosee.

A member of Smyrna's Greek community, Rosee could converse in several of his city's languages and pick his way through local manners and customs, and he seems to have aided his master in striking some of the bargains that built his business. With all that, he possessed one skill his employer valued perhaps above all else: he could make coffee.

Most of Smyrna's devotees took their coffee in one of the city's more than forty coffeehouses, described in a wonderful phrase by a French visitor, Jean de Thevenot, as "*cabarets publics de cahue.*" Edwards slipped into such local habits—but soon found himself more than commonly hungering for his dose, so much so that it was later recorded that he required as many as nine cups a day to satisfy his hunger. Edwards ordered Rosee to buy all the necessary apparatus so he could prepare the essential first cup in his employer's home every morning.

Given his compulsion, it's no surprise that when Edwards returned home he brought Rosee and his equipment with him. Making coffee in the Turkish manner was an involved process: the raw beans had to be roasted, then ground into a powder. Water would be poured over that powder and brought to a boil, then allowed to relax, then reboiled as many as a dozen times until a thick, bitter brew remained. In London, Edwards introduced his habit to his friends and family, and they rallied to the strange new drink. It got to the point that he couldn't get his work done because of the stream of coffee-seeking visitors "drawing too much company to him."

The solution to his predicament was obvious: turn the constant clamor for coffee into a paying business. Smyrna had its coffeehouses. Why not London? Edwards and his father-in-law backed Rosee as the front man for the new business. The partners opened their operation in 1652 (there is some dispute about the date) in a shed in St. Michael's Alley, a narrow passage running south from the Cornhill. The Royal Exchange, the grand old building that Thomas Gresham (of Gresham's law) had built with Queen Elizabeth's blessing, stood perhaps two hundred yards away. That was home turf for the Levant traders—Edwards's people, those merchants doing business across the Ottoman Empire, as well as many others engaged in England's domestic and international commerce, exactly those who might be enticed by a taste of something luxurious and exotic. Within a few years, according to the historian Markman Ellis, the shop's annual revenue approached or topped £450—a very nice take indeed.

That a small business could be nurtured to profitability three and a half centuries ago wouldn't usually inscribe much of a mark on time. But the opening of the first coffeehouse in the English-speaking world did more than expose Londoners to an exotic drink. London's coffeehouses were the meeting places in which patrons discovered new ways of thinking. Within a surprisingly short time it would become clear that they had become both emblems and drivers of change.

Rosee's itself was lost to the Great Fire of 1666. It was hardly missed, given that six more coffee sellers had already opened in the streets and passages around St. Michael's Alley, with scores more spread around the City of London. Some of Rosee's imitators may have managed to approximate an authentic Smyrna brew. More likely, though, most produced a very suspect product. At its worst, a seventeenth-century cup of coffee in London would be a hideous stew of burnt breadcrumbs, hickory, and, one hopes, some fraction of actual ground coffee beans. One satirist described the typical draught as "Burnt break and pudle water," smelling of "old *Gehenna, Lucifers* deep furnace, a stench to stifle virtue and good manners." But for the growing crowds gathering in the new cafés, the point was less what they drank than what happened when people met to consume it.

Crucially, the coffeehouse temporarily suspended the usual social hierarchies. A broadside printed in 1674 presented a tongue-in-cheek list of the "Rules and Orders of the Coffee House," the first of which was that "gentry, tradesman, all are welcome hither / And may without affront sit down together." The routine was much the same: a penny bought a seat; a dish of coffee; access to another novelty, the newspapers; and, as one chose, conversation with friends or with strangers, sometimes organized, often not.

Such radical informality catalyzed by an exotic new concoction was catnip to Royal Society men. John Houghton captured this eagerness in his report to the Society meeting in 1699: "[The] Coffee-house makes all sorts of People sociable, they improve Arts, and Merchandize and all other Knowledge"—which was to say that the Fellows found themselves returning over and over again to favored establishments, many within a few hundred yards of Rosee's first location. The polymathic Robert Hooke, for example, traveled daily around a series of regular perches, with Jonathan's Coffee House in Exchange Alley vying with Garraway's, a few doors farther down the Alley, as his most frequent destination. There he wrote up experiments to present to the Society and handled odds and ends of business (a pound of tea changed hands on January 17, 1680). He would pursue deeper Society matters too, as on the day he found himself "at Jonathan's, [with] Sir Chr[istopher] Wren." The topic? "Planetary motion"—the question that would evoke Isaac Newton's *Principia.* It's even possible, perhaps likely, that the ultimate encounter between baristas and men of science reached its climax at Jonathan's—if, as is plausible, Hooke there read the letter from Antoni van Leeuwenhoek in which the great microscopist first described the microstructure of the coffee bean. Leeuwenhoek's report even extended to instructions to make his vision of the perfect cup.*

Jonathan's wouldn't remain as hospitable to Hooke and his philosophical friends for long. As coffee culture evolved in London, its sites grew more specialized. Theater people had their haunts, scribblers too, and so did a group engaged in a business that had burst into prominence suddenly after William and Mary had gained their

* Leeuwenhoek preferred a light roast to avoid the "bitter and burnt taste" of overcooked beans. He would grind his beans "so fine, and sifted through a silken Sieve, that one cannot with the Finger feel the slightest hardness in it." "A certain quantity" of that powder would go into a pot, onto which he would pour water "that has boiled for some time and is still boiling." He then put the pot back on the fire, taking care to keep it just below the boil, and then would let it stand for a short time—just as with the coffee-water ratio, he doesn't specify how much—until finally, Leeuwenhoek told Hooke, "I drink that Coffee beverage."

throne, the men who traded in money. They found each other and their customers in the coffeehouses, along or near the few hundred yards of passageways of the Alley, and especially at Hooke's former haunt, Jonathan's itself. There, these latest in the line of England's new men—including the infamous stockjobbers—dealt in a variety of the forms money was beginning to take, especially in the shares of one of the most fertile of these inventions, the joint-stock company.

AT BOTTOM, THE joint-stock structure was simple: a group of investors provided capital to pay for some action—in the early days, usually a trading venture. In return they received shares in proportion to the amount each invested. This was more than a simple pooling of resources; partnerships and more temporary alliances existed long before the joint-stock company emerged. Unlike such arrangements, a joint-stock approach shifted what it meant to own a business. Instead of indicating part ownership of discrete chunks of actual stuff—a ship, a stand of timber, a machine—shares in a joint-stock company carried with them the right to claim some portion of profits to be earned over time. While, again, partnerships and other arrangements included provisions to distribute the gains of any venture, the joint-stock structure made it possible to pool capital from a wider base and to operate indefinitely—a more flexible and more scalable approach to doing business. As it matured, however, this new commercial form became more complicated, and more powerful: a way to translate a process in the world into numbers on paper, the idea of a thing and not the thing itself. The joint-stock concept in use by the late seventeenth century thus echoes the essential insight of the natural philosophers: bringing an ever-growing range of different phenomena—buying and selling instead of planets—into a form that could be analyzed, compared, quantified, and, most important, readily bought and sold.

That was the theory. In practice it took more than a century for the power latent in the form to fully materialize. The first joint-stock company in England launched in 1553: the Muscovy Company (originally known, marvelously, as the "Mystery and Company of Merchant Adventurers for the Discovery of Regions, Dominions, Islands, and Places unknown"). Its approximately 250 shareholders invested in expeditions in search of the rumored Northeast Passage to China and launched a series of trading ventures with Russia. Almost half a century later, Queen Elizabeth granted a charter to what would become the largest and most famous of the early trading concerns, the East India Company. The East India Company itself illustrates the difficult process of change in commercial practices as people experimented with the joint-stock form. For example: for decades, it did not create a permanent fund of capital, the kind of financial structure that enables continuous operations. Instead, it used short-term agreements to support its ventures as late as the 1650s.

As the East India Company evolved into its more potent form, a few other large trading companies began to copy its example. In 1660, the Royal African Company began to trade in gold and slaves; eight years later Hudson's Bay Company trapped its first pelts in the far north of what would become Canada. Taken together, while these early joint-stock ventures carried Englishmen around the globe and hauled the world back to London (sacks of coffee included), there weren't very many of them, and those few were owned by a very small circle of rich people. Just eighteen people founded Hudson's Bay Company, and as late as 1688, after nearly a century in business, the East India Company listed only 551 owners of its stock. Shares could be traded on Exchange Alley, but for the most part they weren't. Between 1682 and 1684, for example, there were just sixty-seven sales of Royal African Company stock, and shares in the giant East India Company changed hands just 537 times.

All that changed in the late spring of 1687, when Captain William Phips, formerly of the Massachusetts Bay Colony, brought his ship *James and Mary* to anchor in the Thames estuary. In March and April, Phips's ship had loitered above a reef off the island of Hispaniola. There her crew raised more than thirty-four tons of silver, gold, and jewels from a sunken Spanish treasure ship, the *Nuestra Señora de la Concepción.* The haul was valued at just over £200,000.

Phips's personal reward was impressive enough: £11,000, well over a million pounds in twenty-first-century currency. The greatest shares, though, went to the seven men who had funded the treasure hunt as a joint-stock enterprise. Each hundred pounds invested returned £10,000, and the adventure's biggest backer, Christopher Monck, Duke of Albermarle, got £43,000 for his pains. Unsurprisingly, then, the almost instant response to the news of Phips's exploit was imitation, in the form of a boom in joint-stock speculation.

The year before *James and Mary* returned to home waters, fewer than twenty English joint-stock companies existed. By the mid-1690s, there were about one hundred. Some of those new enterprises were direct copycats, get-rich-quick schemes born more of desire than of any rational calculation. But over the next decade London saw the start of a boom in offerings of more plausible concerns. They ran the entire range of human imagination: a company making wallpaper managed to raise capital, as did two glassmakers (one manufacturing plate glass, the other bottles), along with a lead-smelting operation and three gunpowder manufacturers (war has long been good for business). The "Society for Improving Native Manufactures so as to keep out the Wet" announced itself, as did another company that produced "German Balls"—lumps of wax and other ingredients that could, it was claimed, "preserve Leather from damp." There were companies formed to hunt whales, catch cod, or dive for pearls, along with several planning to support colonies in Pennsylvania, New Jersey, and Tobago. Water companies sought cash, and so did mining operations, along with a company that planned to gild metals and another "for lacquering after the manner of Japan." A stagecoach operator attracted funds, as did a concern producing whalebone whips. If someone had an idea, almost any idea, it seemed that there were people out there with money who could back them.

Most of these were small operations, and many failed to survive into the new century. But the sheer range of the opportunities they offered to Englishmen and -women who might support them combined with one other crucial development of the 1690s. The new national debt was itself a potential investment. A share in the Million Act or a ticket for the Million Adventure could be sold to third parties. One could trade in Bank of England stock too. All those who had supplied the Bank's initial £1.2 million in capital received shares that carried with them the right to whatever profits the Bank might gain in proportion to their initial investment. And just like those of the East India Company, Bank shares could be offered up for private sale. By the end of 1694 that meant that alongside the dozens of exuberant joint-stock adventures, over £3 million worth of financial paper—the two national loans and the Bank's stock—could flow between buyers and sellers, if only the two sides could find each other.

SHARES HAD CHANGED hands before the coffeehouses existed, of course, and the brokers who handled those sales had tended to gather in the Royal Exchange. Each type of trader had an assigned location or "walk" on the trading floor. The concentration of exotic figures attracted the attention of seventeenth-century travel writer Ned Ward, who described a cross between a kind of "nations of the world" exhibit and a simulacrum of hell, filled with demons. Strolling through the Exchange, Ward wrote first of "a parcel of Swarthy Buggerantoes," then a "crowd of Bumfirking-Italians"—employing the rhetoric of sexual crime to condemn whole nations. He then crossed to the Dutch walk, there to confront "a throng of strait-lac'd monsters in Fur and Thrum-Caps." The bawling merchants he encountered, Ward claimed, were "the Water-Rats of Europe, who Love no Body but themselves & Fatten upon the Spoils, & Build their own Welfare upon the ruin of their neighbors." As he roamed deeper into the throng, he balked at the Spaniards and the French and, inevitably, at "the Lords Vagabonds the Jews," the "Hawks of Mankind." His scorn wasn't purely xenophobic, though: for Ward, the true strangeness of the place was the unanimous obsession with profit. True-born Englishmen were just as bad as any foreigner, no matter how suspect, and perhaps worse. Ward described "Coasters and English Traders" rejoicing "in Out-witting one another as if Plain-Dealing was a crime and Cozenage a Vertue."

There was one group missing from Ward's lineup. His rant singled out merchants, but by the time he took his stroll at the end of the '90s, one claque of traders was gone: anyone dealing in joint-stock shares. Stockbrokers were, it seems, too loud, too bumptious, the wrong sort, and just generally ill-behaved, even by the standards of the shouting scrum of all the other buyers and sellers in the building. Licensed brokers were formally ejected from the building in 1697, but the exodus had begun years before because their true crime had been to grow too numerous: the boom in joint-stock companies and government notes had drawn in waves of aspiring jobbers, and the old structure couldn't come close to accommodating them all.

The opening to Exchange Alley lay (and still does) just steps away from the Royal Exchange itself, which made it an easy destination for those in the stock trade. Its coffeehouses were natural gathering spots, and those dealing in financial papers flowed to Jonathan's—closest to the Royal Exchange's address on Cornhill—and then deeper into the Alley, to Garraway's, Tom's, the Jerusalem, and still more. As they came, what had been a small "d" democratic home where "gentry, tradesmen, all are welcome hither" began to shift in character. Where Hooke and his friends had explored any kind of novelty, Jonathan's and the other caffeinating holes became the increasingly exclusive habitat of the jobbers and brokers—and hence of the infant English financial market.

The earliest surviving image of the inside of Jonathan's is an engraving from 1745, after it had been rebuilt following a fire. The viewer peers into an infernal chamber where (as the caption explains) "those Exchange mountebanks" put all honest men "to the mercy of their mercenary tongues." The room is filled with a ravening horde of men, reaching, grasping, grinning, passing paper, plotting. Greed is clear on most faces. Some can't hide their sidelong, too-clever-by-half glances. Everywhere the hunger shines through: a desperate desire for more.

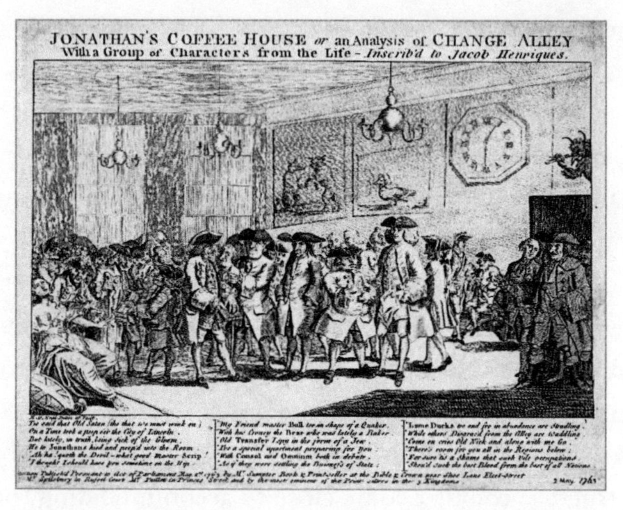

Jonathan's Coffee House in its second incarnation

Three large paintings depicted on the back wall of the room complete the scene. On the far left, a pair of foxes stand on their hind legs, most likely colluding against the complacent, easily plucked duck of the center image—a very dignified fowl beneath its handsome wig. To the right, presiding over the whole, stands Satan himself, peering through his spyglass at the melee below and, his words captured in the caption, exhorting his servants: "Come on cries Old Nick and along with me Go. / There's room for you all in the Regions below. / For sure 'tis a shame that such vile occupations / Shou'd suck the best Blood from the best of all Nations."

That's how many Londoners saw those who made their living off moving bits of other people's money. They were an alien species to their contemporaries, different from and much more suspect than those merchants who handled actual stuff, cloth or coffee and the like. To Daniel Defoe, speculators in financial paper were the dregs of London, "Bankrupts and Beggars" who "rais'd themselves to vast estates . . . by the sharping, tricking, intreagueing scandalous Employment of Stock-jobbing." The stockjobber's "whole business is tricking," Ned Ward agreed. Such a man was "a kind of speculum [mirror] wherein you may behold the passions of mankind and the vanity of human life"—laughing, grinning, maddened, driven by "those twin passions, hope and fear."

Those knaves came to life in Thomas Shadwell's play *The Volunteers; or, The Stock-Jobbers,* published in 1693. In it, a jobber tempts an equally unlovable pair of rascals with shares in a patent-pending mousetrap "that will invite all Mice in, nay Rats too, whether they will *or* no." Buy at fifteen pounds a share now, the jobber urges, before its patent arrives, or who knows how much afterwards? Not interested? Fine. For a mere twenty pounds per share, take a chance on a company claiming to know a method for walking underwater. No? How about "a great Vertuoso" who claims to be able to fly—fast enough to beat "any Post-Horse [by] five Mile an hour." Shares in that project are available too.

Shadwell's scene turns on the tension his audience would feel between the absurd and the possible. In an age when men had just learned how to lift treasure from the sea bottom and keep time to the second, what was so impossible about a better mousetrap? By the time we get to flying virtuosi, of course, everyone, audience and characters alike, know what's going on. But then the jobber gives his pitch its final turn: what about shares in the business of a traveling company, "some Chinese Rope-Dancers, the most exquisite in the World." It's a good play, he says: "Considerable men have shares in it." But he and his targets both know what's truly on offer. One says the only point of such a venture is to "use it whereby we may turn the Penny." The beauty of the dancers (and hints of negotiable virtue) don't matter: "The said shares will sell well, and then we shall not care, whether the aforesaid Dancers come over or no." That is all that matters: flip the shares, grab the cash, and never mind what comes after.

No one would have called Shadwell subtle, but he caught the temper of his times. Most of his audience believed that Exchange Alley was hostile territory and that the trade in financial paper was fraudulent to the core. Those who read their Defoe (or any other typeset scold) knew satire when they heard it. But Shadwell's scene worked because it was plausibly an exaggeration, not pure fiction. Fake news was a daily fact of life in Exchange Alley, as tipsters for another novel phenomenon, the newspapers, prowled the coffee-houses in search of choice nuggets. Information—or what could be passed as such—had a recognized price, "a shilling, or a pint of wine," paid to "some little Clerk, or a conference with a Door-keeper." It was easy enough to plant the seeds of outright frauds. As the historian Anne L. Murphy has pointed out, the incentives for stockjobbers to cheat were obvious to all, and there's no doubt that some in the market succumbed to temptation. Murphy uncovered the case of Estcourt's Lead Mine Company, floated in 1693 at ten pounds a share by a group of stockjobbers and wealthy Londoners. Over the next two years the company's stock price behaved suspiciously—with big jumps associated with large sales by insiders to naïve buyers along the Alley. It likely helped that those running Estcourt's business had in Charles Blunt, cousin to one of the principals, John Blunt, a licensed stockbroker—someone legally allowed to trade shares on behalf of clients—to push their deals.

TAKE A NOTE of that name: John Blunt. If he hadn't existed, Defoe could have invented Blunt as the archetype for his catalog of Exchange Alley's sinners, "*Jobbing* their Stocks about" without remorse, "raising and sinking them at the Pleasure of Parties and private Interests." His father, Thomas Blunt, had made shoes in the country town of Rochester. He was a well-established tradesman, but there was no escaping the fact that he had worked with his hands. John, born in 1665, had ambition above that station. In his late teens or early twenties, he made his way to where the money was, London

John Blunt in an undated engraving

and the neighborhood around Exchange Alley. He apprenticed himself to a scrivener, learning how to draw up contracts and other commercial documents. That was, in effect, training in the legal technology of credit and company structures, evolving around him as the joint-stock boom took off.

By no later than 1689, Blunt began to deal on his own account. The 1693 venture with Estcourt's Lead Mine Company showed how swiftly he'd mastered the stock market's habits, exactly the behavior Defoe condemned. Blunt and other insiders managed to get in early on Estcourt stock, curried interest in the scheme, and cashed out as knowledge of the company's potentially declining prospects gave them the edge over whoever was on the opposite side of the transaction—or, as Murphy puts it in the dispassionate language available to the long view, "The only reasonable conclusion, is that the price of Estcourt's Lead Mine stock was manipulated by its managers to their own advantage."

Blunt would hone these skills over the next quarter of a century as he accumulated what became, briefly, one of the greatest fortunes of the age. But even in this early stage, he embodied both the obvious perils of London's infant financial markets and their resilience. Estcourt's Lead Mine Company was a vehicle for some very aggressive stockjobbing. But it was also a real company, not a fiction. Lead did emerge from its patch of Welsh dirt. Similarly, for all the real evidence that Exchange Alley could be a very dangerous place for any but the most sophisticated—or exceptionally canny—there is this fact: it worked.

Partly this reflected a fact of life that still holds in financial markets: the easiest cheats occur at the margins of the marketplace. Pump-and-dump schemes (talking up a stock, then selling out before reality hits) in Welsh lead mines or dot-com fantasies are much easier to execute when one's marks can't discover much about the security being manipulated. For all the new issues promoters were trying to sell, most of the money in the market went to the largest propositions: the East India Company, the Bank of England, slices of the Million Act loan, Million Adventure tickets, and a handful of others. These were all either large, well-understood trading companies or financial shares backed by the government. Such securities were not altogether immune to ups and downs, certainly. Bank of England shares fell when the government briefly suspended interest payments on its loan; Million Act paper would fall in value for similar reasons. But the kind of national and international happenings that could bolster or threaten the large stocks were hardly secret; the nefarious wiles of Shadwell's jobber had few levers to pull on such well-known merchandise.

That is: England and Britain's financial revolution turned on the themes most closely associated with the parallel scientific transformation: the application of number to experience; the discovery of the mathematics of risk, both in the moment and over time; and the incorporation of both those leaps into money expressed as credit, a form of wealth whose value evolves over days and years. But one more step remained to turn the Alley, and Jonathan's in particular, into the first recognizably modern financial market, and not the innocents' slaughterhouse Shadwell's actors depicted. There had to be a way to figure out how much each piece of paper on offer was actually worth at any moment in time. It took a series of trials over the better part of a decade before a broker named John Castaing finally realized that he didn't need to trade in shares—not when there was knowledge to be sold.

Chapter Six

"John Castaing, Broker, at his Office"

Jonathan Castaing washed up in London an outcast, a refugee from Louis XIV's persecution of France's Protestants. Little is known of Castaing's life before he crossed the Channel, though he was certainly involved in buying and selling. He put his skills to use in his new home, becoming a licensed stockbroker and taking his place on the walk at the Royal Exchange—and he was equally swift in seeing where that situation was headed. By 1695—two years ahead of the jobbers' expulsion from the Exchange—he was already advertising that he "buys and sells all Blank and Benefit tickets [in the lottery-loans] and all other Stocks and Shares" to anyone who made his way to "*Jonathan's* Coffee-house."

Once established in his new premises, Castaing confronted a problem he shared with everyone else doing business in the Alley. How much did any given share actually cost? In modern financial markets, of course, that's usually a trivial question: the price of a share is whatever the last person paid for it, a number visible almost instantaneously to all interested parties. But in 1695, it wasn't so simple. A jobber looking to buy East India stock in the coffee room at the Jerusalem had no way to know what a broker had just charged for the same security a few hundred yards away at Jonathan's. It didn't even require that much distance to run into trouble, as the banker Thomas Martin found when he discovered that shares offered at £920 at one end of Garraway's long room would cost him £1,000 at the far wall.

Such wild fluctuations certainly fueled the belief that outsiders venturing into the Alley were nothing more than sheep to be fleeced, but the danger implicit in such uncertainty ran much deeper. If no one—brokers and dealers included—could be sure how much to pay for anything, the Alley's men could trade in shares, but they couldn't perform the essential function of a market: enabling buyers and sellers to meet and strike deals with confidence that no one was pulling a fast one, cheating on what the rest of the market thought the price should be. Exchanges, stripped down to their essence, offer a way those who use them can express—or, in the jargon of the craft, discover—what value the collective judgment of everyone in the marketplaces on whatever slice of the world is up for sale. In England's early financial markets, dominated by the state's borrowing, that meant settling on the public assessment of the price for shares in each of the loans the Treasury sought from London's bankers and investors. Such prices would vary, depending on the details of each attempt to borrow—the terms of the Million Adventure were different from those of the 1696 Malt Lottery offering, for example. But whatever the particulars, Exchange Alley had to give its clientele a reasonable expectation that the amount paid for a slice of the national debt would be the same at Jonathan's as it was at the Jerusalem. Otherwise, if the men and women crowding the Alley couldn't feel at least fairly confident in their prices, the state's ability to borrow by selling debt to be traded on such a fractured market would suffer. In the worst case, given sufficient uncertainty, a momentary panic could depress the state's credit, inevitably at the point of greatest need.

Within two years of his move onto the Alley, Castaing offered a solution. What was lacking, clearly, wasn't the stuff to sell: there was plenty of government paper to trade, while England's eager projectors could be relied on to come up with an ever-renewed catalog of joint-stock propositions. The dearth lay in information, a clear, simple, and reliable record of the aggregate conclusions of the market at any given moment. To meet that need, Castaing came up with the obvious-in-hindsight response in March 1697: a simple list of prices, updated daily, under the title of *The Course of the Exchange and Other Things*. His operation was at once as basic as can be imagined and also the prototype for so much of the market infrastructure that we now take for granted—the lists of quotes for everything from shares on the New York Stock Exchange to the current levels in bond markets across the world. He enlisted a platoon of clerks to walk the Alley and nearby streets in midafternoon, just as the day's affairs were winding down. They would canvass jobbers and dealers to find out the last prices received for the various investments covered by *The Course*. Final, printed sheets came out twice a week on Tuesday and Friday evening. Some readers chose to have each issue delivered—for the quite reasonable cost of twelve shillings (roughly £100 today) per year. Alternatively, one could simply pay a call on "John Castaing, *Broker, at his Office at* Jonathan's *Coffee-house.*"

(1)

The Courſe of the *Ex-change*, and other things.

London, Tueſday *4th January*, 1698.

Amſterdam	35	9a10
Rotterdam	35	11a35
Antwerp	35	9a10
Hamburgh	35	2a3
Paris	47	1/4
Lyons	47	1/4
Cadiz	51	1/4a51
Madrid	51	1/4
Leghorn	52	1/2
Genoua	51	1/4
Venice	49	1/2
Lisbon	5	7 3/4
Porto	5	6 3/4
Dublin	16	1/2

Gold	4 *l.* 00 *s.* 6 *d.*
Ditto Ducats	4 · 5 6
Silver Sta.	5 *s.* 1 *d.* 1/2 a 2 *d.*
Foreign Bars	5 3 1/2
Pieces of Eight	5 3 1/2

	Saturd	*Monday*	*Tueſd.*
Bank *Stock*	86 1/2 a 3/4	86 1/2 a 3/4	86 3/4
India	53 3/4	53 3/4	53 3/4
African	11 3/4	11 3/4	11 3/4
Hudſon Bay	110	110	110
Orphans Chamb.	53	53	53
Blank Tick.M.L.	6 15	6 15	6 15

No Transfer of the Bank *till* January 7.

In the Exchequer	*Advanced.*	*Paid off.*
1ſt 4 Shill.Aid	1896874	1814575
3d 4 Shill. Aid	1800000	1392377
4th 4 Shill. Aid	1800000	886492
3/4 Cuſtom	967985	764328
New Cuſtom	1250000	655200
Tobacco, *&c.*	1500000	119400
3/4 Exciſe	999815	864260
Poll-Tax	569293	479328
Paper, *&c.*	324114	65512
Salt Act	1904519	73772
Low Wines, *&c.*	69959	11100
Coal Act&Leath.	564700	17162
Births and Marr.	650000	2000
3 Shill. Aid	1500000	601555
Malt Act	200000	163746
Exchequer Notes, ſunk		5850000*l.*

Coyn'd in the *Tower*, laſt Week, 00000*l.*

By John Caſtaing, *Broker, at his Office at* Jonathans *Coffee-houſe.*

The Course of the Exchange for Tuesday, January 4, 1698

The Course of the Exchange wasn't the first stock sheet to appear in London. At least two short-lived price lists preceded it in the first half of the '90s. But Castaing's was the one that stuck, appearing every Tuesday and Friday for well over a century. At its launch and for decades after it kept the same basic look: a single sheet printed on one side. At the top of the page came the latest currency rates, the prices at which English bills of exchange traded in cities across Europe. That was testimony to the reach of England's economic life—and more practically, it showed the advantage reliable information could give an investor. For example, one of the earliest issues (see the image on page 91) alerted its readers that its correspondents had found that the pound traded for a slightly different price in Genoa than in the Tuscan port city of Livorno. A little over one hundred miles separates the two cities, so it would have been difficult to take advantage of the arbitrage opportunity—buying pounds in the cheaper location to sell, preferably very, very quickly indeed, where they commanded a higher price. But the possibility was there, and Castaing's readers would have known it.

Below those numbers came the quotations for joint-stock companies. The early issues listed only the largest and most commonly traded shares, but there they were: in the first edition every week, the Saturday, Monday, and Tuesday prices for the Bank of England, Hudson's Bay Company, and the East India Company, with the Friday issue covering the rest of the week's trades. Farther down the page came the latest quotes for the growing list of government debt instruments, from the Million Adventure forward. Taken together and viewed over time, week after week, the importance of this single sheet of newsprint appears: it distilled the tumult and welter of the hundreds of individual transactions up and down the Alley into a serene table of numbers, snapshots that accumulated into a moving picture of what England's infant stock market was thinking.

This is what transformed that talking shop Jonathan's Coffee House into an institution, London's first stock exchange. By the mid-1690s, it was known that if you wanted a stockjobber you headed for Exchange Alley. But that just made it a meeting place. It took more than just a conveniently localized crowd to create an exchange in the modern sense. That required a common stock of knowledge, the detailed observations that allowed both sides of any trade to have a shared grasp (imperfect, then and now) of what the market as a whole believed about whatever was changing hands. That is: a market is more than a place; it is information, accessible to those who take part in it. When that information is poorly communicated—or doesn't exist—bricks and mortar can do little more than keep the rain off a failing bourse.

The Course of the Exchange was thus much more than a mere price list. It was, rather, a distillation of disciplined observation and the quantification of events—the accumulation of all the individual decisions by independent buyers and sellers—processed into a consistent form that anyone could grasp and use. Of course, such a paper couldn't solve every problem for a nascent financial market. Insider trading—especially in the smaller joint-stock companies that never made it onto Castaing's list—remained a constant danger. Rumors could swamp the kind of hard data Castaing tried to gather. Outright frauds certainly occurred. But *The Course of the Exchange* gave the market its memory and, effectively, reported on the insight of the hundreds of individual traders making bets on what would happen next, tomorrow and for years to come. It powered the development of a financial market that worked tolerably well—one in which buyers and sellers could reasonably expect to deal in prices that were largely consistent across the Exchange.

To steal Winston Churchill's line, this was, if not the beginning of the end, at least the end of the beginning of the financial revolution. By the turn of the eighteenth century it was all in place: the power of mathematics and the habits of observation most closely associated with the scientific revolution had created new ways for late-seventeenth-century Englishmen and -women to think about the future. The near-simultaneous crisis for England's hard currency and the emergence of forms of credit subject to calculation and mathematical analysis created new forms of money. In the nascent stock market both such credit novelties and the newly constructed private enterprises could be priced and traded—thus cementing the connection between ideas about numbers and those about money.

For England's rulers, at least, that was the whole point of the exercise. A functioning stock market created a so-called secondary market for government credit,* both the large, long-term debts Parliament authorized and guaranteed—what became known as the national debt—and so-called floating obligations, which were short-term loans used to raise funds quickly to cover ongoing operations. The floating debt was a common tool of official finance, sums borrowed for a few months or so. Such borrowing was useful to allow the Treasury to navigate problems like needing to pay bills—military pay, suppliers, and the like—at different intervals than the timing of its income, tax payments, for example. Usually, it was assumed that such loans either could be paid off out of ordinary revenue when it finally arrived or could be rolled over into new short-term paper on similar terms. That meant that this so-called short-term debt could in fact stick around for a very long time, but it also meant that the market in government finance had a vital role to play. The daily price quoted on Exchange Alley for the Treasury's short-term paper sent a clear signal to both officials and everyone else about how much confidence the monied public had in the government at any given moment. Castaing's *Course* often showed that these debts traded at a discount. That is: an instrument with a face value of £100 might sell for £93, or £87, or whatever sum reflected the judgment of the clever men at Jonathan's as to whether the Treasury would make its interest payments on the agreed schedule—or at all.

Such discounts were and are the price paid for any uncertainty in

* The primary market for a security is the one in which the creator of that instrument sells its bonds or other types of financial paper to their initial buyers. In these early days that was usually by subscription—a company or the Treasury would offer shares or debt, and those who wanted to invest would sign up for the amount they wanted to buy. The secondary market is where members of the public buy and sell such securities to each other—which happens now in a wide range of public exchanges, and did so then in every nook and cranny along Exchange Alley.

the nation's finances, the market's perception of the risk that the Treasury might not meet its obligations. Any collapse in the value of government paper—a deep discount—would suggest a high risk of default. When that occurred, it was wildly expensive for the state to borrow.* Any new loan would have to offer enough of a return to persuade investors to take the risk; that would make refinancing the short-term debt more difficult and more expensive and would impose decades of increased obligation on the Treasury if it tried to raise any more long-term national debt funds. But until and unless a crash occurred, the persistence of buyers willing to trade in England's menagerie of loans, lotteries, short-term notes, and the rest helped the Treasury borrow, even if Exchange Alley priced government paper a little below par.

That newly enhanced ability to borrow worked like this: the reason many bought shares in the new national debt was to ensure a long-running stable stream of income—all the regular interest payments that Parliament guaranteed the Treasury would pay over all the years and decades for each particular loan. But the deal was that an investor's capital—the money paid to purchase that stream of payments—would stay with the government until the debt was paid off (or, for certain types of borrowing, forever). That could be a problem if someone suddenly needed their stake for some other reason—and that's where Exchange Alley came in: it was a secondary market for the Treasury's paper. In that case, those who needed to get at their capital could head to Jonathan's, where, if they consulted *The Course,* they could be reasonably certain they were getting a reasonable price. This ability to take one's cash out of the world of finance was an es-

* Every drop in the price of a piece of debt raises the effective interest on the loan: if the original rate was 5 percent, then £100 of face value still earns £5 even if that instrument sells for just £80. That's a 6.25 percent return on the £80 the lender actually had to pay for her investment. That, in turn, influences the effective interest a borrower will have to pay on any new debt it issues, as potential lenders will not hand over their cash for a lower rate than they could get on Exchange Alley—or Wall Street.

cape hatch that made it much easier for the government to borrow. The new loans thus depended on the ongoing ability of London's new stock market to buy and sell all the debt the government chose to create.

Create it they did, swiftly, and in astonishing amounts.

FROM THE TRIUMPH of the Glorious Revolution in 1688 through 1697, England spent almost £50 million on its war with France. The Treasury borrowed almost exactly one-third, just over £16 million in all. Of that, about £7 million came in the form of the new "national debt" instruments, long-term obligations guaranteed by Parliament. This emerging system of credit and finance was clearly a success by at least one measure: it raised the capital needed to prosecute national aims that cost more than the state had at any given moment. But was that a good thing? Once the war was over, clever Englishmen began to ask if the nation should continue to borrow, now that the immediate pressure of war had eased. The answer wasn't obvious, even to some of the most sophisticated thinkers in London.

By the end of the seventeenth century, John Pollexfen had become a wealthy man. The second son of a rich parent, he pointedly recalled that he "never had but £200 from his father." From that modest stake Pollexfen built a fortune as a merchant, trading initially with Spain and Portugal. He grew comfortable enough to marry in his thirties, and by 1677 he had accumulated a sufficient pile to build one of London's most notable mansions, Walbrook House, for the impressive cost of £3,351 6s. 8d.

By then Pollexfen had added public duties to his private business, first serving as a member of several official committees on trade and other commercial matters, then, in 1679, gaining a seat in Parliament, where he gave special attention to the evolution of England's mercantile and financial systems. Finally, in 1695, he joined the recently reconstituted Board of Trade. Thus few men in England were better able to think about the risks and rewards of the ongoing experiment in credit and national debt. So, just as England's wartime borrowing reached its peak, he decided it was time to sound a warning. In his *Discourse of Trade, Coyn, and Paper Credit, and of Ways and Means to Gain, and Retain Riches* he wrote that "it may occasion that a Nation that relies much on Paper Credit, may be thought Rich one day and be found Poor another." This discovery would come, Pollexfen noted, precisely "when some great convolution [a sudden twist or surprise] may happen, which only can discover (like Death to some real Trader) whether a Nation be Rich or Poor." Thus exposed, he warned, "such Credit may fail"—and the nation would not be able to raise any cash—"when there may be most occasion for the use of it." Last, he suggested, somewhat cryptically, no one should place too much faith in the ability of any private market to prop up the state's debts: "No Nation can be too cautious with whom they trust their Riches," for the bankers and monied men might be tempted to manipulate the price of government paper as it suited their interests, and not necessarily those of the Treasury. Those "interested with the artificial Treasure, or passing of Notes or Bills of Credit," would, he wrote, face the same "great Temptations to multiply" their fortunes at the expense of the country as a whole.

That sly gibe aimed at the Alley's men was perhaps inevitable but slightly off his main point: promises made by individuals or businesses—or the Treasury—that relied on events that would unfold years down the road would seem perfectly safe up to the moment they were not. A crisis in which the government found it could not borrow new funds—or could do so only at impossible rates of interest—would occur only at exactly the point when the money was most needed, which meant, Pollexfen argued, that it was too risky for any sensible administration to pin its financial fate on uncertain and uncontrollable markets in money. Pollexfen conceded that paper credit could be useful, able "in a great measure [to] supply the want of *Coyn* for the carrying on of Commerce," and perhaps equally so in the face of the extraordinary costs of war. Still, the risks remained, to the point "we may soon experience a great want of valuable Riches, and have only in its room what is imaginary." Ultimately, he concluded, this new invention of credit was a broken reed, "not to be depended on." England could borrow, to be sure, but should it, beyond a very carefully drawn limit? Pollexfen, as sober and clever a businessman as could be found in London, was clear: it should not.

POLLEXFEN PUBLISHED HIS argument in 1697, at the height of the Great Recoinage. The warden of the Mint was thus rather busy, so despite Isaac Newton's urgent interest in any debate about England's currency, he does not seem to have noticed the pamphlet on its first release. It was reprinted in 1700, though, and this time Newton paid attention. A draft memo survives in his own hand that captures what was so transformative in England's newly developed ability to turn mere paper to national ends.

Newton began softly, broadly agreeing with Pollexfen that "too much paper credit in proportion to money is not safe." But when one looked at England's accounts as they actually stood, no such danger seemed imminent: "In examining . . . whether this credit has been of more advantage to England by increasing its trade or disadvantage by increasing its luxury [i.e., wasting borrowed money on unproductive consumption], I do not find that it has hitherto done us any great damage." That is, Newton agreed with Pollexfen that credit, by allowing a borrower to spend more money than he, she, or it possessed, could produce a false sense of wealth and thus feed an urge to fritter away paper wealth on fine living. But Newton performed an exhaustive study of the history of England's currency—how much metal money the Mint had coined over the preceding century and what the current state was of the worldwide trade in precious metal—noting that gold was much cheaper in China and Japan than in London. He concluded that government borrowing, with its creation of forms of credit that did some of the jobs performed by ordinary currency, had played a crucial role in helping England deal with the collapse of silver coinage and hadn't produced any significant rise in reckless spending. Credit helped the nation, Newton concluded, much more than it hurt—if it caused any harm at all.

To this point, the two men were answering the same question: How risky is debt for either private enterprise or government finance? Pollexfen said, "Very"; Newton responded, "Not much." Such differences in degree, not kind, reflect shared assumptions, a common view of what they were talking about. But there was more. As he thought back over the crisis he'd just helped resolve, Newton wrote, "But as for paper credit, that was so far from hurting us that the want of it during the recoinage brought us into the greatest difficulties." The shortage of ready money, Newton wrote, "made the Interest of money very high which . . . put the greatest damp upon trade." That observation led directly (for Newton, though not for many of his peers) to the next step in the argument: "If interest [on government borrowing, and for private loans as well] be not yet low enough for the advantage of trade & the designe of setting the poor on work & encouragement of all such business as is profitable to the nation, the only proper way to lower it is more paper credit till by trading & business we can get more money."

Newton made it clear that he had no desire for any direct government control of how much individuals could charge for money—the interest rates at which the state or ordinary men and women could borrow. Rather, imbibing the spirit of the classical liberalism of his friend John Locke, he noted that "to lower it by Act of Parliament is a violent method, & . . . is apt to put trade & business out of humour." Better to let the market work—including (by implication) those rowdy louts along Exchange Alley. "The law should rather follow & comply with the free & voluntary course of interest then attempt to force it." But—and here he leaves behind the most orthodox Enlightenment liberal economic argument—"Let it be considered therefore what rate of Interest is best for the nation & let there be so much credit (& no more) as brings down money to that interest."

This was a view of official borrowing as a kind of currency, a way for the government to affect the total supply of money available to the nation. Newton was not a deep financial theorist, but he leapt ahead of many of his contemporaries in this argument, that the official obligations, divided up into easily traded shares, could be an instrument of policy. The Treasury's borrowing was thus much more than just a way to deal with a moment's shortfall. It was more useful than that. Credit touches the future, as it makes a prediction about the economic life of the nation in years to come: the interest on a loan is a claim on future revenue, on the taxes a government can extract from those it rules. A loan today, then, pays for stuff and deeds in the present that can generate more economic activity in the days to come, which (it's hoped) will produce more profits to tax . . . which both service the loan and (again, it's hoped) encourage yet more trade and production and all the different things Englishmen and women do in their working lives. Hence Newton's conclusion: credit was an abstract representation in the present of coming years' flow of silver and gold, of profits—and it could cure whatever ailed England's economic life. "Credit is a present remedy against poverty," he wrote. He acknowledged his earlier concession to Pollexfen. The use of credit was never risk-free; it was powerful medicine, and its effects depended strongly on the dose. "Like the best remedies in Physick [it] works strongly & has a poisnous quality. . . . For it inclines the nation to an expensive luxury in foreign commodities." But, again, Newton emphasized, wise men running a sober government could be trusted to handle such powerful medicine. "Good Physitians reject not strong remedies because they may kill but study how to apply them with safety & success. Let that expense be corrected by good Orders & laws well executed & then credit becomes a very safe & soveraign remedy."

Correct the spelling and update the syntax a bit, and Newton's argument starts to sound suspiciously modern, a call for what we would now call fiscal policy: if the economy languishes, if businesses are struggling and employment lags, well then, the government should issue new credit—seek new loans—and spend that borrowed money until trade swells and the poor can set to work. That's not quite what he meant. Newton was no proto-Keynesian. He wasn't suggesting that the Treasury should choose to create a deficit—deliberately pay out more than it took in to help out in hard times—nor was he suggesting that the government should manipulate interest rates to stimulate economic activity. There was no real way to think such thoughts at that moment, in fact, as the infrastructure of modern financial policy did not yet exist. You can't call for a central bank to do something when the concept of such a bank lies centuries in the future. This was the period in which notions of what money could be and do were still being worked out, and Newton did not have a clear idea about how quantities like the money supply—how much money, coin and paper, exists within an economy—interact with fiscal questions or how government spending affected interest rates or other prices. The master of the Mint was the most accomplished thinker of his (and perhaps any) age, but he wasn't a twentieth-century macroeconomist. He wasn't even much of an early eighteenth-century political economist. His draft response to Pollexfen was a sketch, never developed into anything that can be seen as a financial theory.

But even if Newton shouldn't be viewed as an economic pioneer, his thoughts on credit do hint at what was happening on his watch. He was, after all, a senior member of England's state financial apparatus, and he saw both national and private credit as at once safe and useful. Others, knowledgeable and worldly, disagreed, but Newton's argument shows how those around the center of power were thinking as they worked out a first, recognizably modern approach to government finance. It's a matter of record that Newton's view won the argument with Pollexfen in the year of our Lord 1700. England had weathered its experiments of the 1690s. Given that, the Treasury's moves in the next crisis were clear. Good ministers need not reject strong remedies. If circumstances demanded it, then "more paper credit" it would have to be.

Chapter Seven

"some way to answer these demands"

Those circumstances came swiftly. In 1702 King William died—eight years after his wife, Queen Mary. Mary's sister, Anne, took power, first as queen of England, Scotland, and Ireland and then as monarch of Great Britain after the Acts of Union merged Scotland and England into a single nation in 1707. Louis XIV of France refused to recognize Anne's claim to the throne, supporting instead James Francis Edward Stuart, son of the monarch that William and Mary had deposed, King James II. That, along with several other disputes, sparked the next conflict between the major European powers, a globe-spanning struggle that began after the childless Charles II of Spain died. The conflict, known as the War of Spanish Succession, pitted Britain and allies against France and its coalition. The war lasted for more than a decade, ending as much in exhaustion as in resolution in 1713. It proved just as costly, or more so, than the previous one. The huge (for its day), ninety-thousand-strong army had to be paid and fed, clothed and armed, while the fleet in its turn gulped down its tons of beer and beef every day. As the war dragged on and on and on, the cost of this massive military effort rose steadily until by 1710 the cost of the war had climbed toward 10 percent of England's income—that's the product of the whole nation's working life, not just the government's revenue.

Faced with such outsize demands, the national debt, whose growth had paused in the brief peace after 1697, swelled again. The government borrowed £8 million from 1704 to 1708. More obligations followed as the Treasury negotiated large loans from the big joint-stock monopolies, the East India Company and the Bank of England. Then the Treasury turned back to the public at large with even greater hunger: £2.4 million offered in 1710, followed by an astonishing £7.1 million over the next two years. By 1713, as the books closed on the war, Britain's long-term national debt, zero at the beginning of 1693, roughly £7 million by 1697, stood at over £40 million. Added to that total came millions more in short-term loans—that floating debt that could, in theory, be maintained indefinitely ("floating" as long as the government kept up its interest payments or raised new loans to pay back each older obligation).

The Alley groaned. The huge costs of the war strained the capacity of London's market in money to come up with all the cash the government demanded. Clear signs of trouble came in 1711, when the Treasury sought buyers for the first of four new lottery loans. It took a heroic offer to draw the punters in: 8 percent interest to be paid for more than three decades. That was at least 50 percent more than private borrowers were paying—a premium that reflects just how nervous the smart money at Jonathan's had grown about the nation's credit. Supporters of the government and the war could argue that peace would restore fiscal prudence. But that was hardly reassuring, given that England and then Britain had fought France for all but five years since William had come to power back in 1689.

Such financial woes added to a political crisis that was already threatening the rule of Newton's credit-happy friends in the ministry (the collective label for the cabinet and other senior leaders who performed the executive functions of government). Sidney Godolphin's administration had been a thoroughly Whig affair, dominated by the party associated with London, a measure of religious tolerance, and the kind of financial invention the Bank of England exemplified. The Whigs were bitterly opposed by the Tories, the party more associated with the country, the landed interest, and orthodox religion—and who feared the great Whig companies' monopoly on government finance, especially the role played by the Bank of England. It is a mistake to see those two labels, Whig and Tory, as marking broadly consistent political commitments. The two groupings had clearly different views of kingship and governance in the years leading up to and just after the end of the Stuart dynasty. But twenty years on, such associations were less about coherent beliefs and more about networks and the interests and patronage power of one's friends.

Thus, though Britain's governing elites had genuinely varied stances about national policy, including the conduct of the war, Godolphin's difficulties had as much to do with ephemeral passions—including who was to enjoy the perks of governing—as they did with any grand disputes of statecraft. This time, it was an infuriated clergyman who tipped the scale. On November 5, 1709, Henry Sacheverell, a clergyman with a loathing for any lapse in orthodoxy, climbed the curving stair leading to the pulpit in St. Paul's Cathedral. Brimstone showered down, a vicious condemnation of lapses within the Church of England. The sermon was later printed under the title *The Perils of False Brethren in Church and State* and was immediately understood as an attack on the doctrinally squishy Whigs. Following a second, even more incendiary sermon three weeks later, Godolphin's Whigs brought him to trial in the House of Commons, convicted him, ordered his sermons burnt at the Royal Exchange, and barred him from the clergy for three years. These were profoundly stupid moves. Sacheverell became an instant martyr and, given the affiliation of those who'd opposed him, a hero to the Tory cause. Riots followed, an election was held, and Godolphin lost, to be replaced by the urbane Tory leader Robert Harley.

Harley inherited a mess. The bloated debts—long-term national obligations, short-term floating loans, every pension and annuity and overdue bill issued over the last two decades and more—were all still there. The war still rumbled on, which meant that more borrowing was likely to come. And perhaps worst of all for Harley's purely political calculus, the major new institutions embroiled in national finance, especially the Bank of England, were clearly linked to the Whigs—and thus had to be viewed as a threat to Tory power.

It was, in fact, a predicament like the one that England had faced in the years immediately after the Glorious Revolution: leaders engaged in tense, high-stakes political gamesmanship at court and in Parliament as the nation first overextended itself in a European war and then plunged toward insolvency. A host of men, gamblers like Neale, theorists of money like John Locke, Defoe's "projectors," and more, had used the moment to launch one experiment after another to extract the nation from its troubles. Two decades later, Harley faced a similar risk of a collapse of Britain's ability to borrow. Trying to navigate that crisis within a financial milieu dominated by his political opponents, he was met by a round of offers similar to those proposed twenty years before, a renewed push for financial innovation that would, if its proponents could be believed, both save the nation and make those riding to the rescue very rich indeed.

ONE SUCH SCHEME stood out. It had several moving parts, but it turned on a single, simple idea: a forced marriage of two of the financial revolution's signature inventions, official debt and the joint-stock company. The plan was to transfer government obligations to a new, private company. Those who had lent their money to the Treasury would now receive shares in that joint-stock operation. The company in turn could use its new capital—all that government debt, with its market value and guaranteed stream of interest payments, now the property of the fledgling concern—to fund whatever business it had been set up to do. This was one of the first examples of what is now called a debt-for-equity swap. Creditors become owners, exchanging the promise of regular payments on their money for a stake in a new, hopefully profit-making concern.

It was a radical notion but not an entirely unfamiliar one. The Bank of England had tried something similar in 1697, offering its own shares to holders of certain loans on which it appeared the government might stop paying the interest due. For those who owned the suspect paper, the bet was that the Bank would be much better placed than any individual to force the crown's officials to pay up, if not on time, then eventually. This was a popular view—investors swapped about £800,000 in obligations for the Bank's stock—and it turned out to be correct. At the same time, the whole affair had a Rube Goldberg quality to it, something thrown together to deal with one particular loan gone bad. It was a way to create an asset worth more in the Bank's hands than in anyone else's, and as such it didn't really establish a clear link between debt and company equity: the Bank's stock could be seen as a kind of currency, and the idea that investors were buying into its ongoing business was secondary. Even so, the transaction did hint at what might be possible, and it seems to have stuck in the mind of some of London's most canniest men, a group led by three Exchange Alley veterans: George Caswell, Elias Turner, and Jacob Sawbridge.

These were men Daniel Defoe would call out by name in his most biting assault on stockjobbing, 1719's *The Anatomy of Exchange Alley*—there recalling this early scheme. Caswell was "sufficient for much more business than he can be trusted with . . . the head Class of the Fraternity called *Pick Pockets,*" Turner was "a gamester . . . all Grimace," and Sawbridge was "as cunning as C (Caswell) is bold." They had combined forces during the 1690s, Defoe wrote, to form "a true triumvirate of modern thieving." The trio's first move was a kind of financial grave robbing: they exhumed the corpse of a near-dead enterprise, the Company for Making Hollow Sword Blades in the North of England. Better known as the Sword Blade Company, it had been founded in 1691 with the stated intention to produce "hollow" or grooved sword blades in the French style—deemed superior to the traditional English flat-bladed rapier. The sword business went well for a few years but collapsed by the end of the decade, leaving behind nothing but a charter as a joint-stock company. These financial buccaneers recognized that such a charter had real value—as did their newest coconspirator, the still-young cobbler's son named John

Blunt, last seen growing fat off insider trades in a Welsh mining scheme.

THE RESURRECTION OF the Sword Blade Company was the making of John Blunt. Few details survive from his earlier years on Exchange Alley, just sketchy remembrances of his twenties and thirties as he scuffled within a scrum of similarly urgent, hungry men. It's clear, though, that he was a more complicated figure than his enemies would later credit. He was certainly viewed, as the historian John Sperling put it, as a perfect avatar of Exchange Alley villainy, combining "the most lethal qualities" of his new partners "in a hard driving, brutal character." By another account, he was unlovely, inside and out, "burly and overbearing, glib, ingenious," and (most damning in the eyes of his social superiors) "determined to get on." Yet even his worst detractors noted that he avoided some of the typical vices of the age. He was a genuinely religious man, a Baptist, strong enough in his views to dissent from the Church of England. Blunt led an apparently happy married life with a woman perhaps slightly above his station—and by all accounts he was a devoted father to his seven children. As ruthless as he was known to be in business, he seems never to have cheated his own partners. There's no contradiction there, of course; Blunt was hardly the first person, nor would this grasping scrivener be the last, to love his family while bludgeoning his foes. The one constant recollection, though, reported by both his enemies and his allies, has nothing to do with questions of sin or grace. Whatever else he was, it was clear that John Blunt was very clever indeed—and thoroughly determined to use his intelligence to get as rich as it was possible to imagine.

Given that pedigree and those of his partners, it's no surprise that this partnership pursued a plan that was expertly designed to conceal how little economic value was left within the Sword Blade Company. The first bargain struck under its umbrella was a dodgy land deal—an attempted grab of Irish estates owned by supporters of the deposed Stuarts, which had been confiscated after William and Mary took the throne. The Sword Blade proprietors offered £200,000 for property in Ireland that returned rents of £20,000 a year—a bid for a very healthy 10 percent return on their money. The only problem? They didn't have the cash.

Their solution lay with that surviving charter as a joint-stock business. With that, the Sword Blade Company could create new shares in the company. In the ordinary course of business the company would sell those shares in the open market (at Jonathan's and Tom's and Garraway's and the rest) to raise the funds needed for their Irish adventure. Instead, in a move that plausibly originated with Blunt himself, the new stock was to be used in a debt-for-equity swap, with shares offered for what were called army debentures, floating notes sold by the Treasury to cover the army's petty cash needs, day by day. Unlike the headline loans of the national debt, these debentures weren't backed by any guaranteed funding stream—no malt tax for them. That made them risky—and prices for the army's paper would often fall below face value on Exchange Alley. The Sword Blade partners saw an advantage in that risk. As they put the last touches on their plan, the army's debentures were trading as low as £85—£15 below their nominal value. By the terms of the proposed deal, anyone with army paper to sell could exchange it for Sword Blade stock issued at £100 per share—an instant gain of that same sum, £15, as long as Sword Blade shares held their value on Exchange Alley.

Naturally, the Sword Blade Company found plenty of people willing to make that trade. Even less surprising, the Sword Blade principals worked their insider knowledge for all it was worth, buying deeply discounted army paper in Exchange Alley before they announced the swap. By one estimate, Blunt and his associates made at least £25,000, about £4 million now, by such self-dealing. In fairness—if that's the word—this sort of front-running was common enough along Exchange Alley. Bank of England insiders had done much the same in 1697. And for all that Defoe and other stockjobbing skeptics scorned the way those in the know took advantage of the uninformed outsider, Blunt and his colleagues could defend themselves with the fact that the Sword Blade partners delivered on their promises, at least at first. The government rid itself of a substantial chunk of debt; the Sword Blade Company gained command of some Irish land that generated an above-market return; and holders of risky government paper gained both a share of that revenue and a chance at further profit if Sword Blade shares rose over time—a win-win-win!

The problem was that this delightful outcome held up just long enough for the transaction to be completed and for the company insiders to bank their profits from their preswap grab of cheap army debt. The setbacks came swiftly, the most significant being a common problem for better-to-apologize-than-ask-permission market traders: the Sword Blade partners were selling something that it wasn't clear they would ever own. Across the water, Irish claimants to the forfeit lands emerged, and Irish courts proved sympathetic to the challengers. Sword Blade stock values fell until, in the spring of 1708, stock acquired at the effective swap price of £85 (the discounted value of the debentures) now brought just £51 per share—a loss of 40 percent over just five years.

But while investors suffered, the company's directors took one lesson from the deal. Their failure was due to a bad—or at least overly aggressive—bet on a specific business proposition. But the transaction seemed to confirm its underlying logic. It was too bad that their investors lost money, of course, but Blunt and his colleagues recognized that swapping government debt for shares in a private company was a trick they could use again.

ENTER ROBERT HARLEY and his fellow Tories, desperately in need of debt relief. One of the first acts of the new administration was to catalog the full extent of Britain's borrowing. Alongside the national loans authorized by Parliament, the floating debt alone was still terrifying: £5 million due on navy accounts, another million for the army, £1.3 million more just to cover general government expenses, and so on. Most of these funds were unsecured—there were no specific sources of revenue dedicated to paying them off—making this among the riskiest forms of government credit. In all, the floating debt amounted to almost £9.5 million, and as far as Exchange Alley was concerned, it looked less and less likely that it would ever be paid off in full: by 1711, stockjobbers were offering only a little over £65 in the hundred for that paper. That was a clear signal to the Treasury that it would grow ever harder, perhaps impossible, for the nation to borrow for its daily needs. This was a true crisis, so much so that in her speech to open the newly elected Tory parliament, Queen Anne almost begged: "I must earnestly desire you," she said, "to find some way to answer these demands, and to prevent the like for time to come."

Under a Whig administration, the Bank of England would already have been recruited to answer Anne's plea. Harley, unwilling to risk dealing with his enemies' bank, looked for another way out—which was all John Blunt and his associates needed. A few months after Harley took office, they delivered their plan, by far the most grandiose scheme the Sword Blade crew had yet attempted. At bottom, the new idea was just another debt-for-equity swap, similar to the ill-fated Irish adventure of a decade earlier. But again, there was a twist: the Sword Blade men wanted to wrap that swap around another project, a commercial monopoly, like that of the extremely successful East India Company. The crown would grant this new business the exclusive right to trade with Spain's South American ports. The capital needed to create this transatlantic mercantile network would come from investors in British unsecured notes—floating debt—who would be forced by an act of Parliament to swap that debt for shares in the new concern. Finally, ensuring that the new trading company would have an adequate supply of working income, the Treasury would pay an agreed but reduced rate of interest on the swapped-in obligations.

Just as in the Irish land deal, the idea was presented as a win for everyone. Britain would consolidate what it owed and pay 6 percent of its debt—less (but hopefully more reliably delivered) than what it currently owed. Those compelled to hand over their risky, maybe even worthless paper would receive shares in a going concern. The company itself would make its profits (and boost its share price) as it pursued what was imagined to be fabulously lucrative commerce across the seas. Best of all, at least for Blunt and his circle, the whole scheme would float on other people's money. They did not have to come up with a single shilling of their own. What would be (on paper) one of the largest economic institutions in the world was to rise out of nothing more than the ever-expanding pile of promises Britain had made to its creditors.

Harley told the House of Commons about the scheme in March. The measure creating the new venture passed two months later, and the charter was granted at the end of summer. Formally, the new enterprise was called "the Governor and Company of the merchants of Great Britain, trading to the South Seas and the parts of America, and for the encouragement of fishing." It was known then and is remembered now by a much simpler name: Blunt and friends had just given birth to the South Sea Company.

Part Two
Money's Magic Power

All *Britains* rejoice at this Turn of the State,
Which rescu'd from Plunder the nation;
From this happy year you for ever may date
Of Credit the Restoration

—Arthur Maynwaring,
"An excellent New Song,
call'd Credit Restor'd"*

* Maynwaring, a skeptic, did not actually believe this.

Chapter Eight
"an Exquisite Management"

The late seventeenth and early eighteenth centuries were a great age for public dispute—and no one mastered the tactic of combat by printing press better than Daniel Defoe. At each major turn in politics and national finance, he was there, with something to say and a brilliant turn of invective in which to say it. The famous novels would come later—he produced *Robinson Crusoe,* his first, in 1719, when he was fifty-nine. For most of his life he wrote more or less anything that could pay the bills—he was one of the first true professional journalists, but he also wrote topical poems, manuals on proper conduct, political tracts, popular history, works on trade, and more besides.

As circumstances dictated, he was also what we would call, anachronistically, a spin doctor, producing copy in support of power for pay. He never necessarily argued against his own convictions, but he was able to trim his sails to the needs of the patron of the moment. One consequence of such works for hire is that over the years Defoe created a polemical, ferociously argued, often ad hominem record of how Britain's leaders attempted to solve their problems of debt and credit. Defoe is thus a not always reliable but emotionally acute witness to what his clients—Britain's leaders—thought they were doing when they tried something new along Exchange Alley.

In 1711, though, Defoe found himself in a very tricky position. His problem? He'd bet on the wrong side. Though the Tory grandee Robert Harley had been one of his first backers in the new century after his imprisonment for debt, for several years before 1711 he had followed power over to the Whig side, serving as a reliable mouthpiece for Godolphin. The Tory triumph caught him out, and for months he struggled to regain Harley's confidence. John Blunt's South Sea maneuver gave him his chance, and his tract in support of the deal reminded the new leader what Defoe was capable of. That pamphlet expressed both what he believed and what his patron hoped to sell the public: the South Sea Company was no mere financial scheme, designed solely to ease the pressure on the Treasury. Its commercial monopoly was expected to be a valuable business. It was good in itself, an engine of wealth to come, and not simply another desperate move by a nation living beyond its means.

Such public propaganda was necessary because not everyone was convinced. Defoe faced some confounding issues in his argument: after all, the proposal relied in part on what he disdained—Exchange Alley trickery. But for Defoe, an ever-hopeful projector, the element of trade seemed to overcome any doubts. So when South Sea skeptics emerged, he was among the first to reply—and, in an illustration of how the pen-for-hire life goes, in defense of some of the same people he had previously scorned.

Daniel Defoe in an undated portrait

He did so in the usual form of such public disputes. He first raised the key argument he intended to refute, that "it falls among People unacquainted with Trade," who, however clever they might be with money, had "no Occasion to venture to Sea, understand nothing of Merchandizing." Which meant, as Defoe restated his foes' claims, "to talk to these of a South Sea Trade is to talk *Hebrew* and *Arabic*."

This was perfectly true, and Defoe knew it. As the historian John Carswell wrote, it was striking that among Blunt and his founding colleagues "not a single one had any experience of the South American or even the West Indian Trade." Given that inconvenient fact, Defoe did his best. The problem, he argued, lay with the company's split personality: it was both a sort of bank and a trading enterprise. Those able to manage debt weren't merchants, he wrote, and those deeply immersed in trade were hardly experts in manipulating money. That one company had to tackle both jobs, he said, was "the Misfortune of the Nation." But "an Exquisite Management"—presumably including hiring some folks who knew about maritime commerce—could avoid disaster. Be smart, be clever, and, Defoe wrote, "This Trade is not only probably to be Great, but capable of being the Greatest, most Valuable, most Profitable, and most Encreasing Branch of Trade in our whole British Commerce."

Again: it's important to remember that Defoe definitely wrote what Harley and the Tory ministry wanted to hear. But it's also true that Defoe himself believed in Britain's trading prowess. The East India Company was perhaps the most obvious exemplar of that skill, but the throngs at the Royal Exchange bore witness to the importance of commerce for the nation in peace and war. And there is always the problem of historical hindsight. We know now what happened. Defoe and a very great many of his compatriots did not, of course, and sincerely believed that the risks the new company faced were small and its potential gains enormous. Given the outsized fortunes East India merchants had gained and the almost two decades of successful operations at the Bank of England, such confidence wasn't crazy. Still, even if it was reasonable to believe in Britain's mastery of both commerce and finance, Defoe and fellow boosters assumed that the new company's plans would survive first contact with the real world. They did not.

THE FIRST TRAP the Company faced was the most obvious: trade with Spanish possessions required Spain's consent. In 1711, the shooting war between Queen Anne's men and the French-backed forces of His Most Catholic Majesty was still going on—meaning that at launch there was exactly zero chance that any legal commerce would take place. From Blunt on down, the directors certainly knew that, and they had an answer ready. In February 1712 they asked the ministry for an armada: twenty ships of the line, forty more transports, and four hundred of Her Majesty's soldiers bent on conquering enough of Spanish South America to create a defensible base in the New World. They believed, apparently, that the new ministry would authorize a new front in the Whig war it was trying to escape—and that any of Spain's possessions would welcome trade offered at gunpoint. They were confident enough to collect a cargo of 1,100 tons, worth £200,000, with a firm date set for the start of the voyage: June 26, 1712.

April passed, then May.

Merchandise was stowed in the Company's two ships. June came, then June 26. Both vessels swung at anchor. Summer passed; autumn came; then winter, when no sane traveler would cross the Atlantic by choice. Finally, in March 1713, the South Sea Company informed its contacts in the government that the goods within the holds of the waiting ships had begun to rot.

No officials responded. No soldiers embarked; no ships sailed. The fate of the trade stuffs is unknown. Harley and his colleagues in the ministry had no intention of allowing the Royal Navy to poke the eye of the Spanish Empire in the midst of the long and difficult peace negotiations. The government had got what it needed from the Company once the floating debt passed onto its books. Still, it was important to the ministry that the South Sea scheme avoid total collapse, if only to prop up a Tory-leaning financial center to counter the Whiggish Bank of England. So, as negotiations continued on ending the war, Britain's diplomats sought some concessions the Company could use.

Harley's negotiators seemed to have won that point in the treaty finally agreed in 1713. Spain granted the South Sea Company a set of enumerated, specific commercial privileges. The terms of the agreement allowed the Company to send one ship each year to the fair at Cartagena (in present-day Colombia) or to Vera Cruz (in Mexico). The British also won something potentially more lucrative, called the Asiento: the right to trade in slaves—4,800 Africans whom the Company could sell each year at any port in Spanish America. The Company formally received these privileges in June 1713 and almost prepared its first significant trading voyage, finding ships, buying trade goods, and selling £200,000 in Company bonds to pay for the whole exercise. Then someone read the fine print: the King of Spain, Queen Anne of England, and some other grandees claimed 58 percent of the profits of each permitted trading journey. Company officials may have been novices at long-distance commerce, but they could do the arithmetic: this was a bargain to pick their pockets clean. They took all the risk of each transatlantic trip and still had to surrender over half of whatever they might make on happy voyages. The 1713 expedition was canceled, and the Company and its supporters devoted the next year to clawing back some of Queen Anne's share.

In practical terms, this amounted to a loss of two years of trade—so much for that "Exquisite Management." Everyone who had been compelled to swap claims on the Treasury for shares had to wait, to be satisfied with the hope that the government's reduced interest payments would yet fund the Company's expansive plans. But there was, at last, a glimmer of hope. In 1715, the first South Sea trading ship crossed the Atlantic. Repeat voyages followed until, by 1718, the Company was able to book about £100,000 in profits.

The slaving side of the business seemed even more promising.

Here the Company had gotten off to a better start. In the last half of 1713, South Sea agents established seven slave-trading factories along the seaboard of Spanish America, north to south from Vera Cruz to Buenos Aires. (The trade in humans was allowed in more locations than the Company's general merchandise.) Way stations in Jamaica and Barbados were set up so that those who survived the journey could be preserved long enough to reach the major markets in Spanish Central and South America. A deal was struck with the Royal African Company to supply "healthful, sound negroes of all sizes, in such conditions as to be able to go over the ship's side." Vessels were found and journeys charted.

Yet for all the Company's preparation and the impressive-seeming volume of its trade, it did not profit from its transport of suffering across the ocean.

The reason? Early critics had been right when they pointed to the Company's lack of experience in any form of trade. Slaving was always a grotesque, brutal, murderous business, but the South Sea Company was more inept—and hence even more wasteful of human life—than the long-established Royal African Company. Its ships lost more slaves on their journeys than those of its rival. At the same time, the Company men were unprepared for competition at the point of sale. British planters in Jamaica resisted the South Sea Company's monopoly of the market in bondage, complaining both that it raised the cost of their own labor force and (less publicly) that it interfered with their own off-the-books slaving operations. Once a Company ship landed its cargo, new obstacles appeared. Local officials did everything they could to block or milk every sale, including outright dishonesty: overcounting the number of slaves being sent ashore when payment of import duties was being calculated, and somehow losing some—to accident, disease, or any other pretext—to undercount the final figures of those sold. The Company couldn't even get paid properly. A line in the Asiento allowed buyers to purchase slaves with the "fruits of the country"—goods like cacao, sarsaparilla, tobacco, and other commodities. Instead of a clear price and a bankable profit, slaves that had been bought with cash and sold for a stack of dry goods to be shipped back to England represented a delayed and uncertain reward at best, and, often, a trading loss on every "sound Negro" borne across the sea. Selling stolen lives took skill, it seems, and local knowledge the newcomers did not possess.

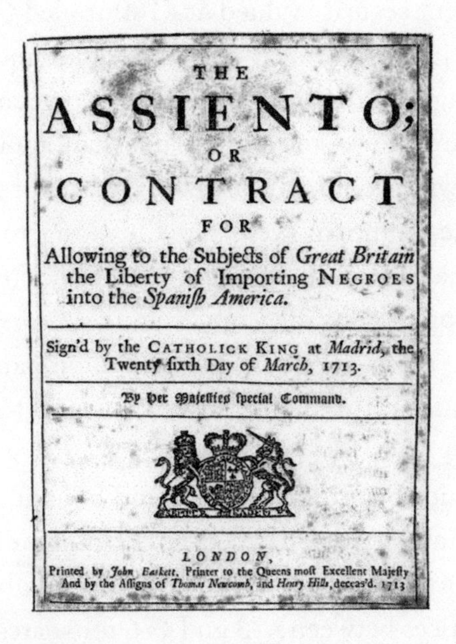

THE

ASSIENTO;

OR

CONTRACT

FOR

Allowing to the Subjects of *Great Britain* the Liberty of Importing NEGROES into the *Spanish America.*

Sign'd by the CATHOLICK KING at *Madrid*, the Twenty sixth Day of *March*, 1713.

By Her Majesties special Command.

LONDON,
Printed by *John Baskett*, Printer to the Queens most Excellent Majesty. And by the Assigns of *Thomas Newcomb*, and *Henry Hills*, deceas'd. 1713

The Asiento Contract gave the South Sea Company the right to sell slaves in Spanish America.

That left the hard reality: the flow of goods and bodies from English ports to Latin America from 1712 forward through the decade never yielded anything approaching wealth for those who'd offered their government paper for equity in a private concern. The South Sea Company, it turned out, was not a twin to the East India fortune-making machine. Cynical observers could be forgiven for believing it had never really been meant to become one. It still had work to do—but that was dull bookkeeper's fare: just taking in the flow of (reduced) interest payments from the Treasury and then passing payments through to its shareholders.

BORING DIDN'T MEAN bad, though—or so Exchange Alley concluded. South Sea share prices quoted at Jonathan's in these early years show that the stockjobbing world didn't actually care about Spanish fantasies. The original debt-shares conversion had been at par: a government security valued at £100 traded for £100 of South Sea stock. But just as the deal was to be announced, that floating debt was trading at a discount of about 32 percent. That meant as long as Company stock sold for no less than about £70, those involved in the swap would still break even. The promised 6 percent payments to the Company were actually closer to 9 percent, measured against that discounted market valuation. To put that another way: the share price published in Castaing's *Course of the Exchange* reveals how the smart money valued the Company purely as a financial play—and in that light the deal seemed to be working just fine.

The first South Sea stock was sold and bought in October 1711, well before any investor could expect any trading profits. Those trades landed just above the break-even territory for government paper, commanding prices between £73 and £76 for shares with a par value of £100. The Alley warmed to the stock over the next two years, buying and selling shares at £94 by the end of 1713. That was a decent bump: roughly a 35 to 40 percent increase over the value of the debt that had been swapped for shares two years before—and crucially, it did not turn on any false optimism about the Company's commercial prowess. Instead, shareholders saw the value of their stock continue to rise slowly, reaching their par value of £100 in 1716. At that point, the Treasury was reliably making its payments—6 percent per year, as promised. This was reckoned a solid return, nothing special, certainly, but not terrible either, just the six pounds on the hundred that London's market in money then believed any relatively safe security should deliver. In other words, at par, £100 per share, the clever men at Jonathan's reckoned the South Sea Company to be worth exactly the value of its financial assets and nothing more. It was a quiet business, sound and slow. If Harley's goal had been to use a veneer of mercantile derring-do to provide cover for a scheme to reduce the immediate pressure of Britain's debt, then by the middle of the decade it was clear that he'd succeeded.

Not that Harley reaped the rewards of such sound judgment. His hold on power had always been a balancing act, managing rivals within the cabinet, the demands of a parliamentary majority eager for vengeance on the Whigs, the Whigs themselves, and his monarch, Queen Anne, who hugely resented any hint that she was being managed by her ministers. Perhaps more important, his dive into the bottle after the death of his daughter in 1713 offended Anne. Her Majesty added that lapse to the tally of Harley's sins until, on July 27, 1714, Anne fired him, she wrote, because he neglected his duties, was careless with the truth, unpunctual—this was all in a public statement—not to mention that "he often came drunk" and "lastly, to crown all, he behaved himself towards her with bad manners, indecency and disrespect."

All true, no doubt, or close enough to serve the needs of the moment. But in fact, Harley was toppled as much by the familiar course of factional warfare as by any swipe of royal displeasure. Another Tory had passed him in Anne's confidence, while his base of support in both houses of Parliament had grown brittle as each contested decision exacted its cost. His Tory rivals had no time to enjoy their triumph, though. Days after dismissing Harley, the queen rose from her bed, reporting that she felt better than she had earlier in the week—but then fell down, convulsing. The fit carried through the morning, and when she regained consciousness she could barely speak. Anne died two days later, just forty-nine years old. She had been pregnant seventeen times but had given birth to just five living children. None survived her.

So, at four in the afternoon after Anne's death, Georg Ludwig, elector of Hanover, was proclaimed King George of Great Britain and Ireland. George was Anne's second cousin and a great-grandson of the first Stuart monarch, James I. He had dozens of cousins with similar or stronger ties to the British royal family, but he possessed one attribute that placed him ahead of them all: he was a committed Protestant. Thus a relatively remote German prince became the center of London's political life—and almost immediately began to make his authority felt. The new king had his grievances with the Tories, not least that some of them retained some sympathy for the deposed and Catholic house of Stuart. George reached England six weeks after Anne's death, and in the election of 1715 the Whigs, to their new king's evident satisfaction, returned to power in resounding fashion.

Enter, again, the South Sea Company. Though the Company was wholly a Tory creation, in just three years it had become so much a part of the financial landscape that it survived Harley's fall. There was just one concession: it agreed to "forget" payments due from the state—about a million pounds—that the Treasury had somehow failed to pay. From there, while its trading luck didn't improve under the commercially minded Whigs—no one could fix the lack of a willing partner on the Spanish side—it continued to do its best to be useful.

IT DIDN'T TAKE much time for the incoming ministry to discover how urgently it needed the Company's services. On regaining power the new ministry found that in spite of Harley's efforts to bring the nation's debts under control, Britain's finances remained both dire and desperately confused.

The tangle of official obligations stretched back at least forty years to the bad behavior of the Stuart king Charles II, whose "stop" of the payments from the Exchequer lingered as an obligation of more than half a million pounds still due from the Treasury. The long-term loans from the 1690s remained on the books. That money, the price of the Nine Years' War, had grown into over £12 million due, carrying interest rates of up to 10 percent over terms that extended for up to a hundred years. These were the "irredeemables," so called because they could not simply be paid off during their run. More than £11 million more in long-term debt had piled on during the Tory years—much of it in redeemable lottery loans that could be paid off whenever the government could find the funds. And then there was another £16 million in floating debt that had been absorbed by the Bank of England, the East India Company (each holding just over £3 million), and the South Sea Company itself, which owned £9 million of government paper—a sum that made it the largest financial company in the world.

Put all that together and the results were grim indeed: £40,357,011 owed, bearing an annual interest charge of £2,519,808, roughly half of the nation's annual revenue. And that was in a time of peace, with no need to pay for an army and navy on campaign. Worse yet, from a survey of the whole miserable list of debts it was clear that there was no obvious way to bring the problem under control. Though some loans were linked to distinct streams of revenue—like the old Malt Lottery of 1697, secured by what was in effect a tax on beer, cider, and perry (cider made from perry pears)—the enormous number and variety of Britain's money-raising schemes over the last twenty years threatened future borrowing. What would happen should the government run out of obvious income streams to pledge—or if potential lenders decided that there wasn't enough revenue left in the realm to keep up with payments due?

As far as the Treasury was concerned, the final insult was that by 1714 the cumulative impact of all the crises in which it had been forced to borrow meant that too often it ended up paying top rates for the nation's money, much more than private borrowers faced. Taken together, Britain's debts carried a roughly 6.25 percent interest charge, with the highest rates for the longest-term, least tractable loans. Meanwhile goldsmith-bankers charged their customers just 5 percent, sometimes less. That difference added up to about half a million pounds in extra cost per year, not enough on its own to solve the nation's difficulties but still a very useful sum. Thus the question for the newly installed ministry: how to cut costs, simplify the debt, and still borrow as needed.

It took the Whig leadership almost three years to come up with a plan, which was mostly the work of one of the ministry's youngest and most clever members. That man was, it turned out, both good at his job and lucky, as fortunate in his setbacks as in his victories.

His name was Robert Walpole.

CHAPTER NINE

"many Conferences and Considerations"

Over his long life Robert Walpole would play every part his times offered an ambitious, brilliant, avaricious, and calculating public man. One role that he surely never anticipated, though, came in one of the most macabre games generated by Britain's financial revolution: as the object of strangers' bets on the time of his death. Gamblers gamble—and here the play came as an unanticipated consequence of the birth of life insurance.

For all that London's insurers paid little attention to Edmond Halley's paper about death in Breslau, life insurance itself was readily available in Britain from the early decades of the eighteenth century. The key historical detail is that until 1774 it was perfectly legal to take out insurance on a stranger. That meant that much of the point—the fun—in such insurance was the chance to place a bet on the fate of a stranger. Such wagers could take many forms. In one notorious example, two wastrels proposed a wager on which of their fathers would die first, for revealing stakes: the one whose dad shuffled off first would be the loser, obligated to pay off all the debts of the one still encumbered by a breathing parent. The two men ended up before a judge when the first to inherit refused to pay up on the grounds that the news that his father was already dead had not yet reached the club when the bet was struck. The court did not agree.

It was hardly a revelation that useless twigs of aristocratic family trees might squander their estates. But the rise of quantified approaches of human experience yielded both new ways to bet and a new culture of gambling, one in which any turn of the human comedy could evoke a flutter. There were wagers on which gentleman might get married first, which gentleman's wife would produce a child next (and how long it would survive!), even, notably and notoriously, another private flutter on the gender of an ambiguously male French diplomat—or spy—then in London. That one also prompted a couple of lawsuits when bettors tried to collect on the evidence of alleged witnesses. Lord Mansfield heard at least two cases on the matter and finally ruled from the bench that all bets on this particular subject were invalid—dismissed for the sin of discussing the glaring impropriety in open court.

Given that mere scandalous gossip could spark a genteel flutter, gambling on the lives of strangers gave little pause. From the early eighteenth century on, using insurance to bet on the fates of prominent men (and some women) became routine, for good reason. If an important figure died at a crucial moment, the resulting disruption in statecraft or business could prove costly to all kinds of ventures. An insurance policy that paid off in the event of such death was a hedge against the twists of politics, war, or insurrection—or the vagaries of Britain's official financial management. Between 1713 and 1717 one insurance concern, the Amicable, was involved in dozens of bets on the great and the good—on the Tory High Church firebrand Henry Sacheverell, for one, along with one of his chief political antagonists, Charles Townshend, a peer and one of the leaders of the Whig ministry that swept away Sacheverell's friends after Queen Anne's death. Notably, the members of the Amicable saw the South Sea Company as important enough to insure John Blunt, almost certainly without his knowledge—and they saw the same opportunity in a still-young, still-rising, not-yet-dominant politician named Robert Walpole.

Walpole would go on to become almost certainly the most frequently insured figure in British history. The cost of a policy on his life even became a kind of index for trouble: the premium rose whenever he faced a political challenge, or the nation he governed faced a threat—the threat of another Stuart invasion, for example. At the same time as policies on lives began to spread, so did commercial underwriting, protecting commerce by insuring ships at sea and individuals and companies against the risk of fire. All of these transactions advanced the broad idea that contingency, risk, and human choices could be measured in money and analyzed in ways that anticipated—laid bets on—the future.

In time—a century, at least—insurance would become routine, simply part of the commercial world. But in its infancy there was something marvelous about making a wager on which famous person would die next. This was a new world, in which numbers on paper could multiply and fortunes rise. That same faith propelled the whole panoply of novel ways to speculate along the Alley, including the newest possibilities provided by the South Sea Company. Its rise would coincide with Walpole's; his career would turn on a series of encounters with it and its leading man, Blunt. Their interaction would climax in a collision that would settle the fate of that enterprise, of Walpole himself, and, in some measure, of the nation as well.

ROBERT WALPOLE'S ORIGINS were sound enough, though hardly suggestive of greatness. Born on August 26, 1676, he was the third son and fifth of seventeen children born to the wife of a well-off Norfolk landowner also named Robert. Growing up, he anticipated a familiar fate for younger sons of the well-to-do: a respectable career in the church. But one older brother was killed in 1690 at the naval battle of Beachy Head, and the other died in 1698, just as Walpole completed his second year at Cambridge. Suddenly the heir, he was called back from university to be trained up as an estate manager, county notable, and hereditary Whig.

The younger Walpole never lacked for confidence. He would later say that had his brothers lived, forcing him into the church, he would have ended up as archbishop of Canterbury. But nothing in his interlude as a country gentleman suggested future consequence. His father delegated tedious chores—overseeing the Walpole cows at weekly markets was one that stuck in memory—and made sure his son fully mastered the habits of a local magnate: "his mornings . . . being engaged in the occupations of farming, or the sports of the field, of which he was always extremely fond, and his evenings passed in festive society." Festive indeed—the elder Robert "supplied his glass with a double portion of wine, adding, 'Come Robert, you shall drink twice, while I drink once: for I will not permit the son in his sober senses, to be witness to the intoxication of his father.'"

The older Robert's management of his son went beyond the dinner table. It was already a truth universally acknowledged that the heir to a respectable estate must be in want of a wife, and so, within two years of his return to his father's house, a suitable mate was found for Walpole: Catherine Shorter, daughter of a wealthy timber merchant and former Lord Mayor of London. Measured by looks alone it was a mismatch: Catherine was accounted "a woman of exquisite beauty and accomplished manners," while Walpole, though not yet the heroically fat man he would become, "bordered on the absurd, for he was short and plump." But marriage in Walpole's class was a rational business, and the fact that the groom was a landed heir while his bride brought a dowry of £20,000 settled the matter. The two were married in July 1700. That November, his father died and Walpole became sole master of an estate with rents that brought in £2,000 a year. He immediately stood for his father's parliamentary seat and, to precisely no one's surprise, won election to the House of Commons.

Walpole and Catherine promptly set up housekeeping in London, to the new member's delight. He was done driving cattle to a village market; there would be no more double bumpers of wine while his neighbors retold stories everyone at table could recite from memory; no more commanding parent calling him to his duties. Instead, both halves of the young couple threw themselves into the worldly—and expensive—pleasures of the capital.

THE KIT-CAT CLUB—the drinking and talking shop that welcomed Walpole shortly after he arrived in London—had gotten its start in murky circumstances. It most likely took its name from Christopher Catt, who ran an inn on Shire Lane. London's satirical chronicler Ned Ward praised him as an "Engineer, so skill'd in the Fortification of *Cheese Cakes, Pies* and *Custards*" that when a Whig wit wished to entertain his friends, he "Invited them to a Collation of Oven-Trumpery" featuring Catt's pastries. His signature lamb pies had already been dubbed Kit Kats, and the name stuck to the growing band of devotees that moved from Catt's establishment to a nearby tavern (as did the tradition of serving Kit Kats at each weekly meeting).

The Kit Kats were never a particularly erudite crowd: a Whig stronghold, the club was famous for well-lubricated "literary" contests in which Whig wits bandied verses at each other, mostly having to do with gossip and scandal. Another of the club's traditions was to offer not-quite-openly-rude toasts to reigning beauties, among them Isaac Newton's niece, Catherine Barton. After moving in with her uncle when he moved to London, she encountered Charles Montagu, the Whig patron who had gotten Newton his Mint post. She became Montagu's mistress, and he immortalized his admiration for her charms in a Kit-Cat toasting cup inscribed in her honor: "At Barton's feet the God of Love / His Arrows and his Quiver lays, / Forgets he has a Throne above, / And with this lovely Creature stays."

That was rather more discreet than much of the club's usual wit, and it marked about the limit of the poetical heights its members could achieve—but such doggerel does give a whiff of Walpole's new world: not overly virtuous but very useful for an aspirant to power.

There were connections to be made with other political comers, a chance to brush shoulders with the occasional grandee (the membership included half a dozen dukes) and the company of clever men, like John Locke himself, who added a bit of polish to the drink and the fast talk.

All this was catnip to the country-born Walpole, entering into his true habitat. There was just one problem: he couldn't afford the life. Between the expenses for a member of Parliament rising within a faction and the cost of life in London with a wife trained in extravagance, Walpole's finances mimicked the nation's: he dove ever deeper in debt and was always reaching for the next expedient. He borrowed money from shopkeepers, then dodged his creditors until his Norfolk agent, Charles Mann, wrote, "They tire my heart out." One lender "rages like a madd man," he added, and two others "have been with me severall times and make great complaint." He borrowed from a man who married his cousin; he shorted his mother on the money due under his father's will so frequently that she finally threatened to take Walpole to court. He mortgaged family land and sold some.

None of this seems to have worried Walpole much. He lacked money, not bravado, and so simply ignored most of those dunning him. His agent, Turner, told him that "some reports spread about your extravagant way of living very much to your prejudice." He may have hoped to shame his client into better behavior. If so, he failed. Throughout his early years in London, Walpole's appetites consistently exceeded his means, and at no time did he tailor his life to his purse.

THERE WAS ONE sure way out of Walpole's predicament. A mere seat in Parliament did not yield major political spoils, but some offices that were in the gift of the crown—which is to say, of the king's ministers—could reward their occupants handsomely. So, to place himself in contention for an appointment up the greasy pole, Walpole threw himself into the business of the House. He regularly served as a teller—a vote counter; he did committee work; he spoke on the floor of the Commons, over time gaining skill at the oratorical knife-work of parliamentary debate. Notably, months after taking his seat, Walpole found his way onto a committee appointed to examine the national debt. Such labor had the desired result. In 1705 he joined the council advising Prince George, Queen Anne's husband and titular head of the Royal Navy. The perks of that office served him well, placing an Admiralty boat at his disposal, which he used to land a useful quantity of un-dutied claret, burgundy, and champagne, smuggled in from Holland.

He revealed a knack for administration there too, even in that minor job. Naval finances were in their usual mess, which meant that naval logistics were too, and as one of Prince George's advisers Walpole took on the task of taming the convoluted mess of navy accounts. That helped earn him his first taste of actual power with his appointment in 1708 as secretary of war. Despite the title, that office had nothing to do with strategy, or with the direction of Britain's ongoing campaigns; rather, it involved managing organizational matters for the army. Given the habits of early-eighteenth-century politics, that made it a potentially very rewarding post for any young statesman willing to make discreet use of any money that passed through his office.

Throughout these early years, Walpole demonstrated the capacity that would serve him through four decades in Parliament: an aptitude for public business that few, if any, of his rivals mustered, married to an enthusiastic appetite for factional warfare. His rise halted during the political crisis sparked by Sacheverell's preaching. As one of the Whigs' most feared tacticians on the floor of the House of Commons, he became one of the leaders in the move to punish the cleric. He helped draw up the articles of impeachment, and at the ensuing trial Walpole sharpened his attacks, arguing that Sacheverell had flirted with sedition: the plain meaning of the sermon, he said, attacked "the very foundations of our government"—even to the point of feeding the old dream of restoring a Stuart king to the throne.

That trial was an uncharacteristic misstep for Walpole, and it contributed directly to Harley's triumph and the seeming collapse of decades of Whig dominance. Walpole himself became a target of the new regime, which responded with a trial of its own, charging Walpole with having accepted bribes while secretary of war. This was (as Sacheverell's had been) a wholly political case. There is no doubt that Walpole grew much richer in office than he had been as a mere landowner and member of Parliament. But the Tories were unable to prove that he had personally gained from any of the transactions they brought up at trial. No matter. Eliminating Walpole's opportunity to torment the government from the floor of the House was all that mattered, so the Tory majority voted to convict him of "a high breach of trust and notorious corruption." He was expelled from Parliament and marched off to the Tower of London on January 17, 1712.

Walpole, never the most sympathetic of men, was thus, like Sacheverell before him, turned into a martyr, to his immense, if slightly delayed, good fortune. He was a pampered guest at the Tower, with servants, ample supplies of pen and paper with which to vex his enemies, and a door that swung open to admit a steady parade of friends, from the Duke and Duchess of Marlborough down. The actor Richard Estcourt produced a ballad, grandly titled *On the Jewel in the Tower*. In verse perhaps slightly more polished than the usual Kit-Cat fare, Estcourt ridiculed Walpole's opponents—"With thousand methods they did try it, / Whose firmness strengthen'd every hour; / They were notable all to buy it, / So they sent it to the *Tower*"—then warned those scoundrels what was surely to come: "The day shall come to make amends, / This Jewel shall with pride be wore, / And o'er his foes and with his friends, / Shine glorious bright out of the *Tower*."

The Tory overreach became a public embarrassment a few weeks into his imprisonment—a replay of the Whigs' self-inflicted wound in the Sacheverell trial—when Walpole won the by-election to fill

Robert Walpole in 1715—in his younger and still fairly svelte incarnation

the seat from which his enemies had just ejected him. Ultimately, Walpole was released from the Tower, but the damage had been done: it had long since become clear to the more acute Tories—like Jonathan Swift—that his return to power only waited on the fall of Harley's government. When that day duly came in 1714, Walpole reaped the spoils. He took the seemingly minor but exceptionally lucrative post of paymaster general. The job came with a fine house and garden backing on the Thames and, even better, a way to match his style of life to his means. From his desk he oversaw a river of cash—and between 1714 and 1717 he managed to accumulate over £60,000 from investments of government funds under his direct control—on the order of £10 million in twenty-first-century terms. Meanwhile, alongside the usual pursuit of the rewards of office, the ongoing problem of paying for British power remained. That fell to Walpole too, following his appointment as chancellor of the Exchequer and First Lord of the Treasury in October. Figuring out how to pay for Britain's wars was now his job.

THERE HAD BEEN some progress on the debt in the year before Walpole took over, but mostly a series of minor fixes. Even with Harley's moves, Britain still owed huge sums at high rates on the guaranteed national debt. It owed yet more in the hodgepodge of floating obligations. Interest payments, even in a time of peace, still consumed more than half of the nation's annual takings. The nation remained, as ever, on the edge of being flat broke. It took Walpole nearly a year to assess the full extent of the damage, but when he emerged he was ready to present a comprehensive plan to fix the problems once and for all.

Walpole began by defending the need to be able to borrow, arguing that the national debt was the price paid to defend "the Religion, Laws and Liberties of *Great Britain*," which, he declared, "I believe, no Man will confess that he thinks were dearly purchased." The total owed, he conceded, "was become large and burthensome," but given that private borrowers were paying just 5 percent on their loans, compared to the rates the Treasury faced—6 percent, 7 percent, or more—now, he said, was the moment to act: "The Circumstances of Time, high Credit, and low Price of Money" led him to try "to make Use of that favourable Opportunity, that the Publick might share in the common Advantage." In plain language, if the state could take advantage of the prevailing interest rates, this was the moment to refinance the nation's debts at a lower charge.

There was one great obstacle to Walpole's plan: a hefty chunk of the national debt, the loans explicitly guaranteed by Parliament, had been issued with the legally binding promise that they could not be repaid ahead of schedule without, as Walpole put it, "the absolute Consent of the Proprietors." These "irredeemables" made up almost one-third of the total owed.* Then as now, people don't give up an advantage—those high-interest-rate payments—for nothing. Walpole understood that the Treasury had to give something to get what it wanted, better terms from its creditors. To that end, Walpole wrote, "it cannot be doubted but many Conferences and Considerations were had . . . among the monied Men, and Money-Corporations."

If anything, Walpole underplayed the agonizing, ponderous pace

* The distinction between redeemable loans—ones that could be paid back at any time the government chose—and the irredeemables, which could not be retired before their specified due date unless the lender agreed, didn't depend on the length of the term of a given loan. Some long-term national debt was redeemable, some was not.

of those meetings. There's no record of what was said, or of how many times the politicians and the monied men gathered to go over every offer, clause by clause. It took more than a year to construct the mere outline of a deal that addressed all the faults in the nation's finances. Some months more passed before everyone—the Bank and the South Sea Company, key private lenders and the ministry—all came on board. But by March 1717 Walpole presented the program that, if fully implemented, would put the Treasury back in control of the full range of Britain's borrowings. Those who held redeemable long-term debt (any loan the government could pay off whenever it chose) would see their interest payments drop. (If they didn't like it, the government was given the authority to borrow as needed to pay off the total loan to any who refused such revised terms.) Owners of lottery tickets (those hybrid long-term instruments that offered both a chance at a prize and bond payments for some fixed period of years) faced the same choice. Most important, those who held the irredeemable obligations, with their guaranteed, decade-after-decade stream of payments, faced one crucial restriction: the kind of debt that is known technically as illiquid. It couldn't be transferred to another person, and it couldn't be sold or bought in Exchange Alley's secondary market. That meant if owners of an irredeemable needed the capital tied up in their investment in government debt, and not just their yearly interest payments, they couldn't get it. So Walpole's deal included an offer that, it was hoped, would be persuasive: trade their higher but fixed and immutable interest payments for liquid investments, financial paper that would still pay interest (albeit at a lower rate) *and* that could be freely traded to third parties, bought and sold at will along Exchange Alley.

The mechanics of the plan were somewhat complex, but the underlying concept was simple enough: the proper price, the number at which government debts would be traded in, could be found with the help of the previous decades' work on present-value calculations. That was the math developed to put a number on the value today of a flow of payments over any span of time. Such calculations originated as ways to value a piece of land producing an annual yield. The same mathematics used to price a pasture could be applied to finding the present value of a bond based on its future yields. In this new setting, the goal was to find a price for shares that carried both an annual interest return and a market value in terms of the long-term payment stream of an untradeable public debt.

People might disagree on details of a model—how much to multiply a year's income to set a price, for example, or how to estimate the demographic risks in, say, an annuity on three lives.* But that was the stuff of ordinary negotiation. The key to Walpole's plan was that by the second decade of the eighteenth century the basic mathematics of investment, of returns over time, was well enough understood to make the whole negotiation possible. Throughout their endless conferences, the Treasury and its creditors knew they could reason their way to a mutually acceptable price that could turn each high-interest, illiquid, irredeemable loan into a reduced-rate, redeemable, and tradeable one. In other words, this was precisely the kind of empirical and quantifying reason that the scientific revolution had both deployed on its own problems and turned into a way of thinking that could be used—and prized—by Britain's elite society as well.

Over the summer of 1717, Parliament got to work on the bills to implement Walpole's plans. One handled the lottery loans, converting more than £9 million into lower-interest-rate instruments managed by the Bank of England. Others reworked various long-term obligations held by the Bank and the South Sea Company, achieving Walpole's aim of dropping the interest charges the Treasury faced. Some of the short-term obligations were retired, and more became cheaper, thanks to another rate cut. Taken together, this was, by any

* Some long-term national obligations were raised as annuities. In return for a given payment, the Treasury promised to pay interest on that sum as long as the lender lived. Some such annuities were offered not just on the life of the initial lender but on the lives of one or two more people. Once the last person designated on the loan died, payments would cease, and that would be the end of it; the purchase price of the annuity remained in the Treasury's vaults.

standard, an impressive achievement. The wild menagerie of ad hoc borrowings, desperate moves across three decades of war, had been reduced to something like simplified order. The nation's debt had just become cheaper, and, for a subtle benefit, the accounting had been simplified to the point where it no longer took months of work just to discover what was owed to whom.

But for all those genuine successes, the biggest task was left undone. The irredeemables were left untouched: their holders weren't to be forced to swap their overmarket returns for paper the government could retire—pay off—whenever the Treasury chose. In the end, compelling or coercing those rich enough to lend to the government to give up above-market returns was seen as too risky for Parliament to tackle. These were the people the government would need to turn to the next time Britain needed to borrow, so for the moment the ministry chose not to attack their interests. That might even have been the right choice politically, given how intimately connected the Whigs were to Britain's monied men. But fiscally, such timidity had its cost. More than £15 million of the most expensive obligations, or over one-third of the nation's total debt, remained a running drain on the Treasury. Combined with the fact that the plan as passed cut the cost of Britain's borrowing but did little immediately to reduce the total amount owed, the brute reality remained: Britain could still run out of money—and the possibility of raising more—at any moment.

THAT WAS CLEARLY the next major job for the chancellor of the Exchequer. But it had suddenly ceased to be Walpole's problem. Labels like "Whig" and "Tory" evoke an image of more or less coherent political parties. That's not how it worked in the early days of the party system. To identify as either Whig or Tory was in part a matter of family ties; Walpole himself had inherited his estate, his seat in Parliament, and his party from his father. Class and faith mattered too: the Tories were generally linked to landed wealth and High

Church orthodoxy, while to be a Whig implied a more tolerant religious sensibility, an affinity for a web of smaller landholders, and an alliance with town and city men of a certain amount of property. But even such categories were at most very loose guides: there were Whig dukes and Tory bankers. Political labels conformed to malleable coalitions—often with no more than "the enemy of my enemy is my friend" allegiances. Such alliances were always vulnerable to personal animus, pure competitive frenzy, and, on occasion, naked plays for power.

Thus, in early 1717, Walpole and Charles Townshend, an old Norfolk friend and one of the two highest-ranking members of the government, were ambushed by fellow Whigs. Their antagonists, Earl Stanhope and Charles Spencer, Earl of Sunderland, had traveled with King George to Hanover in the summer of 1716 while Walpole and Townshend remained in London. In Germany, Stanhope maintained the facade of what had once been a genuine friendship with Walpole, while both he and Sunderland managed it so that the king came to see that the ministers left behind in London were unreliable at best. When Parliament convened in early 1717, the two earls were able to drive Townshend and Walpole from the cabinet, with Stanhope seizing Walpole's jobs at the Treasury, while Sunderland occupied Townshend's offices.

The two defeated men planned revenge but could do little in the face of their rivals' coup. Walpole voted for the financial measures he'd done so much to craft, but after those passed he became, in effect, a member of the opposition, using all his skill in parliamentary maneuver to discomfit his rivals. Over the next few years, he regularly allied with the Tories and in 1719 even worked to block the long-held Whig priority of tolerance for Protestant religious dissent.

Walpole lost that battle but won others. Behind all the skirmishing between the two Whig factions, the irredeemables continued to loom over every attempt to bring the national budget into some kind of stable state. On that question, though, the one on which he had more knowledge than any other leading member of the House of Commons, Walpole was pushed to the sidelines, denied any influence at the precise moment the South Sea Company presented itself to the ruling ministry, proposing an experiment that would, if it worked, solve Britain's money worries for good.

CHAPTER TEN

"true to, and exact in, the performance of the General Work"

Dreams linger, perhaps none more tenaciously than those of wealth beyond avarice. For the directors of the South Sea Company, duly and dully turning receipts from the Treasury into payments to shareholders, their first hope was still that despite every setback trade might yet produce truly epic fortunes. In March 1717 the Company's first purpose-built ship, the *Royal Prince,* with Captain Baynham Raymond commanding, departed English waters carrying cargo worth over £250,000. Raymond made landfall at Vera Cruz that October and sent printed lists of his trade goods throughout New Spain. The *Royal Prince*'s lading likely exceeded the agreed total of 650 to 700 tons, but the ship was cleared on arrival by the local commander, Antonio Serrano, who was likely well compensated for his conveniently blind eye.

Shipping label for South Sea Company trade goods

Such local corruption irritated those Spanish traders whose cargoes followed the British into port, as Serrano's swift decision allowed the Company ship to sell off its goods ahead of the competition. The South Sea men were, of course, thoroughly pleased, and the rich haul of the *Royal Prince*'s first voyage led them to order a second ship, the *Royal George,* to be completed ahead of the 1718 sailing. She was launched early in the year, fitted out, and then loaded with goods bound across the Atlantic.

She never sailed.

Spain had never reconciled itself to the Treaty of Utrecht, the agreement that in 1714 marked the end of the War of Spanish Succession. The "Peace of Utrecht" forced Madrid to surrender a number of territories, including several Mediterranean islands. It was little surprise, then, that in the summer of 1717, Spanish soldiers returned to the battlefield, invading Sardinia while its adversary, Austria, was distracted by yet another conflict, this one with the Ottoman Empire. The Austrians ignored the provocation for a time, but once they reached a peace with the sultan in July of 1718 they turned their attention back to their old adversary—just as the Spaniards invaded Sicily. Other powers rallied against an overambitious Spain, which threatened the balance of European power. Austria, the Netherlands, France, and Britain, in a rare concord with their traditional rivals in Paris, created the Quadruple Alliance. Shards of war struck several fronts (including Florida, where French forces destroyed the Spanish outpost of Pensacola and thus prevented the invasion of South Carolina).

The various campaigns shuddered to a halt in 1720, but in the meantime the predictable effect was felt in London: trade between any British enterprise and any Spanish territory instantly and completely collapsed. The Company's slave trade evaporated, and though the *Royal George* would resume its South Atlantic passages in the 1720s, its first cargo of mercantile goods festered in its hold.

Alongside that private loss, the nation was again at war, with all that this implied for Britain's financial condition. The South Sea Company was once again thrown back on its purely financial role, meekly passing Treasury payments through its books. In that context, the endless pressure of the national debt became too good a crisis to waste, certainly for the clever men in John Blunt's inner circle.

THE SOUTH SEA Company had never stopped being a company in love with financial tricks. For example, by 1715 the government had failed to make almost a million pounds of the interest payments due to the company and its shareholders on the debts that had been converted into stock at the Company's founding. There was a simple solution to the problem: more financial alchemy, transmuting the interest arrears into South Sea capital—which in theory could have funded more trading adventures. Under the deal, the Company's capital increased to the satisfyingly round number of £10 million—the original subscription plus the interest arrears owed by the Treasury. That capital could be held as new shares with a face value of £100 or sold at Jonathan's and along the Alley for whatever the market would bear. The Treasury's payments to the Company would rise as the Treasury faced interest due on this higher total, the original debt plus the missed payments. In sum: the Company received more paper obligations from the government, a higher regular payment due, and the right to treat the missed interest payments as a store of capital that increased the total value of the company—and hence the number of shares it could issue. On the government side—in modern terms—this deal was very much like what happens when you skip a payment on a credit card: next month you'll owe your original balance, plus the interest on that balance accumulated over the prior month, *and* the interest on that new total, compounded out of both sums.

It's rarely a good idea for a person or a family to run their finances this way, of course, but—to step ahead of the story for a moment—nations are different, for several reasons. As early-eighteenth-century treasury officials knew to their sorrow, realms and individuals alike may make lousy deals when trying to borrow in an emergency. But it was and remains true that states have ways you and I don't that allow them to lay their hands on more money. King George's ministers could (and did!) hike taxes, raise excise and tariff charges, sell licenses (like the monopoly on trade with Spanish America granted to the South Sea Company), and so on. As Defoe put it, this meant that as needed, "it is easy to propose sufficient Funds . . . without any such Taxes, as shall either be burthensom to the Poor or scandalous to the Nation"—by which he meant that the Treasury could borrow at will because it could identify enough sources of revenue to persuade lenders that their returns would (mostly) arrive on time. For all the various examples of hiccups in interest payments on the new national debt, this wasn't the idle boast of a pamphleteer in the ministry's pocket. The first phase of the financial revolution in the 1690s had demonstrated that Parliament could indeed raise loans at almost any moment and use its powers to persuade investors that such debt would in fact be covered by one new stream of income—a fee on malt!—or another.*

The second advantage that the British government possessed was the inexorable passage of time. The funds it borrowed at any moment became bets on the nation's economic life year over year. The wager was that the ongoing work of every new enterprise, each voyage, everything that Britons did to get and spend in the future, would create

*To step well ahead of this historical moment, in modern, postmetallic-money states, there's a more direct way to deal with government shortfalls: states can simply create money through a variety of central bank operations. One controversial form of such money creation was used during the Great Recession that began in late 2007, so-called quantitative easing, in which central banks buy financial assets—bonds and other forms of debt—and thereby inject money created by the central bank into the broader financial marketplace; that "new" money is then available for use in the economy as a whole. The same money-creation power can in principle be used to pay back any official obligation expressed in that nation's currency; the risk of paying off the entire US debt in this way would be inflation and a rapid depreciation of the value of the US dollar against any other currency.

enough wealth to support the debts being incurred. The chancellor of the Exchequer didn't have to treat every expense as a pay-as-you-go imperative. Whole nations, as London's monetary thinkers had discovered, need not perform the virtue embodied in the very good advice to pay off a credit card balance in full every month. Rather, the task was to balance the needs of the moment with an analytical picture that could be drawn of Britain as a whole, all its getting and spending and accumulation, integrated over years to come.

From this realization—that time matters—flows perhaps the most significant advantage states have over families: what a government does, how much it chooses to spend and on what, can directly affect the way a nation's economy will perform going forward. When Parliament approved the Longitude Act in 1714, for example, the British government authorized spending significant amounts of money both as prizes and as research funding for anyone who could solve the problem of finding longitude at sea—a vital invention for a maritime power with global ambitions. This was at once yet one more example of the scientific revolution intertwining with the daily working life of the nation—both Newton and Halley advised Parliament on the measure—and an instance of the way official spending produces both direct and indirect effects on economic activity. Over the next several decades funds spent under the act paid for several people's working lives, while the eventual invention of a timepiece that could keep accurate time at sea yielded uncountable rewards, both to maritime trade and to the nation that fostered it.

The recognition that debt is not simply an encumbrance on a more or less fixed trove (a way of thinking that resonated with the older notion of the nation's budget as a royal family's revenue) was in some sense the central realization of Britain's financial revolution. It would take decades, at least until the mid-eighteenth century, for this view to be fully internalized. But already, in the early 1700s, it lay at the heart of every attempt to address the national debt. Defoe expressed this insight in an image his audience would have immediately recognized. "Credit," he wrote, "is not the Effect of this or that

Wheel in the Government, moving regular and just to its proper work"—no one person governs the process, and the nation's borrowing does not depend on the integrity or intelligence of a single minister. Rather, Britain's capacity to borrow more or less at will turns on "the whole Movement, acting by the Force of its true original Motion."

None of his readers would miss the reference to Newton's clockwork universe, nature moving in strict order, governed by fundamental laws that produced sure and certain outcomes each succeeding hour. Defoe's theory wasn't as precise as Newton's. He wrote that it was "the Honor, the Probity, the exact punctual Management" of Britain's finances—the commitment to deliver each succeeding year what was promised to those who supplied it with credit—that drove both "the Great Wheel in the Nation's Clockwork" and "the Great Spring that turn'd about that Wheel." But the metaphor was clear. The strength of Britain's credit, he wrote, lay in the fact that proper, honest administration of the debt "keeps it true to, and exact in, the performance of the General Work, (viz.) the equal and punctual dividing the smallest measures of Time."

To be sure, Defoe was at best an amateur financial theorist. But the broad idea he conveyed was correct. Neither the specific deals Defoe defended nor the broader idea of manipulating Britain's debt seemed to either the author or his audience to be particularly dangerous, and certainly not inherently corrupt. It was simply the normal business of government, one that British officials were fully, perhaps uniquely qualified to perform: maintaining the machinery of state to ensure the smooth advance of a meticulously arranged mechanism.

Any number of arguments could be and were laid against particular decisions and transactions, of course, and vicious party and factional politics could derail even clearly sensible moves. But while hindsight makes Defoe's happy confidence seem almost unbelievably naïve, it's important to remember that he and his contemporaries did not see the South Sea Company's new plans for Britain's finances as anything radical, unfamiliar, or dangerous. Rather, its proposal turned on precisely the kinds of deals that had worked in the past—which themselves mapped onto the same kinds of reasoning that had so recently brought order to the entire universe.

THE FIRST MOVE was familiar, essentially the same idea Walpole had proposed just before losing office two years earlier. In January 1719, the Treasury and the directors of the South Sea Company agreed to a kind of experiment on a fragment of the national debt—a proof of concept that could, if it worked, expand to the whole pile of outstanding British obligations. This was to be the first head-on attack on the problem of the irredeemables. The instrument chosen for this trial was a £1.5 million lottery loan from 1710 that carried a staggering rate of 9 percent interest over thirty years—roughly double what a new loan would cost—and that, by its original terms, could not be called in by the government before its term had run.

Strictly speaking, the 1710 lottery wasn't a loan, as the principal, or the purchase price of the lottery tickets, wasn't to be repaid. Instead, it was a term-annuity, a promise to pay a certain amount of money for a fixed number of years. The idea was to perform a present-value calculation—how much those yearly payments of 9 percent on the remaining years of its original 30-year term would be worth as a single amount of money right then—to set a fair price for that income, to be paid in South Sea shares (which could potentially pay dividends for as long as someone chose to own them, and not just the last twenty-one years left on the lottery paper).

Tables for such calculations were available to buyers and sellers of land in England as early as the 1610s. What had changed since those early attempts to price the future of a fixed, material asset like land was the emergence of the menagerie of types of borrowing the British government now used. If the government's credit, in the full sense of the term, was to thrive, then it was necessary to find a persuasive mathematical apparatus to capture the full promise of times to come and the full range of risks. For this particular lottery obligation, the price finally agreed upon was eleven and a half times the annual return, to be paid in Company stock at its par value of £100 per share. This would transform the lottery annuities into actual debt held by the South Sea Company on which the government would pay just 5 percent—thus cutting the annual cost to the Treasury almost in half—and that, crucially, it could redeem, pay off, at any time they chose, with cash or, if interest rates fell further, with a new pile of cheaper borrowed money.

As the deal emerged, it was as usual billed as a win-win-win proposition. Those who held lottery tickets got a great deal because of the provision that their holdings would be swapped for South Sea shares at par value—£100 of the face value of Company stock for every £100 worth of lottery paper at the agreed eleven and a half years purchase price. Given that the stock was selling for between £112 and £117 in the secondary market on Exchange Alley when the lottery-for-shares offer opened, anyone making the swap was instantly ahead of the game.

Over the longer term the original lure dangled by the Company would still be there too. New shareholders would command their fraction of Company dividends funded by the government's interest payments, plus their fair proportion of any gains from the trading side of enterprise. Perhaps most important: in the language of finance, the proposed conversion allowed lottery holders to transform an illiquid asset into a liquid one—the kind that lets you get your hands on your capital whenever you want.

The advantages for the Treasury were equally clear. They got rid of an obligation that cost way more than any other borrower faced, year after weary year. The deal simplified financial administration too: instead of having to account for hundreds or thousands of ticket holders, there would be just the South Sea Company, and a single payment every six months. And, best of all, the deal promised to achieve the previously impossible: through an act of financial manipulation, an instrument the government had been stuck with for decades became something that could be modified, or even paid off entirely, whenever an opportunity presented itself.

Finally, the Company did very well for itself indeed. The most obvious change was that the full value of the lottery obligation swapped for its shares was added to its capital—just over £1.5 million. The £168,750 in overdue interest payments on its preexisting debt holdings was converted into capital as well, as was a new loan of £778,750 that the Company agreed to provide the government. Put all those pieces together and the result was £2.5 million added to the Company's books—the formal measure of what it actually owned—all of which was earning interest from the Treasury.

That in itself didn't offer extraordinary gains, as any earnings from the deal would be passed on as dividends to Company shareholders. This was the same constraint that had frustrated John Blunt and his allies from the beginning: the financial side of their business remained fundamentally boring, and hence not wildly profitable. But this time, lurking within the seemingly innocuous machinery of the deal, there was indeed real money to be made. The new loan to the government promised a sum the Company did not actually have. Accordingly, the agreement allowed the company to raise that money by issuing new shares to be sold to the public.

As straightforward as that seemed, there was this twist: the Company was allowed to value this new stock at a par value of £100 per share. Every £1,000 to be loaned would thus add ten shares to the Company's accounts. But the money that would actually fund that loan would come from selling shares at whatever the market would bear. So if Exchange Alley quoted prices over par, the Company would pocket the difference—in essence, enjoying the same windfall offered to those being enticed to swap their lottery annuities for shares. At the moment the first new stock was sold, each share brought in £114. Of that total, £100 would fund the loan, leaving £14, a 14 percent gain, for the Company itself. Not bad for a day's work—especially if you were an existing shareholder, reaping the benefit of that extra cash provided by the newcomers.

To emphasize: nothing in this arrangement was seen as either dangerous or venal. Quite the reverse. In principle, at least, this was nothing more than sound management, a clever and equitable approach to a long-standing problem. As Defoe boasted, such financial manipulation was just one example of the "Methods which made our National Credit rise . . . to such a height," methods that made it easy to raise any sum of money at a price of 6 percent, or 5, or even less. This was, Defoe noted—suspending for the moment his antijobber crusade—a crucial advantage in the never-ending struggle for power: "Foreigners had been heard to say," he wrote, "that there was no getting the better of *England* by Battle. . . . That while we had thus an inexhaustible Storehouse of Money, no superiority in the Field could be a match for this superiority of Treasure." The ability, seemingly at will, to shape the market for money at the Treasury's word was, Defoe concluded, "the Honor and Advantage of *England*."

THE SALE OF new shares to the public began in the spring of 1719, when the South Sea Company opened its subscription books to anyone holding a lottery annuity. They responded briskly, though not quite as enthusiastically as either the Treasury or the Company had expected. Roughly two-thirds of the lottery tickets were exchanged for shares, which meant that the Company added about £1.7 million to its capitalization instead of £2.5 million. The deal with the government had tied the loan the Company would provide to the amount of lottery debt it absorbed, so that sum shrank as well, to roughly £544,000. At those numbers, the deal gave the Company an instant profit of almost £270,000—the gap between the open-market price its stock commanded and the loan it handed over to the Treasury, along with the overdue interest it had converted into shares. Crucially for all concerned, the moving parts of the deal spun smoothly, suggesting that this type of transaction could be repeated—potentially on a larger scale.

To be sure, some were unconvinced. For all his earlier evangelizing for the South Sea Company, maneuvers like those powering this deal—the willingness to exploit the difference between the par and market value of shares—lay behind Daniel Defoe's deep distrust of anything to do with stockjobbing. On July 1 he published *The Anatomy of Exchange Alley*. There he argued that a secondary, unofficial market in financial instruments put the nation in danger by subjecting Britain's finances to the incorrigible dishonesty that shot through the entire trade in money. In *The Anatomy* he ripped a story from the news of the day to illustrate the damage malign stock dealers could do.

That spring, there had been yet another in the series of risings in support of the House of Stuart, this time in the Scottish Highlands. The rebellion was crushed at the Battle of Glen Shiel, on June 10, a nasty little fight with roughly a thousand men on each side. It is remembered as the clash in which the to-be-legendary Rob Roy MacGregor was badly wounded—and as the last time British forces faced invading troops on home soil: two hundred soldiers from a Galician regiment had been sent by Spain to aid a rising in the hopes that civil strife might destabilize the British monarchy. The Spanish were disappointed in their ambition—but once the rising began, and until news of Glen Shiel reached London, anyone trading on Exchange Alley was vulnerable to any reports, fake or not, of the rebels' progress. The scare had passed by the time Defoe wrote, but he used its recent memory to demonstrate just how the sharp men of Jonathan's and Garraway's would behave—even in the midst of national crisis.

Defoe's cautionary tale begins as a young man with more money than sense walks into the coffeehouse. As soon as this "cull" appears, a pair of confederates sweep him up and set their lure: "Sir, here is a great piece of News, it is not yet publick, it is worth a Thousand Guineas but to mention it." The patter continues nonstop: "I am heartily glad I met you, but it must be as secret as the black side of your Soul, for they know nothing of it yet in the Coffee-House, if they should . . . I warrant you *South-Sea* will be 130 up in a Week's Time, after it is known."

Who could resist such a lure? Not Defoe's patsy: "Well, says the weak Creature, prethee dear *Tom* what is it?" Would the man reveal this key to riches? Why yes . . . yes he would: "Why really Sir I will *let you into the Secret,* upon your Honour to keep it till you hear it from other Hands; why 'tis this, *The Pretender is certainly taken* and is carried Prisoner to the Castle of *Millan,* there they have him fast; I assure you, the Government had an Express of it from my Lord *St——s* within this Hour."

The Pretender, the Stuart claimant to the throne, eliminated forever as a threat! Obviously, such intelligence would matter to the markets. As long as James II's son roamed free, he would be a threat to the German interloper on the throne—and to anyone trading in debt incurred by the interloper's ministers. After all, a restored and vengeful Stuart monarch might well refuse to honor his predecessors' promises. A captured Pretender would cheer the Alley enormously and thus prompt a sudden boost in the prices for government paper. This is the kind of inside knowledge from which dreams—and fortunes—are made . . . as Defoe's "cull" knows full well.

But not quite so fast! The innocent prides himself on his due diligence: "Are you sure of it, says the Fish, who jumps eagerly into the Net?" Oh, indeed, Tom is sure. The two Alley men send their Fish to speak to a third member of the conspiracy, who confirms the story. With that, writes Defoe, "Away goes the Gudgeon with his Head full of Wildfire, and a Squib in his Brain . . . so that without giving himself Time to consider, he hurries back full of the Delusions, dreaming of nothing but of getting a Hundred Thousand Pounds, or purchase Two."

The rest of the story unfolds with utter inevitability. Defoe's eager innocent "meets his Broker, who throws more Fire-works into the Mine, and blows him up to so fierce an Inflamation, that he employs him instantly to take Guineas to accept Stock of any Kind, and almost at any Price; for the News being now publick, the Artist[s] made their Price upon him." But, alas and of course, the news is false; the stock falls, and the slick gents at Jonathan's make out like . . . well, bandits: "The Jobber has got an Estate, the Broker 2 or 300 Guineas." As for the "Gudgeon"? He now owes "Fifty Thousand Pounds more than he is able to pay" and "remains at Leisure to sell his Coach and Horses, his fine Seat and rich Furniture, to make good the Deficiency."

THE
ANATOMY
OF
Exchange-Alley:
OR,
A *Syftem* of STOCK-JOBBING.
Proving *that* Scandalous Trade, as it is now carry'd on, to be Knavifh in its Private Practice, and Treafon in its Publick:
Being a clear Detection
I. Of the Private Cheats ufed to Deceive one another.
II. Of their Arts to draw Innocent Families into their Snares, underftood by their New Term of Art (*viz.*) (*being let into the Secret.*)
III. Of their Raifing and Spreading Falfe News to Ground the Rife or Fall of Stocks upon.
IV. Of their Joyning with Traytors in Raifing and Propagating Treafonable Rumours to Terrify and Difcourage the People with Apprehenfions of the Enemies to the Government.
V. Of their Improving thofe Rumours, to make a Run upon the *Bank*, and Ruin publick Credit.
VI. Of the dangerous Confequences of their Practices to the Government, and the Neceffity there is to Regulate or Suppress them:
To which is added,
Some Characters of the moft Eminent Perfons concern'd now, and for fome Years paft, in Carrying on this Pernicious Trade.
By a *JOBBER.*
The Second Edition Corrected.
LONDON: Printed for *E. Smith* near *Exchange-Alley.* 1719. Price One Shilling.

The title page to Defoe's blistering *The Anatomy of Exchange Alley*

The point for Defoe, abandoning his poor fish to the sharpers, was not that credit and finance were in themselves terrible ideas. He remained a booster of both private projects and public ones, all funded with borrowed money. But he was unable to make the connection between Exchange Alley and the government's ability to borrow at the levels the secondary market in credit made possible. The risk that clever bad men could turn useful and necessary loans into a commodity they could manipulate scared him. He never offered a real alternative to the Alley when the Treasury needed to raise money on the scale required to pay for Britain's wars. Such demands, and the role of an exchange in making enough private capital available to the public purse, were still recent enough that even a dedicated observer like Defoe found it hard to distinguish what was needed from what was simply dangerous. And certainly Defoe wasn't alone in his reflexive loathing of those infesting Jonathan's and the rest. He and Jonathan Swift agreed on little, but Swift fully shared his rival's disdain for those who made their living playing the market in money, writing, "There is a gulf, where thousands fell, / Here all the bold adventurers came, / A narrow sound, though deep as Hell— / 'Change Alley is the dreadful name."

Such lamentations cut to the critical issue at this early stage of Britain's financial revolution. What still-uncharted risks lay hidden within all these new conceptions of money? Could the kingdom reap the benefits of a system in which the national debt became the "Honor and Advantage" of the realm without giving up too much in a marketplace in which private interests acted—and misbehaved—for their own reasons? And if not—what then? Could public credit survive the wiles of men who made London's money market "dance attendance on their designs, and rise and fall as they please"?

The South Sea Company's 1719 deal put that question to the test. The proposition to lottery holders—exchange your assets for something that paid a bit less but offered more flexibility—was wholly dependent on Exchange Alley. That was where loathed jobbers and brokers would turn this new asset, shares in a joint-stock company, into cash. The market's reaction, then, was a measure of how well this early example of financial engineering actually worked in real life.

So how did it go?

Just fine!

Throughout the summer of 1719, none of Defoe's dire warnings about Exchange Alley came true. Quite the reverse, in fact, as the progress of the Company stock for lottery bonds demonstrated, revealing a basic flaw in Defoe's grasp of the new markets in money. Official Britain was able to raise new loans not just because the Treasury could (mostly) be relied on for honorable and exact management of its payments due. But its very ability to extract cash from British subjects turned on what the Alley did to coax those with money into the arena. Then and now, a critical function of secondary markets in securities—stocks and bonds, debt and equity—is to allow investors an easy way in and out of their investments. A central risk in any decision to release your money to someone else's use is: What if you need that cash before your debtor wants to pay it back?

Before the Alley provided a way for buyers and sellers to exchange government obligations at broadly consistent prices, the king's creditors were mostly out of luck: whatever they had lent to the crown would come back to them at the agreed term—or worse, at the monarch's pleasure. By giving investors a guarantee that they could enter or exit the business of lending to the Treasury whenever they chose, the stock traders at Garraway's and Jonathan's and the rest expanded the pool of capital available to the state. Britain's access to credit depended, that is, not just on the existence of those with capital to lend, but on the Exchange, an open marketplace in which, for example, shares in the South Sea Company could easily change hands.

In late summer of 1719, that was the story the Alley told. It appeared that it was indeed possible to serve Mammon and the public good all at the same time. Just after the exchange was completed, South Sea shares traded between £113 and £115 5s. 8d.—a gloriously uninteresting range. Not much changed through October, before a slow rise started in the second week of November. Defoe's story of plundered innocents was, it seemed, paranoia: nothing in the record of prices for South Sea stock that year implied that jobbers manipulated the market while stripping fools of all their worldly goods. Throughout the last half of 1719, Exchange Alley appears to have viewed the deal as nothing more than a successful and relatively minor Treasury move to control the cost of national borrowing. Against that record, Defoe's pamphlet could be seen as the wailing of an unreliable prophet, to be safely ignored while the great and the good planned what was next for the national debt. There was, after all, an obvious next step to follow a small-scale play on a single set of annuities. If most of the 1710 lottery, a million or so in irredeemable national debt, could be made to disappear into Company stock, why not try the same kind of maneuver again—on a much grander scale?

South Sea Company directors certainly thought this was a good idea. So did several leading ministers. The records from both parties are frustratingly fragmentary for the latter half of 1719, but scattered and mostly allusive accounts of who spoke to whom suggest that as early as the summer, and almost certainly by November, the two sides were contemplating a stunningly ambitious second act. The target? Everything. All of it. The whole mountain of debt that had accumulated over a quarter of a century of paying for Britain's ambitions with borrowed money.

Chapter Eleven

"resolving to be rich"

The fifth volume of Alexander Pope's translation of the *Iliad* reached its subscribers in the autumn of 1719—the latest installment in what had become *the* annual literary event for English letters. As the first version of Homer's poem to read like English verse, and not some awkward Greek-lish translation, it entertained a list of subscribers that included "Her Royal Highness the Princess," along with dozens of dukes, duchesses, earls and their countesses, and many more.

This was both a literary triumph and a very worldly one. The project had earned Pope an almost unheard-of amount for a man in the scribbling trade, and by the fifth installment his fortune had grown to the point where he was genuinely well-to-do—and beginning to act accordingly. He started work that year on his famous villa in Twickenham, fronting on the Thames, complete with its notorious grotto—Pope's fantasy palace. The poet had been born poor. In funds now, the cavern he excavated behind his house became his playground. By day, it was a camera obscura projecting river traffic onto its walls, transforming at night into a mirrored cavern lit by "a thousand pointed rays [that] glitter and are reflected over the place."

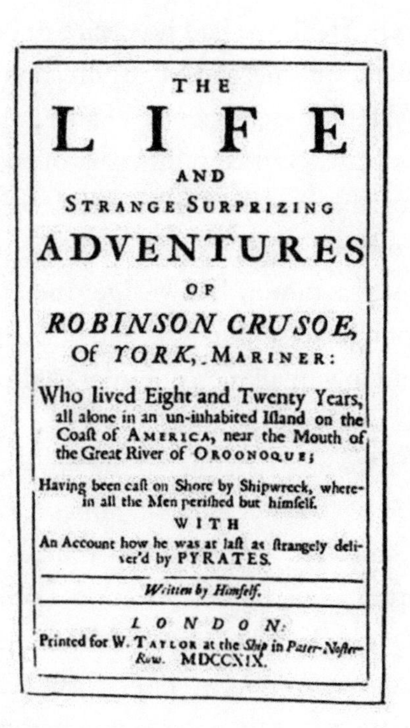

THE

LIFE

AND

STRANGE SURPRIZING

ADVENTURES

OF

ROBINSON CRUSOE,

Of *YORK*, MARINER:

Who lived Eight and Twenty Years, all alone in an un-inhabited Iſland on the Coaſt of AMERICA, near the Mouth of the Great River of OROONOQUE;

Having been caſt on Shore by Shipwreck, wherein all the Men periſhed but himſelf.

WITH

An Account how he was at laſt as ſtrangely deliver'd by PYRATES.

Written by Himſelf.

LONDON:

Printed for W. TAYLOR at the *Ship* in *Pater-Noſter-Row*. MDCCXIX.

Robinson Crusoe as readers first encountered him in 1719

The year 1719 was a good one for ambition, literary and otherwise. That was when Daniel Defoe introduced Robinson Crusoe to London society, while the nonfictitious James Figg won the bout that made him England's first recorded bare-knuckle boxing champion. (He would retire, still champion, fifteen years later, with a stated, though not fully documented, record of 269 wins and one loss.) That same year, Georg Frideric Handel became one of the founders of the Royal Academy of Music—and notably, it was formally structured as a joint-stock company, funded by subscribers buying shares at £200 apiece in what was functionally an opera business. And across the autumn of 1719, a small group of men plotted an epic of their own, though they tried to keep well out of view.

John Blunt made the first move, approaching one of the two leaders of the ministry, Earl Stanhope, with a plan for a repeat on a grand scale of the just-completed lottery conversion. It was a misstep; Stanhope had no desire to dive into the grubby world of arithmetic and speculation. Blunt soon recovered, shifting his attention to Walpole's successor as chancellor of the Exchequer, John Aislabie. Unstated, but certainly part of the calculation, was the thought that Aislabie was no more honest than he needed to be. As Arthur Onslow, later speaker of the House of Commons, put what was apparently common knowledge: the chancellor was clever, with a knack for business, "but dark, and of a cunning that rendered him suspected and low in all men's opinion." He was "much set upon increasing his fortune and did that."

Joining Aislabie on the government's side of the negotiation was the postmaster general, James Craggs. Craggs had his own intimate connection to South Sea House. Over the autumn of 1719 he was often seen in conversation with the Company's cashier, Robert Knight. Knight was, after Blunt, probably the most important of the Company officers. His salary offers oblique insight into Knight's character: at the Company's founding in 1711 he earned a mere £200 per year, enough to keep body and soul together, perhaps, but hardly luxuriously. Over the next several years, that number rose to £250, still a risible number for someone overseeing £10 million in capital. More curiously, that sum could hardly have paid for the landed estate he bought and the mansion he began to build in 1716. Knight did have some family money, but it's almost impossible to come up with an honest explanation for Knight's swift rise to fortune. The millions that flowed through the Company presented any number of opportunities for a clever man, from simply exploiting insider knowledge to making more direct use of funds in his purview. Knight was a man who loved money and, on the circumstantial evidence that survives, was not too particular about how he got it. In Craggs, he found a perfect confederate. Here's Onslow again, recording the consensus view of the postmaster general: Craggs was, he wrote, a tireless, driven man, "restrained by no scruples of conscience," which is to say, just the sort Knight and his colleagues could do business with.

From the start, then, whatever was to come would emerge from a web of long-nourished relationships and the certain knowledge that insiders would do just fine, thanks very much. With the possibility of a deal whose profits would dwarf the already very comfortable returns of the last transaction, it was natural for everyone involved to proceed carefully, and above all discreetly. Blunt drew a few of his closest associates into the discussion, including his old ally George Caswell, while keeping most Company directors in the dark. On the government side, Aislabie and Craggs drew a similar, if not quite as opaque, veil between them and the rest of the ministry—for which the kindest explanation is that the principals wished to avoid any excitement in the market for South Sea stock. (A less generous interpretation is that the two ministers planned to profit by self-dealing along the Alley.) The first records of their discussions date from January 1720, and they omit details from the backroom conversations of the prior months.

The draft agreement contained no surprises, though. With the success of the 1710 lottery conversion, the new, much larger plan followed the same lines. The real question was the scale to be attempted: how much of the nation's debts should be included in the new deal. Blunt seems to have first proposed taking all of it down to the last guinea, even the assets held by the Bank of England and the East India Company. But dispossessing those two great rivals was politically impossible, so negotiators soon settled on a more plausible scheme. In this plan, the conversion would take virtually all of Britain's outstanding debts: the irredeemables, all the debts the Treasury could not pay off before their term; the redeemable portions of the national debt (those that could be retired early, if the ministry had the money); and all of the floating obligations except those held by the two other great monied companies. Adding it all up, that formed a pool of about £31 million in Britain's obligations to its creditors to be converted into South Sea stock.

All this was handled in secret, but inevitably some hints leaked—along with some red herrings. In early January, for example, the *St. James Weekly Journal* claimed that both the South Sea and the East

India Companies had offered debt-conversion schemes to the government, which must have surprised the East India directors. Other London gossips were better informed—and assumed the most cynical explanation for the veil of secrecy that had shrouded the negotiation. "The town," as one member of Parliament put it, had learned that "the bargain with the South Sea Company was agreed," after which "Mr Aislaby [*sic*] has bought £27,000 stock." In this instance, as the historian P. G. M. Dickson uncovered, "the town" was right: Aislabie had started to buy Company shares in January and would continue to do so for several more months. In any event, now that the details of the scheme were slipping out, it was time to push the plan through.

Blunt and his inner circle relayed the details of the agreement to the rest of the Company's directors on January 21. Aislabie told the House of Commons the next day that this deal would enable the nation to pay off the entire national debt within twenty-six years. With notable stage management, Craggs stood next, praising the chancellor for "the clear and intelligible manner in which he had explained the business" and then laying it on even thicker, congratulating "the nation on the prospect of discharging the debt sooner than was generally expected."

The announcement stunned the Commons. According to a later, quite possibly embroidered recollection, no one spoke for almost fifteen minutes. But when at last an MP rose to reply, he shattered the hope of a rush vote. Thomas Brodrick, member for Stockbridge, was a Whig but no friend of the ministry. He began piously, telling the House that he was perfectly happy to get rid of the national debt, declaring that until this was done "we could not properly speaking, call ourselves a nation." But if such sentiments encouraged Aislabie and his allies, what came next was a disaster. Why should the South Sea Company be the only possibility for the task at hand, Brodrick asked. The two ministers had "appeared to recommend this scheme exclusively," but, he argued, "It was to be hoped that with a view of obtaining the best bargain for the nation, every other company, or any society of men might be also at full liberty to deliver in their proposals."

That phrase, "the best bargain for the nation," showed that Brodrick had penetrated to the nub of what made this plan different from the 1719 trial run. Then, the South Sea directors had agreed to a straight-up swap in which the only elements to the deal were the firm's stock and lottery tickets on which the Treasury was obligated to pay interest. This time, Aislabie and Craggs aimed for something more: cash on the nail, a fee the Company would pay for the right to do the deal. As the nation's creditors were not going to be compelled by law to trade their debt holdings for Company stock (just as in 1719), the final sum was to be calculated on a sliding scale, but the maximum possible price topped out at a genuinely impressive number: £3 million. Clearly, the deal's proponents hoped that it was enough to stun any potential opposition into silence—and it had done so until Brodrick reminded the House that they did not know if that was the most that could be squeezed out of Blunt's shop or any other.

This was exactly what Aislabie had hoped to avoid in his haste to shove the deal through (and, perhaps, among other considerations, protect his early speculation in South Sea shares). He fumbled his reply, waffling that the business "must be carried on with spirit." As that shambles of a statement hung in the air, the nemesis of the sitting ministry intervened. Out of power though he remained, Robert Walpole had lost none of his command of debate on the Commons floor. Suavely, smoothly, he too praised Aislabie's fiscal prudence. But then he noted, gently, that perhaps an aspect or two of the plan might need to be amended and that "a few others" were "unreasonable." It was thus, Walpole argued—and who could dispute it?—Parliament's duty to invite the full range of ideas that this one proposal might inspire.

He had the measure of the House, much better than his factional rivals. After a brief further exchange the Commons agreed with Walpole. Competing proposals were welcomed, and the members would return in a week to consider whatever might have arrived in the interval. It was at once the obvious move and the first, fateful swerve off the path Blunt and friends had mapped.

DESPITE WALPOLE'S PIOUS suggestion that anyone could offer a counterproposal, there was really just one challenger that could disrupt the pitch. The East India Company did not put up their own bid. It was rich enough to try, but it was the inverse of the South Sea Company: a trading business first and a financial one only as a by-product of what it took to run a mercantile empire. That left the Bank of England. The Bank had, from its origins as a Whig-leaning private business, served as the government's lender of first resort. Over the first quarter century of its existence it fought ferociously to maintain that position. There was no chance that it would let an upstart rival usurp its position without a fight.

The Bank started shaping its offer within a day of the first parliamentary debate. The proposal was presented to a meeting of shareholders two days later, on January 27. There was one detail missing: how much the Bank was willing to pay to play in this particular transaction. Instead, the directors asked for a blank check, the right to wage a bidding war as far as necessary to secure the Bank's interests. The assembled shareholders said yes with no apparent qualms. The Bank's allies in Parliament transmitted the offer to the House of Commons that day, with the South Sea Company's revised proposal coming on its heels. Its shareholders, like the Bank's, authorized the same unlimited license to top their competition.

So far, Brodrick's and Walpole's stratagem was working exactly as intended. Both firms promised the Treasury a bigger upfront payment for the right to do the deal. The Bank pledged £5,547,500, almost double the Company's original offer, and added other sweeteners: the Treasury could begin to redeem any of the debt taken in by the Bank that it could afford to pay off as early as 1724, three years earlier than the South Sea bid had promised, and it agreed to handle some other, minor financial tasks for the government.

The Company countered by raising their proposed payment for the right to the deal to £3,500,000. That was, obviously, less than the Bank's offer, but the plan included a new major concession: in 1727, the interest on the converted debt would fall from 5 percent to 4 percent—a potentially huge savings for the Treasury. Members of Parliament debated both offers but couldn't come to resolution, while Blunt himself shot down the possible compromise of splitting the work with the Bank. "No, Sir," he said, "we will never divide the child." To settle the matter, the Commons sent out for one more round of offers, which, as Aislabie put it, amounted to "setting the nation to auction."

Auction fever can be deadly. The Bank upped its pay-to-play fee only slightly, while matching the Company's interest rate cut. The South Sea men, though, were determined to win, no matter what. At a directors' meeting, the Company's negotiators were ordered to "obtain the preference, *cost what it would* [italics in original]." Blunt and colleagues did so, settling on an offer of a truly mountainous sum: up to almost £7.6 million—just over £1 billion in twenty-first-century currency, and, incidentally, money the Company did not then possess—would go to the Treasury, not as a loan, but as a pure payout.

That made the decision easy for the Commons. The choice of the Company was generally, though not universally, seen as a victory for the government. Brodrick, writing to his brother, boasted that it might have been doubted that the public "should be fortunate as to more than double the sum intended for them, butt thus it has for once happened."

ONE STEP REMAINED. The deal had to be translated into legislation that could be voted on. This was no simple matter. In its final form the South Sea Act filled thirty-five printed pages with each party's rights and obligations laid out in meticulous detail. The Company was given the choice to exchange official debt held by the public for either cash or shares. The act also specified the conversion formula for the long- and short-term debts. The long obligations were valued at twenty times their annual return—£2,000 for every £100 of yearly interest—and those with shorter terms went at fourteen to one. Given those figures, the total of outstanding British official debt eligible for conversion would be priced at £31.5 million, and the Company's capital—the sum on which the Treasury would pay interest—would increase by that amount.

Within this meticulous, comprehensive, tedious litany of numbers, duties, and powers, though, was one conspicuous omission. The nation's debts had been valued—the legislation specified the conversion rate to capture ongoing payments in a single number. But they had not been priced: nothing in the legislation required the Company to declare how much stock each creditor would receive for his or her holdings. This was a shift from the previous year's lottery conversion. In that transaction, the exchange had a fixed price: the present value—that twenty- or fourteen-to-one conversion of the annual return—would be paid in stock at its par value of £100. The Bank of England's latest offer in this competition worked the same way. Those who agreed to swap their pieces of the long-term national debt for stock would receive seventeen times the return on their holdings, £1,700 in Bank shares at their par value for every £100 of annual interest earned, or about a 6 percent return.

But in the legislation before the Commons there was no such provision. Nothing specified how much stock (or cash, or anything else) the Company would be compelled to exchange for each unit of debt. Rather, it was required only that the bargain would take place "at such Prices . . . as shall be agreed."

This was critical, the hinge on which all of what was to come would turn: the bill allowed the Company to create new shares at par up to the amount of the national debt exchanged under the deal—the full £31 million. But if the South Sea price in Exchange Alley's secondary market rose above that par value of £100 per share, then the Company was free to offer debt holders a swap payment based on that higher market valuation for its stock.

For the South Sea men, this was the beauty of the deal. Given a valuation at seventeen times its annual payment (the figure ultimately offered by the Company for short-term irredeemables), the legislation allowed the Company to create seventeen new shares for each slice of debt with £100 in yearly income. But if its shares were changing hands at £170 along the Alley (just to make the arithmetic easier), that would mean just ten shares were worth (for the moment) the £1,700 to swap for that £100 of income. As long as the Company could persuade the credit holders to accept stock at that market price or something close, it could keep the surplus shares from the par-value creation authorized in the South Sea Act. In this example, that would be seven new shares, with which it could do whatever it wanted. It was a magic trick: wealth conjured out of the jobber's price in the long room at Jonathan's.

There was no mystery as to why the Company and its allies built the deal this way. It rewarded existing shareholders at the expense of those making the swap. These newcomers would, in effect, pay more for their shares than their predecessors had. The Company holdings owned by those who had bought before the swap would now include a fraction of all those new shares that the Company got to keep, while later arrivals would get less of the business than they would if they had received their stock at par. These outcomes would grow ever more unequal the higher the prices on the Alley climbed and the fewer shares it would take to buy a slice of the national debt. The last in, that is, would always be the most vulnerable, their investment, their fortunes most at risk.

Such an asymmetry of interest did not go unnoticed. As the South Sea Act was being debated, Robert Walpole again inserted himself between his rivals and their dreams of wealth, arguing that "as the whole success of the scheme must chiefly depend on the rise of the stock, the great principle of the project was an evil of the first magnitude." He was speaking in support of a motion that would have set the missing price—exactly how much debt each share of Company stock would buy. The failure to do so, he argued, would encourage the Company "to raise artificially the value of the stock, by exciting a general infatuation." Walpole didn't say how the South Sea men might pump the stock, but as Defoe had already written, it was common knowledge that stockjobbing men knew how to stoke reckless enthusiasm in the coffee rooms and passages of the Alley. Walpole would come to seem astonishingly prescient. In this still-hopeful moment, he went unheeded.

PARTLY, THAT WAS because too many of those involved were very well paid to remain oblivious. Up to a point, "ordinary" corruption wouldn't have offended most worldly folk in the London of 1720. A genteel level of influence peddling was simply the way things worked, not so much a breach of trust as the familiar rhythm of patronage, reward, and the economy of favors that has been part of human governance as far back as memory can reach. In Britain in that uncertain age, everyone took care of his friends—even Isaac Newton, for example, who as warden of the Mint placed his fellow comet lover and disciple Edmond Halley in an official post. Newton had honored the principle: a man with gifts to give would be expected to take care of his own and, perhaps even more, those who might be useful in the future.

Some of what happened over the winter of 1719–20 was consistent in form with such behavior. The scale, though, was another matter. Over the two months that Parliament debated the move, South Sea Company advocates pursued a systematic, clever, and ruthless campaign of bribery. The most impressive sweeteners passed through the hands of the Company's cashier, Robert Knight. He kept a special account book in which he recorded "sales" of shares to the Company's targets. These were fictions from start to finish, not least because until Parliament actually approved the South Sea Act there wasn't anything to sell. No one actually paid for this ghost stock. Instead, Knight wrote down a number—the shares assigned to each politician or grandee. Then everyone waited. "If the Price of the Stock had fallen," as one witness later told Parliament, "no Loss could have been sustained by them [the sham buyers]; but if the Price of Stock should advance . . . the Difference . . . was to be made good to the pretend Purchasers."

That is: those to be bribed would never have to pay a shilling. The entries in Knight's notebook worked like those modern-day share warrants that companies often use to reward their executives. Holders of such warrants can choose to exercise them once the underlying stock rises; they pay the price specified in the warrant to acquire the underlying shares, which they can then sell immediately at the higher price available on the open market. If share values fall, or just stay level, then they do nothing. In a notably less formal arrangement, that's what those the Company sought to bribe received. Nothing would change hands in either direction until a recipient decided to take his or her profits. That there should be no doubt as to the aim of such gifts, "No Entries of such Adjustments, or the Names of the Persons with whom the same were made," was inscribed "in any one of the Books of the Company." In other words, this was no mere "friends-and-family" subsidized deal. It was a payoff, clearly recognized as such by everyone involved.

The first to receive such rewards were those who had already earned their payola. Aislabie and Craggs appeared by name on Company books, and witnesses would later admit that leading members of the ministry were taken care of as well: the Earl of Sunderland and Charles Stanhope, the junior Treasury secretary and cousin to the other top minister, James Stanhope, were high-value targets. Soon, though, the Company's leaders widened the field, scrambling to buy the loyalty of anyone who could help ensure passage of the South Sea Act. These included "Persons whose Names were not proper to be known . . . Persons of Distinction at Court . . . forty or fifty of the Company's best friends"—and, in perhaps the clearest signal that Blunt and Knight wished to leave nothing to chance, "certain Ladies whose having Stock would be of Service to the Company." Among such helpful consorts: the Duchess of Kendal, mistress to King George himself, who "seem'd well pleased with the Offer, thanked them, and wished them good Success."

Still, though it may have seemed that everyone who could possibly be of use got their taste, there were some conspicuous omissions. James Stanhope was one. Contemporary accounts and subsequent investigation show him to be a sincere proponent of the scheme. With his fellow earl, Sunderland, he pushed the South Sea Act through the House of Lords—but he didn't need to be bribed to support what he seems to have genuinely believed was a good idea. (That's not to say that those who were bribed couldn't also believe; doing well by doing good is always a seductive dream.) One more uncorrupted man: Robert Walpole.

This was no proof of Walpole's probity. None of his contemporaries would have believed he was unbribable. The Tories' failure to prove that he'd dipped his beak into army contracts as secretary of war didn't dispel the cloud of suspicion, and his sudden rise to real wealth in the office of paymaster showed he knew how to play the game. But in February and March of 1720, Walpole remained the odd man out. His disruption of the debate on the South Sea Act was consistent with his broader eagerness to make life difficult for his former allies. As long as he remained out of the ministry his role in the drama was fixed: he was the opposition, to be defeated, not bribed. This was, it seems, an episode of incidental integrity.

It was to be the making of him.

In the meantime, events showed that Knight had worked the numbers in his little book with practiced skill. Walpole and his allies were unable to overcome their persuasive effect. The bill received its first reading in the House of Commons on March 17. On the twenty-third, Walpole and others put forward their motion to force the Company to set its price. Runners were staged in the lobbies in Parliament, carrying word down the river to Exchange Alley. The price for Company stock leapt and fell at every turn of the debate. When the motion was defeated by a comfortable majority, the market responded exultantly, racing to a new high of £195. The House of

Commons voted its final approval on April 2, and Sunderland and Stanhope experienced no difficulties with the House of Lords, which passed the act on April 7. The king assented to the measure that same day.

South Sea Company was now officially empowered to take in government obligations in exchange for its stock. There was still much to do. The Company had to raise enough money to cover the outlandish payment it now owed the Treasury for the chance to absorb the national debt, and it had to handle the tricky accounting involved in creating new shares as that debt came in. It began that process immediately, even before the first swaps. Those would begin on April 28—*after* the Company sold a first tranche of its new shares to raise the money it needed for its vastly expanded financial operations. That sale, called a "money subscription," was set for April 14, with a price to come, based on the context of Exchange Alley's trading range as the day approached. The game was afoot and, it seemed, going exactly according to the Company's plan.

IT'S EASY NOW, when we know how it all ends, to wonder why only a handful of opponents vaguely glimpsed that something might go awry. It's much harder to squint through time to feel that moment as those living through it did. The sleaze surrounding the deal is a fact: dozens of powerful men and women were paid to allow the debts of a nation to become shares in a private company. But it's also true that no one seems to have been bribed into doing something they actively opposed. There was, after all, very recent precedent to show that however ambitious this new move might be, it was not obviously programmed for disaster. If everything had gone as there were plenty of good reasons to expect, the tale of all these gobbling snouts at the trough would be an amusing anecdote, a footnote about some clever scoundrels on the fringes of a triumph of early modern capitalism.

Looking forward from the morning of April 14, that is, rather than with all the advantages of knowledge no one living then could command, the key issue wasn't that the powerful were corruptible. It was that they were persuadable, willing in those sweet days of spring to believe in the revolutionary possibilities of financial alchemy and the prospect of wealth without tears.

CHAPTER TWELVE

"The Humour of the Town"

As 1720 unfolded, Daniel Defoe would find himself on almost every possible side of the question of the South Sea Company, mindful of its good works, its sins, and how it might advance or mar Britain's pursuit of power. But that spring, just before the epic swap would go live, he was, for the most part, entirely complacent. "A Man that is out of the *Stocks,*" he wrote on Friday, April 8, "may almost as well be out of the World."

This was Robinson Crusoe's literary father talking: Who would spurn such an adventure? With seven days to go before the sale of new South Sea shares, Defoe showed nothing but confidence. "The certain test of the Value of such Stock," he argued, was a reasonable dividend—2 percent would do. There was no reason to suppose that the South Sea Company would fail to achieve such returns, so Defoe chided those who thought the price too high. For all the plaints of "some angry Gentleman," he wrote, it could be seen that "the present Rate of Stock is far from being exorbitant."

Yet for all such confidence Defoe's anonymous antagonists could make some reasonable arguments for their unease. South Sea share prices had been ticking up nicely since January and the first public mention of a possible deal. They hit £130 on January 13, up only about £4 from the beginning of the month, and crept up a bit over the next month. Then, from a St. Valentine's Day quote of £138, they leapt to £155 on February 15, a 12 percent gain in twenty-four hours. It is almost certainly no coincidence that this leap came just as the South Sea Act was being drafted and the Company's campaign of judicious bribery began to reach more and more of its targets. The stock touched £181 on February 26, up eleven from its previous close, and then a pause descended on the market. But as the date for the deal to go live approached, that gentle trot became a gallop. The Company's shares stood at £198 on March 29, £220 on the thirtieth, £255 on April 1, £275 on the second, when the House of Commons approved the South Sea Act—and then, stunningly, £350 on the fourth. That was a price too far for the moment, the market bubbling over the rim of sense, and the stock settled into a range between £310 and £320 up to April fourteenth.

That seemingly reckless ascent was what led those critics Defoe had mocked to suggest that South Sea shares had become "exorbitant." At almost triple the price of three months earlier, Company stock represented an explosion of paper wealth that seemed wholly divorced from any material cause in the real world: a successful voyage, a new mill, acres to be brought under the plow.

There was a ready response to such doubts, and Defoe was just sophisticated enough as an economic thinker to make it: any investment makes sense if the market says it does. April 14 was a special date: it was the Thursday when the first money subscription in South Sea stock opened and investors with cash could buy shares directly. In that subscription the Company offered to sell £2 million worth of newly issued stock, priced at £300 per share. Demand was so high that the amount to be sold reached £2,250,000—all of which was, at least on paper, instantly in the black: at the same moment that subscribers jammed themselves into the South Sea Company's office, brokers at Garraway's and Jonathan's just a few hundred yards away were offering £317 to buy South Sea shares, almost 6 percent profit in an instant.

Those crowds and that number were, in essence, the gauntlet Defoe threw down before those who, after the months of maneuvers in and out of Parliament, still talked down a deal that appeared to reward the government, the Company, and the speculating public. Those buying into the scheme, he wrote, "were in a good deal of Pannick, and had now and then some little Convulsive Twitches." Still, "They looked better than those who were out: They had a pale Envy upon their Cheeks, and discovered too plainly, that they wanted either Money or Courage."

In the event, such cowards were vastly outnumbered by the eager among those who were the primary targets of the deal: anyone who owned Britain's debt. The oversubscribed April 14 sale of stock for cash was followed two weeks later by the first subscription offer in which those who owned government debt could exchange their holdings for South Sea shares. *Offer* isn't quite the right word. Investors were invited to surrender their assets in exchange for "such Terms and Conditions as the said Company shall appoint"—that is, whatever Blunt and friends deigned to give them.

The terms were finally revealed on May 19. Holders of the long-term irredeemable debts, the ones the government couldn't call in early, were to receive "thirty two years purchase"—which in the language of eighteenth-century investors meant that the price for those longest-lasting and hence most valuable credit instruments would be £3,200 for every £100 of income such debts paid to their holders. Short-term irredeemables would go for the same price the Bank had quoted: seventeen years purchase, or £1,700 per £100 of payout. Most of that sum, about 80 percent, was to be paid in South Sea stock, with the balance to come in cash and newly issued Company bonds.* Again, the crucial question was not the nominal numbers attached to the deal—that impressive-sounding £3,200—but just

* This was an early form of a commercial bond—debt on which the issuing enterprise promises to pay a set amount of interest but that does not convey any ownership in the underlying business, as shares of its stock do.

how much the shares in the swap were to be valued, as the higher that number, the fewer shares each creditor would receive and the more the Company would retain on its own books. Recall that the Company could value its shares at any price it could get away with. For this first debt-for-equity exchange, that number turned out to be £375: 25 percent higher than the money subscribers had paid two weeks earlier, and almost four times par.

THAT GAP BETWEEN the swap price and par was what made the deal so attractive for those at the center of the action, Blunt and his associates, and their allies in the ministry and Parliament. Each hundred pounds of income from one of the long-term, irredeemable obligations, valued at £3,200, that came onto its books became that much new capital for the Company—thirty-two new shares at a face value of £100. Setting an above-par price for the shares that debt holders received in exchange meant, in effect, that the Company—and only the Company—could buy that debt at a discount. Pricing those shares at £375 instead of par meant that those on the other side of the sway would get the equivalent of a fraction above eight shares of South Sea stock for each £100 of income they surrendered . . . leaving almost twenty-four shares to the Company. Those shares could be held on the books, or sold to raise cash for whatever the directors might choose to do with it, or distributed in some other way, but however they were deployed, they represented a pure, if paper, profit that would only grow as long as the market could be persuaded that more and better was to come.

There was an obvious way such a fortune could evaporate. If all its new owners rushed to turn their holdings into cash, South Sea stock would flood Exchange Alley, lowering prices and hence insider profits. So Blunt and friends simply blocked such inconvenient competition. Instead of transferring stock to its new owners at the time of the exchange, they postponed the exchange for months, all the way to December 30. It was possible to trade the rights to the shares that would arrive later, but the move ensured that the Company could control the absolute volume of shares available to the market, which helped protect the price level of the stock. The subscription process for both cash sales and swaps would recur several times over the spring and summer of 1720. New stock was offered to the public again on April 29, June 17, and August 24—the second, third, and fourth money subscriptions. Debt-for-share swaps reopened over the summer, sweeping up short- and long-term obligations in three separate offers, with similar delays in the delivery of stock to those making the exchange.

Looking back from the twenty-first century, with its long experience of manipulated markets, perhaps the most shocking lesson of the events of 1720 is how little under the sun is truly new. The tricks and tools Blunt and his associates deployed to pump their stock along Exchange Alley haven't changed much in the three hundred years since. Then as now, such devices—the ones the South Sea directors used early in the transaction like the games with delivery as above, and many more that would appear over the course of the South Sea spring and summer—can produce innocuous results in ordinary circumstances but can become gasoline on a fire when deployed maliciously, ignorantly, or some luckless mixture of the two. For instance, offering stock instead of cash can be a perfectly ordinary move, a way to transfer equity, ownership, from an enterprise to its investors. Such a device, and many others in common use in financial markets, can serve useful purposes for buyers and sellers alike. They can also give the unscrupulous (or the merely desperate) the capacity to do real damage.

In 1720, one such device was a variation on what we now call buying on margin—paying a fraction of the market price for a stock, while borrowing the rest from a broker or another party, an utterly routine transaction in modern financial markets. It works like this: when an investor buys on 50 percent margin (the current standard in the United States), she puts up half the price of a financial purchase while borrowing the rest, usually from the broker handling the sale.

That doubles the number of shares she can purchase with the same amount of her own money, allowing her to potentially double her profit, while also doubling the impact of any market drop.

Margin buying is an exercise in what professional investors term leverage—using borrowing to increase the amount of an asset under either an individual's or an enterprise's control. Where whole stock markets are concerned, such leverage is another way to funnel more demand for shares into an exchange, and hence to boost prices, at least for a while. South Sea insiders clearly grasped the concept and exploited that knowledge by offering a payment schedule that allowed cash buyers over a year to come up with all the money due. At the first money subscription in mid-April, purchasers needed to come up with only £60, just one-fifth of the issue price of £300, paying off the balance at intervals for another sixteen months. Two weeks later, the second money subscription opened at a higher price—£400—but this time it took just £40, or 10 percent, to buy into the issue, with the schedule for the rest of the installments stretching out even longer, all the way out to the end of 1722. The third and fourth subscriptions came over the summer on similar terms, allowing individual punters to leverage themselves five and ten times over—that last a rate that meant each £1 invested would control £10 worth of stock—which would provide outsize returns in a rising market, and the reverse as well: potentially devastating losses in any decline.

To be fair, South Sea Company didn't invent the layaway plan for stock purchases, but it used the device with ferocious eagerness—and with a crucial difference from modern practice: in a margin trade today, some third party (usually the broker handling the trade) lends the money needed to cover the full cost of a purchase. The seller of the asset gets all their cash at the time of the sale. In 1720, though, the seller and the notional lender were the same: the Company. As long as it didn't require full payment for its new stock, it could use whatever combination of price and terms was required to hook buyers—and pour demand into the market, thus propping up the prices the jobbers were pushing. The advantage for the Company was clear: all the shares it retained at every debt swap priced above par would rise in value as the market climbed. The risk was less obvious: if the market moved against its shares hard enough, those who owed on their partly paid-up purchases might default—fail to come up with the balance of what they owed for their stock. Those losses would then land on the South Sea books. Eventually that bill might come due. But in the heat of the moment, all that mattered was how high the buzz could push the stock today.

THE MARKET WENT high indeed, for a time. New money had chased shares along Exchange Alley in March, just as Parliament worked out the details of the deal. But it was when the deal went live that Exchange Alley began to live the dream. Shares traded at £328 at the end of April. On May 2, jobbers at Jonathan's could get £339 for the stock. It hit £352 by the end of the next week, £375 on the nineteenth, and £400 on the following day—and at that, the market was just warming up. South Sea stock hit £500 on the twenty-eighth and leapt to £595 on the last day in May.

Near double in a month! Just as earlier in the spring, nothing had actually happened to justify such a move. There had been no change in the numbers for the deal, no land bought or projects launched . . . nothing. This was the triumph of hope, or perhaps a demonstration of just how easy it was to push the unwary along Exchange Alley—with more to come. In late April the Company's shareholders voted to allow the directors to lend money with South Sea stock as the collateral—creating a kind of circular financial engine, pumping more cash into a rising market, where each step up inflated the nominal value of the pledged shares, which could then be used to justify more lending, more cash, and, when brought to Jonathan's or Garraway's, more demand and hence higher prices on the Exchange. As financial chronicler P. G. M. Dickson wrote, everyone, including those voting to give company officers this power, knew exactly what they were doing. "The profit of the Company," shareholders were told, "do's chiefly depend on the price of the Stock,"—and the point of making those loans was to allow more and more buyers to acquire more and more shares, thus keeping the market price high. Other moves followed, most likely including ("though it cannot be proved," according to Dickson) unannounced and unacknowledged purchases of its own shares, the Company itself playing the Alley to prop up its own shares.

Most important: these were no mere one-off gambits to get the deal off to a running start. The directors made it clear that market manipulation was to be the Company's strategy for the foreseeable future. In a decision made on May 20, executive officers were "hereby empower'd . . . to do all such things as they shall think necessary . . . for the Company's Interest." That sounds innocuous enough—What should leaders of a business do but serve its interests?—except that this directive came in the context of granting its governors the unlimited authority to use the Company's cash for any purpose deemed needful, including propping up the market for its stock. John Carswell was a South Sea historian with a fine eye for folly, and as he put it, Blunt and his closest associates had built "a financial pump, each spurt of stock being accompanied by a draught of cash to suck it up again, leaving it higher than before."

Yet for all such dubious (if, so far, legal) trickery, a basic reality remained: even as Company leaders tried to inflame desire for its stock, no spark can start a fire that lacks fuel. May 1720 saw Exchange Alley set ablaze by the dream of seemingly instant and unlimited wealth shared not just by men like Blunt himself, who had spent decades pursuing their fortunes, but, increasingly, by everyone within reach.

AS USUAL, DANIEL Defoe was there to chronicle the frenzy. "The Humour of the Town has been for some Time under a Kind of Visitation or Possession," he wrote on May 6. "Clergy, and Laity, Men and Women; nay Servants, Foot-men, Cook-maids and even Children, they are all touch'd." As the market shot upwards, powered by and inflaming the passions of the crowd, Defoe scowled: "I thought once, that Love and Jealousy were the only Two Things that could make the World mad; but I see now that Avarice and eager Flight of the grasping soul after money is capable of all the Fury and Rage" of any human desire.

Within weeks of the first great leaps in the market, early winners began to flash their cash. Financier and Company director Sir Theodore Janssen had been a founding investor in the Bank of England, a financial polemicist, and, as of 1717, a member of Parliament. He had thus been perfectly situated to get in on the game early. By late spring his Company holdings were said to be worth a million pounds, and he was seen in public sporting a new diamond ring, a gift, it was known, from the Prince of Wales—the reward for services the public could infer. More luxury followed, as he knocked down the old house on his estate in Wimbledon to build anew on a grander scale that matched his purse. He was, at least, kind enough not to keep his good fortune entirely to himself: it became known that he had helped his housekeeper to accumulate £8,000 in Company stock of her own.

Janssen's baubles and building spree were matched over and over in a town awash in paper fortunes. Writing at a fairly early point in the rise, Defoe reported on a Sunday visit to a church he hadn't attended before. He remembered nothing of the sermon, distracted by the unusual clamor at the end of the service. "As I pass'd by one of the Pews, I heard one Gentleman say to another . . . *'Twas Two Hundred Twenty Six last Night.* Ay, says t'other, with some Surprize, *And did you sell?*" More of the same occurred in the next pew, not just among the gentlemen, but "I found the Ladies Whispering too. . . . 'Well, Madam, *Has your Ladyship Sold your Stock?*' "

No, she had not, it turned out, because she had it "on good report that yet more of a rise was to come." Defoe walked on, making his way to a Quaker meetinghouse—in eighteenth-century England usually the very avatar of modest sobriety—and even there it was the same: "While I was in the very Porch I heard a voice . . . cry out loud . . . *How goes the Stock?* Go, says t'other, it does not go, it flies."

To Defoe, this was evidence not simply of folly but of something worse. The obsession with the last price "breaks in not only to all our Conversation, and all our Pleasures," he wrote, "but it breaks into the very Worship of God, and into our Sabbath-Day Duties." Such piety aside, Defoe documented throughout the month a popular obsession with the market, its tokens of riches, with a sequence of numbers that promised the pot of gold, today, tomorrow, perhaps soon. "*How goes Stock, how goes Stock,* was the question before me and behind," he wrote, "within Doors and without, and through the whole Crowd."

Not everyone talking South Sea marvels was actually in the market, of course. But the seemingly unceasing progress of a rising market drove many to an almost feral urgency to get in on a good thing. As shares continued to gain in June, the Duchess of Marlborough, once the most powerful woman in the realm after Queen Anne and still a force to be reckoned with, badgered the Earl of Sunderland to make sure that her friends would be able to sign up for more shares in the latest money subscription to the point that "to make your Grace as easy as I can," he interrupted his work as the head of government long enough to contact the Company's cashier, Robert Knight, on her behalf. He was then able to reassure the duchess that Knight had promised "that they had taken care of these names before and that I might depend on it." (In this correspondence Sunderland ignored—or may not have noticed—the fact that the finan-

Sarah Churchill, Duchess of Marlborough, in 1702

cially sophisticated duchess wanted to reserve stock for her friends, but not for herself.)

Less connected souls dove in just as avidly as those in the noble lady's orbit. James Windham, an official of the Salt Office (administering the taxes on salt), was no nabob. In May, he wrote to his mother, overcome with joy, "I grow rich so fast that I like stock jobbing of all things. Since the South Sea have declared what they give to the annuitants, Stock has risen vastly." Gloriously wealthy—on paper—Windham told his family that he "has a mind to buy land" and would "willingly buy a clever estate . . . if it costs 10 or 15 or 20,000 pounds [roughly £1.5 to £3 million in twenty-first-century currency]."

More signs of the times: coach makers saw their business rocket up—dozens, eventually hundreds of vehicles delivered over the spring, with more on order. With such tangible evidence of new money visible to anyone in London, Exchange Alley grew ever more irresistible, to the point that its narrow passages became a kind of carnival of money—as depicted in the satirical poem "A South Sea Ballad": "Our greatest Ladies hither come, / And ply in chariots daily / Oft pawn their Jewels for a sum / And venter't in the Alley." Those of quality rubbed shoulders with "Young *Harlots* too from *Drury-Lane*" who came "To *Fool Away* the Gold they gain / By their *Obscene Debauches.*" The unnamed poet added the crucial detail: "Our *South-Sea* ships have Golden Shrouds, / They bring us Wealth, 'tis granted, / But lodge their Treasure in the Clouds, / To hide it till it's wanted." No one—almost—allowed this to concern them. Wealth, even if only on paper, piled up on every corner, and those as yet unblessed by such promises of gold dove into the scrum jostling from one end of the Alley to the other.

As those in the stock gamboled in the South Sea surf, others, equally enterprising, trolled for them. There is always more than one way to make money off a boom, and few hunting grounds have been as filled with targets as those few streets and passages in London during that South Sea spring. Pickpockets worked in groups of three or four: one to distract the target, while the real artist of the gang—a nip, who cut purses, or the yet more skilled foyne, who could dip into the inner pocket of a coat—made off with the mark's purse. They would pass on their spoils to another, who would melt away before their victims knew they'd been robbed. Even South Sea House itself offered no sanctuary, as John Carswell reported: "An Irish earl lost his wallet" coming out of the Company offices.

An undated satiric depiction of the South Sea throngs in Exchange Alley

Put such crimes down to the irrational exuberance of the day; it's not surprising that a ridiculous explosion of wealth should render winners vulnerable to traditional predators. More cautious or more canny people took care of their paper fortunes by transforming them into something much more tangible. Land, the ultimate measure of wealth in eighteenth-century Britain, saw its own boom. Among those with new fortunes to dispose of, sixteen South Sea directors bought acreage during that spring and summer—Janssen and Edward Gibbon, the historian's grandfather, among them, as well as Knight, who kept the Company's books, and his assistant. In quieter times, productive real estate went for about twenty times the revenue it could generate each year. The Company's new rich paid from twenty-nine to thirty-seven times real estate's annual income—at the most extravagant, almost double what the same property would have cost a few months earlier.

STILL, NOT EVERYONE was money-drunk, even at peak frenzy along Exchange Alley. One Londoner wrote that he "had a fancy to go and take a look at the throngs" along the Alley, "and this is how it struck me: it is like nothing so much as if all the lunatics had escaped out of the madhouse at once." That echoes Defoe and is plausible enough—but it's too kind a diagnosis. Not everyone was insane. In an echo of the flood of new companies that had hit the Exchange in the 1690s, and again in the next decade, some of the new ventures were led by gamblers, or sometimes mere swindlers searching for yet bigger fools. Adam Anderson, a South Sea Company clerk, wrote in his remembrance of 1720 that "any impudent imposter needed only to hire a room at some coffee house . . . [,] having advertised it in the newspapers the preceding day, and he might, in a few hours, find subscribers for one or two millions (in some cases more) of imaginary stock." But that fraud didn't mean all those who signed up were deceived. "Many of those very subscribers were far from believing those projects feasible," he recalled. When, as happened throughout the year, their subscriptions popped in value, "they generally got rid of them in the crowded alley to others more credulous than themselves."

Anderson was describing those who were buying into what were called "bubble companies"—projects launched as the booming stock market created both new investors and an appetite for anything that seemed like easy money. But his judgment applied equally to those who confined themselves to the South Sea Company's stock itself. Its unbroken rise pulled all else upwards with it—and through the first half of the year one man became the face of what appeared to be an explosive triumph of financial wizardry: John Blunt.

HE AND HIS associates knew this was false. None of the South Sea Company's maneuvers had been a one-man show. But Blunt was the face of the Company, and hence of the carnival playing out in London. He had promised the nation both relief from its debts and wealth equal to that of Spanish America, and as the summer of 1720 approached, it seemed to all appearances that he had delivered.

Honors came in due course. On June 17, King George graciously made the shoemaker's son a baronet, a title that survives to this day. Sir John, as he was henceforth known, received wildly extravagant praise from much less exalted quarters too. Nicholas Amhurst, expelled from St. John's College, Oxford, in 1719, was trying to make a living with his pen as both a poet and a political polemicist. In July 1720, he tried a familiar trick, publishing a poem of praise aimed at getting in on a good thing: his "Epistle to Blunt" is testimony to the public's enthusiasm in the glorious adventure that, as of its writing, had been running for three months. "The Nation," Amhurst wrote, "distress'd of late / With publick Debts," was now "Renew'd in all her ancient Strength." This was a triumph, not just for a few newly and luckily rich, but on an imperial scale. "Thro' every Realm her Credit reigns," Amhurst crowed, "And *Europe* of its riches drains." Oh, and by the way, Amhurst added—could Blunt please send along some South Sea stock.

All through spring and into high summer, the promise of wealth was being kept, as the numbers the old *Course of the Exchange* reported out of Jonathan's confirmed. On June 17, as Blunt knelt for his baronetage, the South Sea Company closed at £755. That month the Company offered a new issue of £5 million in stock at par value of £100, to be sold at £1,000 each. There was no trouble filling the subscription list—which included one Sir Isaac Newton—to the point that many complained (with Amhurst) that there was none left for them to buy.

And yet, for all the noise, all the din on the Alley, each broadsheet paean to Britain's new financial geniuses, the parade of shiny new coaches in the Park, and the crimson river of fine old Hermitage, there remained a few for whom the glorious rise of the stocks on the Exchange was not persuasive—men who looked at the ever-upward climb of the stock exchange and wanted to know what lay behind such cheerful figures. The scientific revolution had made experience subject to number—but that idea, for which Newton remained the most visible symbol, did not turn on any single data point, say, yesterday's closing quote for a particular stock. What was required was to put such quantified information to the test of reason.

That is: the price that South Sea stock could command was not an answer but a question: What lay behind that claim of value? What expectations, what motions—not of planets but of human acts and transactions—could determine what the South Sea Company might actually be worth? In the South Sea year, there were at least some in Britain whose minds had been steeped in the intellectual revolution of the preceding decades. For them, questions of what one should pay for what expected return, and how to incorporate judgments about risk into that calculation, were problems reason and experience should be able to answer. Was it possible to make the choice to buy or sell, not on the riot of the crowd, but after careful analysis? Some, even at the height of the frenzy, believed they could.

Chapter Thirteen

"If the Computations I have made, be right . . ."

The loudest of those to subject the South Sea scheme to number was a notoriously disputative member of Parliament, Archibald Hutcheson. Hutcheson, though a member of the Royal Society, was no natural philosopher. He was instead a thinker drawn to the play of numbers to model human affairs. The stock market of 1720 presented him with an irresistible target. He hit it hard and often enough so that having begun as a lonely voice, by Christmas his contemporaries would credit him with having grasped the year's follies as well as any man in Britain.

Hutcheson had entered Parliament as a Tory in 1714. In keeping with what would become a settled habit, he soon shifted his allegiance, allying with the new Whig ministry. That didn't last either: he was by temperament a man who loved verbal combat, and he seems to have had a reflexive disdain for whoever held power. Also, he was subject to one of the common sins of his day (and ours!): the compulsion to prove that he was the smartest man in the room, which rarely serves to make or keep friends.

That urge to oppose and self-aggrandize may have come from Hutcheson's reaction to his own modest beginnings. Born in 1660 or thereabouts (the surviving records don't quite specify) to a Scottish family transplanted to Ireland's County Antrim, he took a first step toward public life by enrolling in the Middle Temple, one of London's Inns of Court. He emerged as a barrister in 1683, and a few years later he took a job on the fringes of the British legal world as attorney general of the Leeward Islands. Over the course of those early professional years he somehow picked up a new skill. There is no evidence that he ever formally trained in mathematics, but between whatever he picked up at the Middle Temple and subsequent work as man of affairs for his most famous patron, the Duke of Ormond, Hutcheson managed to turn himself into one of the foremost practical mathematicians of his age. In the 1710s, these "calculators," as they were dubbed, sometimes derisively, became the heirs to the founding political arithmeticians William Petty and Charles Davenant, who had turned their attention to questions of national finance—and Hutcheson rapidly became one of the leading quantitative disputants on the matter of the national debt.

Soon after he entered Parliament, Hutcheson published his first major treatise on the question. In the *Proposal for the Payment of the Public Debts,* he used a simple accounting technique to calculate how long it would take to retire all of the nation's obligations, showing the different results from a variety of approaches. He returned to the issue in 1718, when he released *Some Calculations and Remarks Relating to the Present State of Publick Debts and Funds*. There he waspishly noted that he had long sought to persuade Parliament of the dangers of debt even as the government recklessly increased the nation's obligations. Sermon over, he went on to analyze what it would take to pay off each of the different types of borrowings. The result was an intimidating display, page after page of figures, designed to bludgeon readers into submission: even if Hutcheson's audience couldn't follow his calculations, they had to be impressed by the sheer volume and seeming precision of his numbers. Thus persuaded—or cowed—they were presented with Hutcheson's solution: a "sinking fund," used to reduce or "sink" Britain's obligations by levying a tax on land that, Hutcheson proposed, would be solely used to pay down the debt.

There was nothing new about this idea. Walpole had suggested something similar and would again. But Hutcheson's proposal was significant for another reason, as a demonstration of how far Newtonian culture had advanced since the days of the early political arithmeticians. As historian William Deringer points out, from his first effort Hutcheson used a compound interest calculation as both an analytical model and a rhetorical device, a way to persuade his readers of the truth of his argument. Such models are mathematical stories that evolve over time, and as deployed here they enabled Hutcheson to point out that as each payment cut the total amount of the debt, less interest would accrue on what was left. Those savings could then be applied to paying down yet more of the borrowed principal.

This is exactly how a standard real estate mortgage works: a homeowner pays the same sum every month for thirty years. Each payment reduces the amount of principal owed on the house, which means that slightly less interest will accrue over the next month. As the payment remains constant, that difference pays back a little more of the borrowed principal. The house gets paid off faster and faster until at last there's nothing owed. Draw a graph of the rate at which such a loan gets retired, and you don't see a straight line—the picture of what happens if the same amount of principal gets paid each month. Instead, a beautiful, smooth curve appears—the signature of an accumulating impact over time. Thus the rhetorical punch of what Hutcheson proposed. There was no need to attempt the kind of fancy maneuvers of debt conversion into stock—like the original South Sea deal of 1711. If the Treasury would only adopt his calculated wisdom and create a tax-based repayment system, the amount the nation owed would shrink at an ever-increasing pace—the "sinking" motion of the sinking fund—until Britain would be wholly free of debt.

This was recognizably a scientific revolutionary's way of thinking. In the *Principia* Newton had constructed mathematical models that could explore the behavior over time of the moons of Jupiter or could predict the motion of a comet with a track that remained mostly unknown. He published his results both as an exercise in scientific reasoning and with persuasive intent: he sought to persuade his readers that what he had discovered "cannot fail to be true." In his earlier writings, Hutcheson attempted much the same double act. His work focused on the dynamics of budgets instead of celestial bodies, but it spoke in the same, unassailable language of numbers in flux—and thus asserted a claim to like power: just as Newton had declared of his system of the world, Hutcheson's arguments could not fail to be on the money.

THERE WAS, OF course, a key difference between Hutcheson's calculations and the utterly authoritative demonstrations in the *Principia.* When Newton bragged about his work's unassailable accuracy, he could let nature be the judge, pointing to the agreement between his mathematical account of a comet's flight and the track it actually traversed. Hutcheson could not command such certainty. Instead, he used the cultural power Newton and his friends had given to mathematical reasoning to strengthen his political argument. Whatever truth his algebra might contain was contingent on the uncertain behavior of the human actors involved in any financial choice.

Cue the South Sea Company. At first, Hutcheson did not oppose the scheme. In the early debates in Parliament, he had preferred the Company's proposal to the Bank's. But once the final form of the deal was agreed, he saw an opportunity to apply his mathematics to the agreed terms—and then to test his estimates for a rational price for South Sea stock against the fact of what actually occurred on the Alley.

Hutcheson would repeat the same calculation several times during the first half of 1720. Each time, he asked the same question: Given what was publicly known about the Company's actual sources of revenue and potential profit—the amount of money to be divided among the ever-growing pile of South Sea stock—what was a reasonable price to pay for any given share of that income? Behind that seemingly simple question lurked two key assumptions. First, as economic thinkers began to work out the new financial ideas of the late 1600s and early 1700s, there was a genuine uncertainty about what a business actually consisted of—what it did, what its assets were, and what all that might be worth. Was a company simply all the stuff one could see, weigh, and count at any given moment? That would be just the money people had invested in it—for example, its financial capital; its physical assets like ships and buildings and goods; and even, if one were to be very adventurous, any money others owed it. Or was what the company *did* over the months and years important as well—and if so, how should one figure out the value of such doings, now and over decades to come?

An example: What was the best way to think about a simple project to send a ship on a trading voyage? If you were involved in one of the new insurance schemes, it could make sense just to value it as the cost of that vessel and everything it contained. But perhaps a merchant adventurer with a longer view would want to weigh the hopes and prospects for that ship on this one voyage, the profit that would come when the goods were sold—adjusted, perhaps, for the risk of storm or plunder at sea—and then, what about any money to be made on future voyages? If Hutcheson had shared the "what you see is what it's worth" point of view, then his calculation would have been simple: add up its assets, divide them by its shares, and there you have the price. He didn't do that, and so implicitly accepted the idea that there is more to a business than what it owns.

That seems obvious now. It's standard practice, absolutely elementary, to establish the value of any enterprise (and hence a price for its shares, which are, formally, fractions of the whole) on the basis of what its combination of people and knowledge and things can be expected to do and thus earn over time.

In these watershed years of the financial revolution, however, it proved hard to think through how to define and value something like a government promise of a stream of income that was subject to the risks that the Treasury might delay or even default on its obligations. Applied to the growing number of new joint-stock companies whose shares traded on the Alley, the same question grew even more intractable: Given the wild uncertainties that attended so many of these new businesses, how, exactly, was anyone to discover a "true" price based on something other than how much money investors were willing to pony up? Perhaps the truest measure of value was simply whatever people wanted it to be, as expressed in the price anyone was willing to pay for its shares—the Humpty Dumpty method of pricing a stock.

Hutcheson—and others—disagreed. In his pamphlets in the 1710s on the national debt he'd already made a critical leap, recognizing that financial analysis is inherently rooted in time. A static snapshot of the national debt—or a company—couldn't measure how much money any given investment choice would deliver to its owners over a year or a decade or any interval. Instead, the task of understanding the value of any of the new financial instruments had to include both an analysis of its prospects going forward and some explicit accounting for how likely it was that those possible future returns would actually materialize. To Hutcheson and others, there was only one way to do that: such future possibilities and the risks involved could be understood only in the language of numbers; they could be measured, quantified, and expressed as mathematical relationships—and hence analyzed, even critiqued.

That is: Hutcheson may have been—he was!—as biased and partisan as any of his fellow members of Parliament. But he was also a member of the Royal Society, and his work traded on the core cultural value the Society had fostered since its founding: that truth could most reliably be discovered by deep encounters with "Number, Weight and Measure." Most important, such an approach allowed this financial philosopher to make predictions just as Newton and his heirs had done when they weighed the probabilities of a game of dice or foretold the return of a comet.

Those two axioms—that what mattered was change over time and that calculating the possibilities over a range of potential outcomes would provide vital insight—underpinned everything Hutcheson attempted in the Bubble year. Doing so, he lived Cassandra's nightmare: like her, he could see the future plain. And as she had found, (almost) no one listened to him, no one troubled themselves about the coming disaster that he could see, right there, in every column of his figures.

HUTCHESON DELIVERED HIS first detailed analysis of the South Sea Company's economic fundamentals on March 20, after the bargain had made it through Parliament but weeks before the swaps and sales began. That pamphlet, *Some Calculations Relating to the Proposals Made by the South-Sea Company, and the Bank of England, to the House of Commons,* presents a meticulously constructed mathematical abstraction of the Company. As William Deringer found in his reading of Hutcheson's work, the genuine flash of brilliance lay in his analysis of the refusal to set the share price at which the stock would be exchanged for debt. Rather than trying to come up with a calculation of the "right" such price, he did something subtly different: he asked instead how successful the Company would have to be to justify any given price. Remember: each share of a joint-stock company is in essence a claim on a fraction of the profits from the business. Its price is thus what someone would pay for its share of that profit, which can be compared to the return on any other investment (land, lottery tickets, an annuity, and so on). The key measure for Hutcheson was thus the amount of profit a company could make each year that could be assigned to each share. That was the genius of his approach: any South Sea stock price implied a given level of profit, enough to match or beat what investors could earn from owning land, or from simply hanging on to the government obligations that the Company wanted them to exchange into its stock.

That was clear enough, and plenty of early-eighteenth-century calculators knew how to think about annual returns. But the South

Sea deal was a more convoluted proposition than a farm or even a mercantile voyage. Two pieces of information were missing: both the price the Company planned to charge for the shares offered in the swap, and plausible projections for profits from its transatlantic trading ventures. Hutcheson made his distinctive contribution by finding a workaround for supplying that missing knowledge.

His approach could work only for someone prepared for an extraordinary amount of tedious mathematical labor. Hutcheson ran more than sixteen separate calculations testing share prices ranging from £100 to £500. He varied how much debt would be traded for stock; how much money the Company would receive from the government in interest payments; the price of the shares to be sold to the public;* and more—with each scenario requiring its own separate accounting. But through all his weary sums, Hutcheson's underlying argument was simple. He just treated the Company's total revenue less its expenses—its profit—as identical to, for example, the interest payment a piece of the national debt would bring, or the rents that could be extracted from a wheat field.

Framing his question this way, Hutcheson turned the usual calculation on its head. His equations spat out an answer—but it wasn't what the stock was worth. It was, rather, how much money the company would have to make in the future to justify any given price for its shares. The range of numbers thus produced wouldn't be *the* answer for South Sea investors—how much they should pay for this new asset. Instead, it turned the future into a set of clear and easily compared numbers that could suggest whether any price the Company demanded at the moment accurately reflected its true current value, as defined by the return on other potential investments. That is: Hutcheson wanted to know whether the Company was likely to generate enough income to offer each shareholder the same or better

* These were the shares left over when the Company swapped stock for debt at prices above par. There was always a risk that this new supply of shares would hold down prices, but Hutcheson did not try to model that complication.

rate of return they would receive if they'd put the same amount of money into any other asset.

HIS ANSWER NEVER changed:

No. It would not.

As the first of his pamphlets was being printed in March, jobbers along the Alley quoted their shares between £207 and £220. Hutcheson hadn't anticipated that rapid a rise from the levels before news of the deal came out, so early versions of the model tested a series of share quotes only up to a maximum of £200. Even so, the numbers didn't work. The Company's prospects had to justify the extra £100 above the stock's par value of £100—which at least in theory represented the monetary value of the enterprise itself, the capital it possessed: it had to show that the profit to come would make that premium worthwhile.

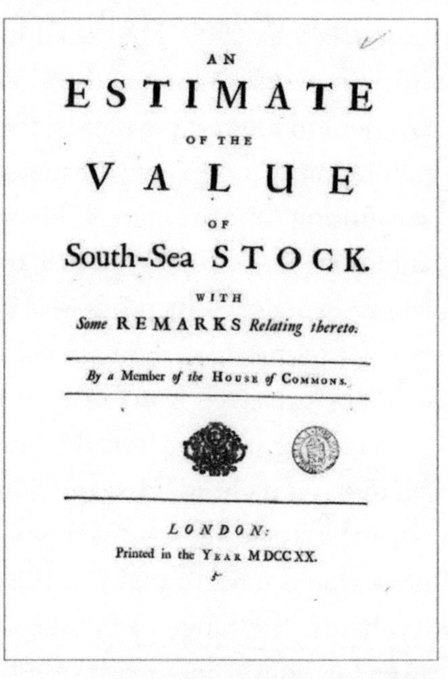
AN
ESTIMATE
OF THE
VALUE
OF
South-Sea STOCK.
WITH
Some REMARKS *Relating thereto.*

By a Member *of the* HOUSE *of* COMMONS.

LONDON:
Printed in the YEAR MDCCXX.

The cover of one of Archibald Hutcheson's early pamphlets analyzing the South Sea Company's prospects and promises

That question set him off on a hunt for both assets and possible future business to cover the difference. At £200, the Company would retain a substantial amount of the shares created as it swapped stock for debt. Hutcheson made the assumption that the Company would take in all of Britain's irredeemable debts, adding £21 million to the Company's balance sheet. Hutcheson divided the value of the shares the Company retained by all the stock held by the public, both preswap investors and those who made the exchange, and found that it represented the equivalent of a one-time dividend that paid £48 0s. 5d. Against that extra £100 over par that the new investors had paid, Hutcheson argued, this still left those who made the swap £52 in the hole.

But of course there was more to the Company's business than its own capital. So Hutcheson looked next at everything else that could produce returns to investors. He handled the known known first, the interest the Treasury would pay to the Company for all the debt it had absorbed. To see how much that income was worth, he performed an absolutely conventional present-value calculation. Dividing the Treasury's contribution among all the outstanding shares, he found that if this banking side of the Company's business stood alone, its present value—what one would expect to pay for such an asset—was just £6 per share. Adding that to the newly created stock "dividend" attributable to each share, that meant that new buyers' investments of £100 per share over par had, so far, bought them an asset worth £54. That left £46 more to account for.

That set up the last stage in the calculation, Hutcheson's unique insight. What the Treasury owed on its debt to the Company was public knowledge. There was some risk that the Treasury might miss a payment, or even default on the debt, but at least all the figures involved were out in the open, from the amount of debt to be absorbed and the interest due on such sums. What remained unknown was how well the South Sea Company would do carrying humans and goods across the Atlantic. So Hutcheson simply asked how much trading income the Company would need to have in order to provide an adequate return for that last forty-odd pounds extra that debt holders had paid for their shares. To make that £200 level—accounting for the £46 remainder—he showed that it would take an annual trading profit of £7 13s. 2d. per share per year to match what investors could get at then-average rates of returns for a wide range of investments. Seven quid and change sounds modest enough. But summed over the total number of shares at large, that added up to a target of £3,335,816 in annual profit from trade with Spanish America.

That was, as Hutcheson knew, almost certainly impossible—and this was before the party really started. April's numbers swamped his early expectations: the debt conversion went out at £375, and the first two money subscriptions, sales to the public, at £300 and £400. At those heights, Hutcheson's model sounded an ever louder siren: to be worth what people lined up to pay, the Company would have to produce trading profits ranging from £5 million to more than £7.25 million.

In his pamphleteering, Hutcheson allowed the impact of those numbers to sink in almost without comment. He knew the Company's recent history, and any sophisticated reader would have too. South Sea voyages had struggled to bring in £100,000 in profits over three years in the mid-1710s, and the Company had made nothing on its trading rights since war with the Spanish had resumed in 1718.

Hutcheson didn't think the slave trade would help either. He had no moral objection to slaving, but he didn't believe there was enough to be made selling African bodies to make the numbers work. "If the Computations I have made, be right, it is then evident, that the Gains of the *South-Sea Company,* in Trade, must be immensely Great"—and for "immensely" read "impossibly." The Company's monopoly over commerce with Spanish America, Hutcheson wrote, would "in all Probability go but a little way towards the Satisfaction of the Loss herein before stated."

In these early calculations Hutcheson did not venture so far as to say that Blunt and the ministry had conspired in a fraud. But he certainly felt that the Company had an obligation to make it clear how current prices could be justified. "Is it not, therefore, reasonable," he wrote, "that the *South Sea* Company should Explain from whence their advantages are to arise which may be a Solid Foundation for the Value of their Stock . . . that Thousands and Thousands of People may not be Undone."

IT'S TEMPTING TO read such a calm and sober suggestion and see Hutcheson as a clearly modern financial thinker. Certainly, he grasped what remains a fundamental concept in asset analysis: to make sense of some economic activity, transform something happening in the real world into the stream of money it pours off each day or month or year. Then run a calculation on that number to transform it into a rate of return over a given term, a dividend—and use that result to inform one's view of a sensible price for any given proposition. Hutcheson's model was far less sophisticated than those used now—he didn't try to explicitly quantify risk, for example—but his habit of mind is recognizably similar to that of someone like Warren Buffett today.

But there is this difference between our now and Hutcheson's then. He wasn't speaking a common language that the public broadly understood. His audience was almost entirely made up of people for whom reasoning about human affairs in mathematical terms was still a strange new phenomenon. That wasn't entirely a negative for Hutcheson. Mathematics, even the appearance of quantitative reasoning, can serve as a rhetorical gambit, and Hutcheson was trying to persuade as much as convince. Few could solve the necessary equations, but the notion that Englishmen had figured out how to trace the stars in their courses by plying equations had become a powerful cultural point of pride. Hutcheson used that cultural authority for his own ends. His pamphlets of 1720 formed part of a long-running polemic aimed at persuading his audience that it wasn't just the South Sea deal that was suspect. National debts were too, and stock market speculation was, if anything, worse. Like Newton himself, he could help create new ways to frame the world without ever being wholly free of his time, place, and deeply held and unexamined assumptions.

But for all those qualifications, and the fact that his analysis was neither pure nor perfect, Hutcheson's writing throughout 1720 affirmed that it was possible to apply rigor to money, even at the height of South Sea obsession. No matter how many contemporary accounts or histories since have portrayed the Bubble year as mass insanity, it wasn't an inescapable madness. Hutcheson showed that the ways of reasoning that had emerged in Britain over the last half century could be brought to bear even on such a radically strange phenomenon as a wild ride on the Exchange. This was the scientific revolution at street level. Anyone could count, think, and use the results of formal calculation to tame his or her emotions and make choices armored by reason.

Anyone could. But for many months, not many did—and it was not obvious that those giddy enthusiasts were wrong. In March, and even in June, the crucial question remained: Were Hutcheson's calculations more than pretty? Was he right?

Perhaps yes . . . but in the moment, no. Through spring and into the summer of 1720, Exchange Alley ignored Hutcheson, and South Sea shares continued to rise. In a second, revised version of his calculations, published on March 31, Hutcheson ran his model on a price that he may have believed was an exercise in absurdity: a price of £500 for South Sea stock. He turned the crank on his model, and out spat its answer: at that price per share, the Company would have to make over £9 million in trading profits every year to produce a merely average investment return. That wasn't difficult; it was ridiculous. There was no reason to repeat the exercise at higher figures.

SOUTH SEA SHARES first hit £500 on May 28. They rose to £530 two days later, which would have seemed a bargain to anyone trying to buy the following afternoon. On June 1, 1720, Company stock was quoted at £720, a gain of almost 50 percent in four days. That represented two hundred and twenty pounds' profit per share: a paper gain equivalent to half a decade of a laborer's toil, more than two years' stipend for the Lucasian Professor of Mathematics at Cambridge, and about five months' salary for that lord of England's money, Sir Isaac Newton, master of the Mint.

Later that year, a deck of playing cards made the rounds in London. Cartoons on the faces of each card made vicious sport of Britain's South Sea follies. The lines on the deuce of clubs pitch-perfectly caught the tone of that glorious first of June:

> A Sharping Min[is]ter bid a Broker run,
> And Buy him Stock, accordingly 'twas done;
> By accident 'twas raised as soon as Bought,
> He sold and got a Hundred Pounds for nought.

From a deck of South Sea playing cards printed in 1721

Money for nothing! And even better, more money for still more nothing if you could bring yourself to wait until tomorrow to dispose of what you had acquired today. As long as the market continued on this happy course, this was a thrill against which even the gods—and certainly the dry and intricate mathematics of that calculating scold, Archibald Hutcheson—would contend in vain. And there was this fact too: some did and would continue to do very well indeed . . . for the time being.

Chapter Fourteen

"the largest honest fortune . . ."

So far, this account has avoided referring to what happened in the first half of 1720 by the name by which it became famous: the South Sea Bubble. That label attached itself to these events as early as 1721, but the term *bubble* wasn't new even then. It had been applied to earlier financial misadventures and was adopted by contemporaries as they lived through the Bubble year. Still, what they had in mind wasn't quite the modern notion of a bubble as a market failure. In London in the 1720s the word had less to do with the mob's inability to grasp what was happening and much more with specific, individual sins: the misdeeds of people who knew what they were doing as they blew up their bubble. That was certainly what the satirist and playwright William Chetwood described in *The Stock-Jobbers; or, The Humours of Exchange-Alley,* written at the height of the public drama of 1720 itself. The work was never performed, and Chetwood may have intended it as a polemic more than a play. The primary villain of the work is a stock trader named Sir John Wealthy, who opens the action by reflecting on a fine day's work: "About 10 shares [bought]," he lists, "and sold 15."

Who were his prey? Wealthy boasted that he'd delivered "Four to Mr. *Noodle,*" most notable for his "fine gilt chariot"—that is, more wealth than wit. Three shares went to that always juicy target, the sailor ashore, a "Captain Sanguine, newly returned from the *East Indies* . . . full of Money and bleeds young Heir." Then there was the "whimsical Musical Sort," just come from Italy who speaks "not one Word without a Tune to't." He took three shares, with two more going to a "young Gentleman out of *Lancashire* . . . just come of Age" and sent by his mother to make his fortune in London. One more went to a merchant on the edge of bankruptcy, risking his last on the South Sea Company "in hopes to give his Fortune the Lye," and the last two went to "a young rattle-brain'd Fellow." He was Sir John's favorite customer: "He ne'er made two Prices [tried to haggle] but as soon as I told him the Value, out flew his Guineas, as tho' they had been but meer Counters."

Chetwood clearly despises Sir John, while ridiculing all those reckless enough to deal with him. Those on both sides of so many trades that spring were risible people: some monsters, some fools, all to be mocked, none pitied and none admired. What was a playwright to do but scorn them all? Still, Chetwood's portrait of Sir John as the knowing cove, a hunter making sport of his prey, captures the nub of what *bubble* meant in 1720. It wasn't just a noun, certainly not as used today to describe market behavior in which prices outrun their before-and-after value. Instead, to someone observing the money trade in these early days, *to bubble* was a verb: to deceive and defraud a person—or a nation. This is the sense that Deringer discovered in the correspondence of an "Eminent Merchant" who in 1718 complained that British traders were being "bubled out of ¼ of the Intrinsick Value of their Money" in currency deals in Paris. This wasn't a complaint about anything so impersonal as the conditions of trade between the two countries. Rather, for the man to whom the merchant wrote, such losses were the result of "artfull Managements"—specific choices made by individual actors aiming to cozen honest subjects of King George.

This is how Daniel Defoe understood the word as well when in 1701 he inveighed against those who "make the *Exchange* a Gaming

Table," who ply "a Trade made up of Sharp and Trick," and who "Jobb, Trick, and Cheat one another"—and feast on those to be "bubbled by them that no know better [morals]." He said the same again in 1719 in his attack on stockjobbers as men "*resolving to be rich at the Price of every Man they can bubble.*"

This early instinct to see the market as moral space—a deranged one, a cesspit of deceit—was never entirely wrong: the invention of new ways to buy and sell money deepened the divide between those comfortable with unfamiliar ways of thinking about finance and everyone else. That in turn evoked an enormous catalog of ways the experts could gain the edge in any deal. That *bubble* had so swiftly become a term of art testifies to the fact that in just a generation the world of money had been transformed, creating opportunities for the active malice of men who knew better and chose to act worse. In the South Sea June of 1720, that meant that many of those thronging the Alley were indeed putting themselves in peril. When such naïfs flocked toward danger, they could, perhaps, be excused, but not entirely. As Chetwood's mug shot of Sir John reveals, if they lost everything, they could blame their tormentor for taking advantage of their collective ignorance of this strange, new, and increasingly complex market in paper wealth. And yet, Sir John's marks would also have to acknowledge their own avarice. Even so, Chetwood's tale implied that it was pretty much impossible for any neophyte investor trying to make sense of the daily passions on the Alley to make a rational judgment about the South Sea Company in the midst of the debt conversion. Hindsight is easy, that is, but in midst of the scrum Sir John's wiles rule the day.

That conclusion—or rather, that excuse—is false. Some investors could and did subject the market to measure, even if such reckonings were usually less sophisticated than a full mathematical analysis. A few who did became very rich as a result.

THOMAS GUY WAS no man's fool. Born in 1644, he was largely but not entirely a self-made man. His father had begun his working life as a collier and carpenter on the London docks. He rose from there until he owned his own barge-building yard. That success allowed the younger Guy to get an education through grammar school, and with it usable knowledge of Latin and Greek. At sixteen he began an eight-year apprenticeship to John Clark, a book dealer in Cheapside, London, and there found his life's calling.

As soon as he fulfilled his obligation to Clark, he set up his own book business, securing key clients, including Oxford University. He branched out into publishing, hitting on a bookman's truth that would be rediscovered many times: if you produce a mass market edition of a classic, you can make a very respectable amount of money. His golden ticket was a compact version of the large and expensive Oxford Bible. Guy priced his volume well below competing editions and so made his first fortune.

Despite being later slandered as a miser, Guy was a charitable man, not just a profitably pious one. He made his first known significant gift in 1677 when he helped fund new buildings for his grammar school. His book trade continued to prosper, and, as the financial revolution generated opportunities for those with surplus cash, he began to dabble in the market. He did well with his first transactions, buying various government obligations in the 1690s, including what were called sailors' tickets, vouchers with which the Royal Navy paid its sailors on those frequent occasions when the Treasury couldn't come up with the needed coins. That early foray into the debt market led to more, until in 1711 he chose to turn some of his government paper into shares in that new creation, the South Sea Company.

Fast-forward to 1720, by which time Guy had become both an experienced investor and, not coincidentally, comfortably rich. He was never a gambler, trading in and out of every market move. He still held his original South Sea shares, and he seems to have acquired more over the decade. By the early months of 1720, as Parliament pitted the Company against the Bank, he owned 54,040 shares that had cost him about £54,000, acquired at prices that average out to just under their par value of £100. The monotonous flow of cash from the Treasury to the Company and thence to him had suited Guy just fine.

Then, after a decade of patient accumulation, he changed his mind. On April 22, the day after the first money subscription was announced at its strike price of £300, he began to sell. He was careful. He unloaded his enormous holdings slowly from April to June, as the stock was steadily climbing. He got rid of the last forty shares on June 14, when South Sea Company stock was being quoted at £700 on Exchange Alley. With that, he was done.

Guy's notes show he took in a total of £250,000—on the order of £400 million now. It was, one historian claimed, "the largest honest fortune made out of the Bubble." That victory left a lasting mark on London when, in 1721, the bookseller Thomas Guy committed the bulk of his wealth to the creation of a hospital in Southwark, to be built for patients "who may be adjudged or called incurable." He visited the building site on December 26, 1724, Boxing Day, returning home complaining of the cold. He was dead within a day. His estate became the hospital's endowment; Guy's Hospital now stretches across nineteen buildings and is one of the world's leading medical teaching and research institutions. An ill wind, and all that.

Guy's Hospital in an early-nineteenth-century print

That largesse was heroic, but the fact that Guy became able to give on such a scale shows what was possible for a cautious and canny—but otherwise ordinary—investor during the South Sea spring. Remember: Guy was no mathematical savant. He didn't bandy equations at Royal Society meetings. He had no insider knowledge of Blunt's schemes or the ministry's plans. Rather, his gift was a readiness to think clearly about what he owned and why. He had held his shares for years. When the Company's circumstances clearly and fundamentally changed, so did his approach, as he recognized that a boring, ordinary business had become something different more or less overnight. Guy kept no diary and wrote few letters, so his reasoning can only be inferred from his actions, but he seems to have reached the conclusion that his shares were worth more as cash than as part ownership of a novel and increasingly uncertain financial experiment.

Having come to that judgment, he made a clear and binary decision. He accepted that he might not be selling at the peak the market might achieve—but recognized that he was still gaining a completely satisfactory return on a long-term investment. Throughout the process of liquidating his holdings, he continued to act sensibly, staging his trades to avoid troubling the market and, along the way, enjoying some of the ongoing rise. Finally, and perhaps most important, he didn't second-guess himself. Once he sold his last shares, he was done. Though the Alley still bubbled and South Sea prices continued to rise after his exit, he did not chase the stock once he'd turned what had been a purely paper profit into sweet, solid, hard cash.

GUY IS THE most famous of the South Sea winners, but he wasn't alone, nor was he the only clearly reasonable one. Some of the others were just as clever and attuned to the market; others were merely lucky, and perhaps a few of those had wit enough to recognize the role good fortune played in the outcome.

The Duke of Marlborough was among the wealthiest men in

Britain. Sarah Churchill, his duchess, was very much a partner in that fortune. She was an experienced investor, tough, with admirably steady nerves. During the 1715 Jacobite rebellion, when much of Scotland briefly fell to supporters of the Old Pretender, James Stuart, she watched as the fear of an overthrow of King George—with its attendant risk that a new monarch might repudiate the debts of the man he deposed—unfolded on Exchange Alley. Like Guy, she had been an early South Sea buyer, and as she wrote to her friend Lady Mary Cowper at the height of the scare, her stock with a face value of £2,100 "is not worth so much as 2 or 300*l*." But, she noted, "I don't find the News from *Scotland* is so bad as some reported." More bad news might come, she conceded, but she, for one, was not going to sell on fear. That, it turned out, was the correct decision.

In 1720, the duchess was likely one of the beneficiaries of the insider gifts brokered by her old friend James Craggs, who, before he became postmaster general and one of the Company's key allies, had been Marlborough's business agent. She got in early and stayed through late April and the celebration of the king's birthday. As the suddenly rich were drinking their loyal toasts, Sarah Churchill took the temperature of the market and chose to take her money off the table, selling out at roughly £500 a share. She missed the early-June rise but, like Guy, didn't chase the market.

Timing such exits from the market is always an imperfect, chancy skill. Robert Walpole, though unbribed in the rush to get the South Sea bill through Parliament, owned a considerable amount of Company stock acquired over the previous decade. It was worth just under £19,000 as of June 1719. He sold about half that year and got rid of the other half in a series of transactions ending on March 18, 1720, at a price of £194 10s. per share. That was a handsome premium over the original value of his holding, but he swiftly became ruefully aware of just how much he'd left on the table—a realization that would gnaw at him over the coming months.

Sir Isaac Newton did better. He too had been a long-term investor in the Company, and he held about £10,000 at par—shares with a market value in early 1720 of about £13,000. He watched as prices rose from there, and, echoing Guy, he began to sell in April, just as the stock price made its first great leap upward. Only scraps of Newton's records survive, but there is a letter, dated April 19, that authorizes his agent (an assistant at the Mint, Francis Fauquier) to sell three hundred shares. By the time Fauquier completed that sale, and, as historians have inferred, much or most of his remaining stock, Newton would have effectively tripled his original investment, booking £20,000 in pure profit.

The common thread is that there doesn't seem to have been any specific incident that pushed Newton or Churchill or Guy to sell, nor did they make any sophisticated quantitative argument (as Newton certainly could have done). Rather, many of those who emerged from the Bubble in the black appear to have shared one key reaction: what had been a dull and long-term investment had changed into something else, a speculation, a gamble. Some were able to decide when they had grown rich enough and then just walk away—from more gains, perhaps, but more risk as well. Those with inside access—Blunt, for example, and those directors who flaunted their new wealth around the king's birthday celebrations—could use private knowledge to decide when and how completely to cash out. But over the course of the spring, outsiders like Guy and Churchill and Newton demonstrated that anyone could walk into Exchange Alley and emerge victorious.

THERE WERE OTHER ways to master the moment too, approaches that did not rely on one's personal guesses. One by-product of the financial revolution was the emergence of a new profession: those who made their daily living dealing in money in all its evolving forms. These were among the first of what we now would call financial professionals. The Bubble year would reveal how potent their expertise could be—and how much it could profit the few who were its masters.

The best example comes from one of the earliest and most successful of the private banks that emerged in the 1680s. Hoare's Bank still exists, and its books provide unique insight into how the cleverest money-men navigated the Bubble year. Economic historians Peter Temin and Hans-Joachim Voth have dissected those records, tracking every South Sea stock transaction the bank made in 1720. They found that the bank held £8,600 worth of South Sea stock early in the year, and, as Parliament debated the enabling South Sea Act, bought another £25,000 at £181 per share. The bank then sold much of its portfolio late in March, after the run-up in prices that followed the final passage of the bill. Hoare's carefully parceled out its sales into different transactions, most likely, as Guy had, to avoid depressing the market.

Hoare's reentered the market at the end of April, buying at £341, rather more than the price at which they'd just sold, and then again in late May, at almost £500, acquiring and selling more in June. As prices peaked and then wavered, the bank methodically set about selling this second portfolio. Over 1720, it completed a total of fifty-four transactions, both buying and selling, with £140,029 changing hands—roughly equivalent to a quarter of a billion pounds now—and by early autumn the bank was about £20,000 ahead on all of its South Sea trades. Though it would suffer some minor losses in Company stock later in the year, still, as Temin and Voth report, by the time everything was settled the two managing partners, Henry and Benjamin Hoare, took £28,000 in a distribution of trading profits based on the bank's success throughout the Bubble year. In total, Temin and Voth estimate, "The bankers earned as much in 1720–21 by buying and selling stock as they had in the twenty years previous."

So, what did Hoare's decision makers know? Why did they make the choices they did? Clearly, they were trading, not purchasing an asset for the long haul. To Temin and Voth, several lines of evidence point to Hoare's principals believing South Sea stock was overvalued (at its peak, wildly so). One key hint lay with how they treated Company shares when these were offered as collateral for loans sought by

Hoare's customers. As early as March, the bank was discounting those shares by more than half, lending barely more than £40 on £100 of stock at then-market prices. To a sophisticated investor, nothing further would be needed to make the meaning clear: Hoare's had no faith in the South Sea Company as a business capable of producing a return enough to justify its then-current heights along the Alley.

That didn't mean, however, that the bankers believed that the trades being made at Jonathan's and Garraway's were incomprehensible. Rather, they made a judgment about human nature. They bought and sold on the basis of their assessment of market emotion, the passions of the mass of investors clamoring for shares (and, later, scrambling to get out). Hoare's partners were gambling, in other words, but never at random: theirs was the smart money at the table. Their strategy was to buy, book profits, and then repeat the maneuver until the Exchange sent them a clear signal that the good times were done. They were steady enough players to maintain a clear, sustained and delightfully lucrative edge over almost every other player in the casino.

COUNT JONATHAN SWIFT among the patsies the hard-eyed types like Hoare's traders could exploit. That's no surprise; almost everyone in any way connected to public life in London risked a flutter along the Alley that year. The dyspeptic writer was still a few years from the lasting fame that would come in 1726, when *Gulliver's Travels* made his reputation as the greatest satirist of his age. But he was already a formidable literary brawler, and it was almost inevitable that he would have something to say about the events of 1720.

He was lucky up to a point: he limited his risk through the first half of 1720 to just £500 in Company stock, most bought years earlier. By late spring he became—on paper—quite comfortable. A year later, with his phantasmal fortune gone, he was still among the lucky ones. He had barely dipped his toe into the South Sea at its height, so he didn't lose much when the Bubble burst. Still, he knew he'd been touched by madness—and he knew exactly whom to blame.

> Conceive the works of midnight hags,
> Tormenting fools behind their backs:
> Thus bankers, o'er their bills and bags,
> Sit squeezing images of wax.

Thus Hoare's clear-eyed professionals! But for all the thump of such bludgeoning verse, Swift ignored the truth on Exchange Alley at the beginning of summer 1720. Individual investors and speculators—however they came to their conclusions—pocketed real gains through the first half of 1720. For a brief time that summer, it seemed as if others, lots and lots of them, would do the same.

Chapter Fifteen

"How goes the Stock"

Alexander Pope had no problem with the idea of making money, with no romantic illusions about the purity of an artist's poverty. Long before his *Iliad* would make him comfortably well-to-do, he thoroughly understood the pleasure of a well-filled purse, as only those who've had to scrape for every shilling do. So when his fortunes rose to make him a man of property, he knew that while enough was pleasant, more was better.

He had been a skeptic early in 1720. Tory in his sympathies, he never trusted the motives of those pushing the South Sea transaction. But once the ride got going, he, like so many with a few pounds to spare, couldn't resist a flutter. He understood what he was about, noting that "there is no gain until the stock is sold . . . and our estate is an imaginary one only," which, he confessed, was a "pretty general case with most of the adventurers." He stayed the course as the adventure continued into June, and his brief verse, preserved as an inscription on a punch bowl, delivers perhaps the perfect snapshot of the passion of the Bubble at its peak:

COME, fill the South Sea goblet full;
The gods shall of our stock take care;

Europa pleased accepts the *Bull*,
And Jove with joy puts off the *Bear*.*

EXCHANGE ALLEY HAD never seen anything like the first weeks of June. The June 1 number, £720, rose to £770 by the fifth. Even more exciting—or terrifying, as one's temperament dictated—its shares had become wildly volatile. Adam Anderson would work for the South Sea Company for more than forty years. He was near the beginning of that career in the Bubble year—still only twenty-seven—but the memory of that climatic month of June remained vivid decades later, when he wrote his magisterial *The Origin of Commerce*. "There happened such sudden fluctuations in their stock, sometimes even in the space of a few hours," he wrote. ". . . For though on the second of June, it got up to eight hundred and ninety per cent (£890), yet that vast price bringing many sellers the day following to Exchange Alley . . . it fell before night to six hundred and forty; and yet the same evening rose again to seven hundred and seventy."

* There is no certainty about where *bull* and *bear* come from as slang names for rising or falling markets, or those betting on one or the other. A common story, which Merriam-Webster supports, gives a nod to the old proverb "It's not wise 'to sell the bear's skin before one has caught the bear'" as having given rise to phrases like "bearskin jobber" to describe the people and the practice of selling in the expectation that the price of a security will drop before the seller has to make delivery. That usage appears in print as early as 1709, and Daniel Defoe uses it in his iconic 1719 pamphlet, *The Anatomy of Exchange Alley*. By then, to be a trader in bearskins, or a bear for short, had clearly left behind its original sense as someone who recklessly deals in what one does not own and had taken on the sense in which Pope uses it: someone who bets on, who hopes for, a market fall. Where *bull* comes from is less clear. It shows up as market cant a little later than *bear* and may have been simply an alliterative and sufficiently aggressive counterpoint to those other impressive mammals prowling the Alley. The significance of Pope's use of terms that had been so recently part of the private slang of the money trade is the speed with which the language and hence ways of thinking within the world of speculation and finance had leaked into Britain's broader society . . . so much so that one of its leading men of letters could bandy them knowing that his audience would understand him. This is what a changing culture looks like when it is in its cups.

So it went, dizzying leaps and heart-crushing falls, but always, it seemed, with the arrow pointing upwards, toward ever more riches to come. Life-changing, generation-spanning levels of wealth changed hands daily at Jonathan's and up and down the Change. As Pope's friend Jonathan Swift wrote, what brave spirit, "Rais'd up on Hope's aspiring Plumes," could hope to resist the lure?

Few did—and before returning to the familiar tales of frenzy, of madmen and madwomen and simple greed, it's important to recognize that even these last, extreme heights of the boom were not simply a matter of irrational exuberance. What was happening to South Sea shares and, somewhat more gently, to other stocks was in part a conventional response to a genuine increase in demand. Foreign money had been moving into London's market since late spring. Dutch speculators had been sending cash to London as early as March. French buyers were in the market by no later than April. By late May, London gossip told of Dutch small craft in the port of Harwich, ready to carry instructions to and from Holland, while hard cash for Exchange Alley was coming back across the North Sea packed in hogsheads, the casks that in more normal times would have carried wine or beer. The sums involved were impressive—hundreds of thousands of pounds—and as prices climbed along Exchange Alley, the London market became ever more attractive to Continental investors souring on investments in rival capitals like Amsterdam and Paris.

Still, even if some of June's rise was a straightforward reaction to more money chasing a given supply of stock, the Company continued to juice the market. That was partly pure self-interest: insiders gained with each new record price. But beyond that, there was a clear personal incentive. Key officers and directors grasped that the survival of the Company itself turned on continuing to attract investors.

Their problem was the same one Hutcheson had identified months earlier. The Treasury's payments on the converted debt were capped by law, and trade in slaves and goods was at a standstill—which left only one other source of Company profit, the stock it retained in the debt exchange. Remember, the key to the deal had always been that the swap wasn't set at par: Blunt and his allies had made sure they would not have to hand over £100 face value in stock for £100 worth of government paper. Instead, the higher the market price went, the more of the newly issued stock would remain in the Company's hands.

It always worked the same way: every piece of debt absorbed became a new capital asset, on which the Company would then issue stock. Using the agreed conversion figures, a government obligation that paid £100 per year was worth £3,200 as its present value, the price to pay for that year-over-year stream of income—and its value as capital on Company books. Under the terms of the South Sea Act, that meant the Company would be able to create thirty-two shares at a par value of £100 to reflect the addition to its total capitalization. But it would take only about four shares at early-June prices to add up to the £3,200 cost of that same piece of debt. The Company thus gained on its own books the other twenty-eight shares, to do with whatever it wished. That was the lead-into-gold transformation in this exercise in financial alchemy: the more South Sea shares rose, the more the Company would be worth—which would make shares in such a valuable enterprise still more desirable. But that only worked as long as that growing heap of shares was valued at the elevated numbers achieved by late spring.

INSIDERS (AND ATTENTIVE unattached investors as well) clearly understood that dynamic. Pumping the stock would keep the whole enterprise afloat. That meant the South Sea Company was in the business of being an expensive business.

In hindsight, the flaw in the plan is obvious. Like many schemes that have found their marks since, the boom in South Sea shares needed new money to chase the stock. Every rise could be justified by the next boost to come—a dynamic that worked until it didn't, until the supply of those willing to take the plunge disappeared.

Some core players grasped the danger—hence the eager land purchases by Company directors in May and June. Even at inflated prices, those deals still turned paper fortunes into useful dirt. But while some cashed in, the implacable logic of the scheme held: the South Sea would remain navigable as long as Exchange Alley continued to bubble. It followed that those in charge would do everything they could to keep the market on the boil.

That was what the 10 percent dividend (to be paid in stock, not hard cash) announced in April was supposed to do: persuade investors that there was ever more money to be made in the Company's shares. That was only the first and simplest gambit. As the market accelerated upwards through the spring, the South Sea men turned to more sophisticated forms of financial manipulation. Ever-easier installment plans to pay for shares brought in investors with minimal cash on hand. Company loans enabled its shareholders to cover the installment payments due on stock they'd purchased in April. The stock itself was the collateral; no money changed hands, and the amount of the loan was simply noted on the share receipts. The virtue of such a trick was that it propagated an illusion: all those shares sold on margin (but not fully paid for) could still be said to be bought and held. If their buyers did have surplus cash of their own, well, so much the better: any such loose change could find its way to Exchange Alley to keep the market humming along. This was utterly circular financial reasoning, just the play of numbers moving through different locations on the Company's account books, but it served its purpose: no one was compelled to sell or abandon shares, which might put downward pressure on Alley trades, merely because of not being able to come up with a few pounds at a given date.

Such loans did not, at least, involve actual cash money; in the most charitable view, the Company was merely extending its payment schedule at a small charge to investors. That restraint didn't last long. Instead, more explosively, it began to lend real cash against the ever-inflating security of South Sea shares. This lending—in effect, pumping money directly into the Alley—accelerated early in June, when the South Sea directors authorized a new set of loans. It was astonishingly easy money: for every £100 of stock, shareholders could borrow £400, up to a limit of £4,000. This was up from the totals originally announced in May—£300 per share to a maximum of £3,000. Thousands took advantage of both offers. The total in play was staggering: throughout 1720, the Company would lend over £9 million on the security of South Sea stock, with at least another £2 million going to cover installment payments.

Some of this flood of credit paid for more excitement in the market. South Sea shareholders borrowed to expand their investment and thus multiply any potential profit—or magnify the cascade of losses that would follow if it all went wrong. Over the centuries techniques for acquiring such leverage have become both more complex and more subtly dangerous, but the basic practice of leveraging an investment has remained constant from the South Sea era to our own, seemingly more sophisticated age—about which more soon.

THE FULL HEIGHT of that spring tide of money came on June 17, 1720. For the first time in six weeks, the South Sea Company offered another money subscription, its third open sale of some of the shares created in the debt swaps to date. In the weeks leading up to the offering, the price to be charged remained a secret. In the April 14 subscription, the stock went out at £300 per share, then, two weeks later, at £400. Both of those prices were more or less in line with what jobbers along the Alley were quoting at the time. Just before this new, third sale was slated to begin, South Sea stock closed at £750.

As that subscription approached, Alley men speculated about the price the Company would seek. Anderson reported that he "remembers distinctly" an incident at this time—"a certain director . . . being asked by a gentleman at Garraway's coffee-house" if it was true "that the court of directors soon intended to open their third subscription at one thousand percent [£1,000]?" The Company man remained coy.

"Truly, gentlemen seem to strive to talk us into some such price, whether we will or no." Finally, as the sale was about to open, the number was revealed, and the Garraway gentleman had got it right: shares were to go for £1,000, one-third—£250—more than the last price quoted along the Alley. For South Sea House, the calculation seems to have been that given the constant clamor for more shares—real, deep, and insistent—why not see what the punters would pay? In the event, the entire issue—£5 million—sold out within hours.

So far, everything was working to plan. This third money subscription had drawn a wide swath of investors. Records from the sale are incomplete (perhaps, as the historian P. G. M. Dickson suggested, to cover up obvious misdeeds), but what survives confirms that Britain's elite remained convinced that all was well: at least half of both houses of Parliament signed up, with 153 members on the list brought in by the younger James Craggs (son of the elder postmaster Craggs, who had helped set up the whole scheme). Aristocratic ladies were still buying too—among them, the Prince of Wales' daughters.

This eager appetite for more shares was driven in part by another round of cannily designed purchase terms. Once again, subscribers had to come up with only one-tenth of the purchase price up front. The rest was to come in nine installments, again following the example of the prior offering, but with one major difference. April's buyers had been given just sixteen months to pay their balances. Those who bought in June could wait all the way to 1725 to cover the full cost of their shares—and if they chose, they could walk away without completing their purchase. This wasn't an explicit right—there was nothing in the terms of the subscription that said in plain language that buyers could jump ship—but there was nothing to forbid it, and given such ambiguity, in practice it was easy simply to stop meeting installment deadlines. That made it simple for sophisticated speculators to place a bet: as long as they were willing to abandon whatever they'd already paid in, what was billed as a stock purchase turned into something else, what we now call an option.

This version gave buyers the *right* to acquire a share of stock at a set price—with the option to abandon the purchase if the market went against them.

That is, those who subscribed to this third Company sale of its shares had actually bought themselves time with their £100: six months to watch what happened to the stock. If they didn't like the view—if the stock slid enough to make the whole idea look bad—they could simply walk away, limiting their losses to that first payment. On the other hand, if the tide was going their way they could hand over the next £100 and buy another six months to watch the market while retaining their right to complete their purchase of that share.

SUCH MANEUVERS DID not originate in the Bubble year itself. In *Politics* Aristotle describes another philosopher's successful option trade based on an upcoming olive harvest. The concept reemerged no later than the early seventeenth century in a form more immediately useful to the jobbers and brokers of Exchange Alley. By the time of the first London joint stock boom, 1690s options were in common use. By 1720, stock traders were familiar with at least two types of options: calls and their inverted siblings, puts. Calls give those who purchase them the right but not the obligation to buy shares at a set price in the future from the person who sold them that call. A put flips that script: it allows the person who buys a put to sell shares at a set price to the person who "wrote" or issued the option.

The South Sea Company's installment plan can thus be seen as a call option that had to be renewed at each installment deadline to maintain control of the underlying shares. But such option trading wasn't confined to the Company itself. In the Bubble year a variety of private call contracts were agreed between speculators. On the buyer's side, paying the call option price is a cheap way to bet on a stock you think might rise while capping your losses if you've guessed wrong. If an enthusiast for a given stock, say, shares in the Acme Company (currently trading at £100), thinks their price may rise by £5 over a month, spending a pound per share for a call on one hundred shares sets up a net gain of £400 (£5 a share on 100 shares, less the cost of the option). That's much better than the £5 to be gained just by buying one share outright, and it caps the amount at risk at the £100 laid out to buy the option.*

These moves were then and remain useful in ordinary investing practice. If you hold a stock that you fear may go down, for example, selling calls for shares you own gives a kind of insurance against the worst of a decline. In the Acme example above, if its shares had dropped to £95, the call seller would keep the call price and thus be down only £4 per share rather than £5 when the buyer decided not to execute the option, while being guaranteed an effective price of £101 per share if the stock did rise and got called away.

Of course, calls can be used to make much more risky bets as well. Selling a call without owning the underlying shares in what's termed a naked option is a fine way to profit on the market's decline. If you've guessed wrong, though, and Acme shoots up to £110, the pain will be all too real: you are on the hook for the difference between your strike price and what it's going to cost you to buy the shares you need on the open market. Thomas Guy, for all of his skill and his steady nerves, seems to have fallen into this trap. When he sold out his Company holdings in the spring, he appears also to have sold some calls for shares he did not own—a bet that South Sea stock would fall fairly rapidly. A few months later, the

* Puts—less important in the South Sea context—allow their buyers to bet that a share will fall below the strike price. If it doesn't they don't exercise their option, but if it does they pocket the difference while capping their risk at the cost of the put. Those who sell puts take on the obligation to accept a security at the agreed price, so they hope prices are on their way up—which is where the risk of options really bites, as they are on the hook for the full difference between the option strike price and the market value of the underlying stock . . . even if it falls all the way to zero.

Alley quote remained above his strike price and, with his entire holding gone, he had to buy shares at a loss to make good on his wager.

There were and still are plenty of other ways to make such naked bets, often demanding little or no money up front. Forward contracts require those who enter into them to complete whatever transaction is specified (unlike an option, which confers a right but not the obligation to execute a trade). Such contracts were common over the first several months of the Bubble, binding one party to buy shares at a given price and another to deliver those shares on a given date. In the rising market of the spring of 1720, such an arrangement could be deeply appealing to those with a taste for gambling in a grand style. At the beginning of the year, the Duke of Portland had been reckoned one of the richest men in Britain, but he could not resist the lure of playing for the highest stakes. From the launch of the scheme in April through the summer, he signed at least twenty-four forward contracts to purchase South Sea stock. His contracts put off delivery as much as nine months—and it was only then that any money would be due. In all, he agreed to buy 507 shares at an average price of £649, for a staggering total of £328,975. This was a straight win-or-lose bet. If the stock rose above his contract price, he could sell them as soon as the contracts came due, making a fortune without having laid out a shilling. If they fell below that number, he would be on the hook for the difference. There could be no mistaking Portland's faith: strong enough to put his entire fortune in play.

FOR ALL THAT such techniques had emerged earlier, the Bubble year still marks a watershed in the formation of modern finance. It's true that ideas like joint-stock companies and the quantification of myriad human actions both preceded and shaped the Bubble. Similarly, the development of financial technology—options, forward contracts, and even more complicated trades—occurred in the decades before the South Sea scheme hatched. But for all the individual developments that preceded it, it was only in the Bubble year that the cumulative impact of those inventions would become clear. Even if they weren't conscious of it in the heat of the moment, jobbers shouting in coffee rooms relied on a concept of money that was becoming ever more abstract, ever more distant from any underlying actions in the real world that yield something to sell.

Joint-stock companies embodied one kind of abstraction. Each share was a number that represented a slice of a profit stream generated by the South Sea Company, the Bank of England, or any other such enterprise. An insurance policy formed another, deeper move toward number and away from the world in its calculation of how much the risk that your house might burn down should command as an annual payment for the policy that would protect you if it did. All such transactions, though, are directly connected to the underlying asset—a piece of land, a debt, a life. Options and similar trading devices took finance one more remove from any connection to the world. A call on its own doesn't buy a piece of a company. Instead, it confers something contingent or conditional—not a share in some venture, but the opportunity or obligation to buy or sell such a share under specific conditions.

There is a name for such financial devices: derivatives. Derivatives are simply securities whose price depends on what happens to another security. Calls and puts are derivatives because their prices shift when the underlying shares on which they confer rights rise or fall on the market. Today, derivatives have become enormously more varied and complex than those in play during the South Sea year. Many of the now-commonplace features of financial exchanges would have been literally inconceivable, built and priced using mathematical ideas that hadn't been invented then. Jonathan's and the Jerusalem and Garraway's and the rest of the Alley's corners and coffee rooms were not yet a completely modern financial market. They were instead places where people and ideas came together in the process of becoming modern. Even so, there are links between their day and ours. For one: derivatives are not inevitably dangerous. They can serve not just as a speculative bet but as a kind of insurance—for example, when someone wants to be sure she can buy a stock at a given price down the road, which is the role of a call option.

Still, there is no doubt that derivatives throughout their history have offered new and often exceptionally exciting ways to gamble without limits. The Duke of Portland wasn't investing when he entered into his forward contracts. He was betting that the boom would continue. His play—clearly, his strongly held belief—was that he'd never need to come up with any actual cash: a rising market would buoy his profits as those contracts came due.

Just by placing those wagers, Portland did his part to make that outcome more likely. Each contract expressed his expectation: South Sea shares would continue to rise. In turn, the market at large registered his perceptions of where the stock would go in the strike price in each deal. Any gain on all those shares would once again be money for nothing: in each contract he signed, Portland did not have to pay a penny up front.

From one such bold (or reckless) investor to the market as a whole: the financial inventions in play during 1720 served the Company's goals. Installment plans, loans on stock, dividends paid in stock—all of these allowed investors and speculators to claim more and more shares with less and less hard cash required, at prices that, for the time being, continued to rise in the face of all that demand. Options, forward contracts, and their kin created leverage: short money, or no money used to control far more assets at a fraction of what it would cost to buy them outright. All these gambits gave more and more people the ability to purchase more and more shares—or appear to buy them, at least—for (proportionately) less and less cash up front. There had never been such a perfect recipe for a boom. It was so successful, in fact, that by June the only question at South Sea House was how to keep this magic ride going.

THE ANSWER CAME in the context of some vexing news. During the first half of 1720, all the major joint-stock companies had shared in South Sea's happy rise. Bank of England stock changed hands for about £139 on April 10. At the beginning of May it topped £155, then hit £200 by the end of the month. The East India Company beat that, going from £230 on April Fools' Day to £420 at the end of June, while the slavers at the Royal African Company saw greater gains still, with their stock more than doubling over the same period. Insurance did well too, with Royal Exchange Assurance stock tripling through May and London Assurance doing even better, running all the way from £14 1s. 4d. to £51 in the same few weeks.

To Company men, such gains were galling. Money buying Bank or East India shares wasn't going into South Sea stock. But the truly vexing problem for Blunt and friends was the sudden appearance of a host of upstart adventurers. The ebullience of the stock market encouraged anyone with a good idea—or a bad one, or, on occasion, not much of anything at all—to form a new company and target anyone with a few guineas who wandered into the Alley. Eager projectors floated twenty-seven companies in April, twenty more in May, and then, over the first two weeks of June, seventy-seven new business propositions emerged, all chasing any money looking for shares to buy. Some were perfectly respectable, including several insurance offerings—the Globe Fire Office, for one, which sought to raise £2 million in subscriptions at the Ship and Castle Tavern on April 23. Similarly, focused plans like those of a "Company for the fishermen in eastern parts of New England, Annapolis, Cape Sable, Newfoundland or any other part of America" at least had a plausible story to tell about how the enterprise would make its money.

But a company with such divided loyalties as "establishing a coral fishery in the Mediterranean, and making calico in England," seems from this distance a little less encouraging, and an investment in a "Company for improving an art, lately found out, for the making of soap, 6 lbs of which is more serviceable than 8 lbs of the best brown soap"—offered at Mulford's Coffee House on June 7—would have required much more investigation than the furious pace of the market would allow. Similarly, it is hard to assess proposals like one in which the projector intended to make what sounds like a kind of plywood by "melting down saw dust and chips."

"The Bubblers Mirrour . . ."—a satire on the flood of joint-stock "bubble" companies in the wake of the South Sea boom

That was the common theme of these "bubble" companies, as their contemporaries called them. They sprouted in the dozens, some plausible, many dubious, others, no doubt, purely fraudulent. There were groups that proposed to extract silver from lead, to buy Irish bogs, to trade in ambergris and ostrich feathers. And while the infamous "company for a project which shall hereafter be revealed" seems, sadly, to be apocryphal, a pretty fair approximation appeared in an advertisement on May 21 "for raising the sum of Six Millions sterling to carry on a design of more general advantage, and of more certain profit" than anything London's optimists had yet seen. History does not relate whether this particular pitch found investors, but as the historian Richard Dale points out, a 1764 survey of the wreckage showed that just four of the companies launched in the carnival days of early 1720 emerged from the Bubble year as going concerns.

Long before that shakeout, the Company's inner circle recognized that every start-up that managed to find investors siphoned off cash that could have bought South Sea securities—in Anderson's account, it was apparent that "the traffic [of bubbles] obstructed the rise of the . . . stock." As the weeks advanced, what had been a minor irritant in the first weeks of the stock market boom became more troubling when the task was to maintain South Sea stock at ever more stratospheric heights. As a result, Blunt and his colleagues grabbed what seemed like the simplest solution to their quandary: turning to their political connections to dispose of their competition.

Accordingly, on June 11, Parliament passed a bill known as the Bubble Act. The act banned the formation of any new joint-stock company without either a royal charter or specific authorization by an act of Parliament. Work had begun on this legislation in February. But as the need to boost the stock pressed on the Company, its officers intervened, ensuring that a late change in the text confirmed that all of the South Sea share transactions would remain valid, as the new law would plausibly have barred them under a provision that forbade even established companies from actions not authorized by their charters.

The rest of the act served Company ends as well. All the so-called bubble companies of the spring and anything inventive Britons might yet dream up could no longer be organized as joint-stock operations, which meant that any cash tempted toward these projects could now be redirected to its proper homes: the handful of already-authorized companies, including, of course, the largest of them all, the South Sea Company itself.

As a practical matter, the Bubble Act had little immediate impact. Several new projects sold shares in June and July without drawing an official response. But it was clear what the measure was supposed to do: sustain the market for the major companies at or above its already unprecedented heights. For the moment, it worked. On June 11, as Parliament acted, South Sea stock stood at £735. By June 24, shares changed hands at £1,050. Share price remained close to that level through the end of the month, pegged at £950 for three days running.

The implication behind that number was staggering. With its

shares hovering around £1,000 apiece, Exchange Alley priced the South Sea Company as a whole at about half of the total value of every business traded on London's financial market—a sum that Adam Anderson, the Company clerk who had witnessed the Bubble, estimated to be about half a billion pounds. For context: that much money could have bought all the fixed property—all the real estate, every acre of useful land and all the buildings on it—in the whole of Great Britain. The wealth of the nation was now a thing of paper and promises—and nothing so mundane as a day's work would do for any Briton with gumption. This was the mood and moment that Alexander Pope captured in this bitter tag of in-hindsight scorn:

> How goes the Stock, becomes the gen'ral Cry.
> Rather than fail we'll at Nine Hundred Buy.
> Instead of Scandal, how goes Stock's the Tone,
> Ev'n Wit and Beauty are quite useless grown:
> No Ships unload, no Looms at Work we see,
> But all are swallow'd by the damn'd *South Sea.*

Chapter Sixteen

"a mighty handsome entertainment"

Those last days of June represented the summit, the very top of the market, the brief, still hopeful moment when, as Alexander Pope wrote, "Stocks and subscriptions pour on every side / Till all the demon makes his full descent."

Not everyone was demonically deceived, of course. Even at such breathless heights it was still possible to count and calculate. Archibald Hutcheson regularly retested the price of South Sea stock against his model and published again in June. The math delivered its implacable verdict: "These New Subscribers," he wrote, "gave £1,000 for that which was worth only £300" (a number calculated in the same way he'd come to his earlier conclusions, adding up all the different assets and sources of revenue the Company possessed).

But still, as he knew all too well, Hutcheson's columns of figures had little impact on "this blazing and astonishing Meteor." In time, he noted, "People, perhaps, may then begin to think more cooly about this Matter, and hearken a little to Reason and Demonstration." But on Midsummer's Day? No, not yet.

Instead, the optimists still smiled. James Windham of the Salt Office, whose earlier paper gains had set him shopping for a landed estate, wrote to his brother on July 12 that he was glad he'd subscribed for more stock at £1,000 and that their mother had as well. Pope's and Swift's friend, the usually penniless John Gay (who would soon write *The Beggar's Opera*), shared their initial enthusiasm for the scheme. In April, James Craggs the Younger had given Gay a few South Sea shares. Entranced by the paper fortune that soon appeared—his stake was worth roughly £20,000 at the top of the market—he refused all advice to cash out and instead (like Windham) put every last shilling of his own meager capital into the June money subscription at £1,000 per share.

William Hogarth captured the moment in his *Emblematic Print on the South Sea Scheme,* published after it had all come tumbling down. Hogarth, as vicious a visual satirist as Pope was in words, transformed the madness of the Bubble's peak into a hilarious, unforgiving parable, complete with lords, ladies of negotiable virtue, Honor being whipped, and the Devil's revels. A verse caption explains all the action, beginning with "See here ye Causes why in London, / So many Men are made, & undone, / That Arts, & honest Trading drop, / To Swarm about ye Devils shop."

William Hogarth, in the single most famous depiction of the Bubble at its height

Twenty-first-century economists still argue about the rationality of this and subsequent bubbles and panics, but Hogarth's cartoon reveals the lived experience of those last weeks, when it was still possible to believe in "what magick makes our money rise, / When dropt into the Southern main." Luck still ruled. Wealth could find its way into a hag's purse as much as any nobleman's. Money begat money, while Trade slept. Any individual may have had a plausible reason for what he or she chose to do that summer. But such reason, the rationale for one course of action or another, was molded by the passions of the moment, shot through, in Hogarth's words, with "Monys magick power."

THAT POWER HELD through the summer of 1720. In early July, the Bubble Act did manage to cool some of the ardor for the tag-along joint-stock projects that had so annoyed South Sea principals, though within a couple of weeks stockjobbers were again pushing new projects. The wild ride in Company stock seemed to be slowing too. South Sea stock had made its grand leap up to the neighborhood of £1,000 in the last week in June, almost exactly double its level a month earlier. But then it steadied and even pulled back a little, trading between the mid–eight hundreds and a bit over nine hundred for the next few weeks. The Bank of England's shares, up about 50 percent in May, slowed as well, hovering in a narrow range between roughly £230 and £245 through July. The other major companies followed similar paths, with no climb to new highs but no substantial slide either.

Even so, Blunt and his allies continued to chase every last shilling. On July 14 the Company offered shares in exchange for the last major pile of official debt in private hands, redeemable loans managed by the Bank of England. This was the money borrowed over the years to meet annual operating deficits, and at the start of 1720 it amounted to over £16 million. To sign up for the deal, its owners had to go in person to South Sea House, and the resulting scrum showed that Company stock had lost none of its shine: "The People came thither in great Numbers," according to the *Weekly Journal or Saturday's Post,* "and great Crowding was there to subscribe." That clamoring multitude signed over to the Company a total of £11 million, or almost double the expected amount.

This urgency seems to have troubled even some of those who had helped set the whole affair in motion. James Craggs the Younger, a junior minister, wrote to his chief, Earl Stanhope, on July 15 that "it is impossible to tell you, what a rage prevails here for South Sea subscriptions at any price." The Bank of England, which had to deliver the financial documents in its care to their customers, faced its own near-riot. "The crowd of those that possess the redeemable annuities is so great that the bank, who are obliged to take them in, has been forced to set tables with clerks in the streets." Their clients' fervor seemed to have infected the Bank's usually sober officers as well. Their institution owned £300,000 worth of redeemables on its own account, and it must have gratified Sir John Blunt to see the humility with which they begged for the chance to put it all into South Sea stock: "Having resolved to write them in upon the present Subscription," a Bank governor wrote, "we desire your Favor to inform us, wither we may depend upon its being effected."

AFTER THAT FLURRY, the market rested. Part of this was due to one of the realities of eighteenth-century stock exchanges. All transactions in a company's shares had to be recorded in its books (to make sure that, among other reasons, dividends went to the correct recipients). Everyone who had subscribed to the South Sea Company's debt exchanges or the sales of new stock had to be tallied and the stock dividends announced in the spring allotted. When shares changed hands on the open market, those transfers of ownership had to be recorded as well. Most of the major joint-stock companies managed to register transactions on a rolling basis, but the sheer volume of business in its stock led the Company to announce that it would close its books for two months, beginning in late June. Any trade in the Alley during that closure would become in essence a forward sale, to be completed once the transfer books reopened. While this didn't stop shares from trading hands, it did add some friction to the process, a tap on the brakes as traders waited for normal business to resume.

Getting its paperwork in order didn't stop the Company's business. In early August there was another offer, aimed at sweeping up every last pocket of official debt. Some holders of the remaining irredeemable annuities that provided their holders with a guaranteed income for life (or longer) hadn't been persuaded to join the scheme in April; now they were to be given one last chance. Among those who took the Company up on its offer? Isaac Newton, who traded annuities worth £650 per year, more than his annual salary at the Mint, for cash and stock in a combination that valued South Sea shares at about £800 per share.

That wasn't a terrible deal given the prices still being quoted along the Alley. But as the Company's leaders knew, it would remain so only as long as the market stayed happy. Thus the Company's next maneuvers all aimed at driving demand for its stock. On July 27, the directors authorized still more loans to cover the next installments due on the shares from the two April sales. Next, the Company ignored one of the few existing market regulations and used its own funds to buy back shares and thus prop up the market. Its largest purchases came in August and September. Some directors made individual bargains also designed to help the stock (and themselves), by offering their own shares to sell in a variation of the forward contracts so favored by the Duke of Portland, taking in "great Sums at Forty and Fifty *per cent* [of the final price] to deliver Stock at six Months for Fifteen Hundred Pounds." Just the rumor of such a sale made the idea of prices at a mere eight hundred or a thousand seem almost a bargain.

Last, the directors authorized one more public sale of new stock, the fourth money subscription. Set for August 24, just after the transfer books were to reopen, it went out at the same £1,000 as the June sale—again offered on a long and gentle payment plan. As in June, spot prices in the coffeehouses ran well below that ask: just £820 on the day the subscription opened. Despite that gap, as pure speculation a subscription purchase might still make sense to sophisticated buyers who would use the small initial installment price as an option. About 2,500 people thronged the South Sea offices on the day the sale opened. Company clerk Adam Anderson recalled their being so eager that "this subscription . . . was completed in three hours' time." There wasn't enough stock to satisfy the demand—and that produced its own mini-bubble and a handful of clear winners, as some successful subscribers turned to the Alley's secondary market and sold "that same evening at forty per cent advance."

AT FIRST GLANCE, this was the same crush, the same expectation, the same passion seen since April—and it now appears absurdly reckless. Were all those swarming the Alley in the last weeks of summer as crazy as they seem now? Among academic economists and economic historians the question of whether bubbles, booms, and crashes in general are "rational" is a fundamental issue.

The belief that markets are usually accurate machines for discovering the "correct" price of anything used to be a dogma of conventional economics. It's now less an article of faith and more a complex argument. It remains deeply troubling within the discipline to think that markets can go mad, that thousands or millions of individuals can act against their own interests. Significant research has gone into the concept of a "rational bubble," in which investors' decisions can be seen to have been reasonable at the times they were made. Such work looks at what happens during an ultimately unsupportable boom to see if, given the available information, ultimately disastrous decisions could be seen to make sense as they were being made (in particular, formal ways to understand what "making sense" means). Are the events of the spring and summer of 1720 best understood as a collective loss of reason? Or is there a more nuanced way to understand why so many would put themselves in such jeopardy?

This isn't a question of whether sophisticated investors could legitimately profit off the Bubble; they could and did. Sarah, Duchess of Marlborough, sold before the last great leap in June and did not choose to swap any of her government annuities in August, noting that "this project must burst in a little while and fall to nothing." Those who acted cautiously are important reminders that money manias aren't all-consuming.

But despite such counterexamples, the deeper question remains: Why did so many succumb in 1720, and why have so many in every bubble since? Throughout the Bubble year Daniel Defoe wavered between disdain for money mania and faith in the underlying approach to the national debt. The Company evoked in him a conflicted set of reactions. He'd been a South Sea booster since 1711; he'd expressly promoted the scheme in early 1720; and while he scorned the year's smaller projects as deceits unleashed on the public by the worst men in Britain—the Alley's jobbers—it took him much longer to accept that the Company's business was equally fraught. As late as mid-August he still expressed nothing but confidence that "the *South-Sea* Company . . . whatever happen'd, the Value of their Stock would receive the least Alteration." Alexander Pope, for all he disdained Defoe's politics, felt the same, writing to a friend, Lady Mary Wortley Montagu, on August 22, "I was made acquainted late last night that I might depend upon it as a certain gain, to Buy of South Sea Stock at the present price, which will certainly rise in some weeks, or less." He was, he told her, utterly confident in the news, "& therefore have dispatched the bearer with all speed to you." Montagu took Pope's advice, not just on her own account, but for at least one soon-to-be former friend—that relationship among the least of the casualties of what was about to come.

And above all, what of Isaac Newton? Newton was, of course, an incomparably more skilled calculator than virtually anyone then living. He had, after all, invented the calculus decades earlier. He could think in numbers evolving over time. There was no man in London—no human in the world—better equipped to figure out how difficult it would be for the South Sea Company to beat the boring, regular returns on his existing portfolio of government debt instruments.

He seemed to have grasped that inconvenient truth in his sober choice to sell out in April. But then he watched the stock he no longer owned move up and up and up. By mid-June, he could stand it no longer. A new analysis of his accounts by the historian Andrew Odlyzko has uncovered Newton's breaking point: his decision to sell about £26,000 worth of government paper he owned on June 14. Odlyzko reports that Newton used the proceeds to buy back into South Sea shares at about £700. (Among those on the other side of his trades: the bookman Thomas Guy, still content to shy away from the Alley's wildest flights.) In July he dove in again, swapping out his last £6,000 in redeemable debt. By the end of his summer-long buying spree, he had reacquired at least 100 and as many as 160 shares of South Sea stock, paying out double or more what he'd received per share in April. Late in August, his bet, and Lady Mary's, and all those placed by the thousands clamoring to get onto the Company's subscription lists were just weeks away from disaster. So is it fair to say that Isaac Newton himself, and all these eager people, had simply lost the capacity to think rationally about the market for South Sea stock? Were they, in this matter at least, actually nuts?

There's a plausible case to be made that they were not. The notion of rationality in a market does not mean that reasoned decisions will necessarily turn out well, just that they make sense given the information available in the market at the moment. Newton in July had traded a stable stream of income for an asset that the market priced at about the same level—and that had, on the evidence of the previous several months, the chance to grow in value. He or you or I can be wrong when we make such choices, but it's not folly on its face to try. The idea that markets are rational, that what happens in human exchange always reflects the reasoned judgment of market participants, has been an article of faith for many economic thinkers. Thus some twenty-first-century observers of the Bubble claim that accounts of Exchange Alley as an open-air asylum are simply wrong. Bad decisions, they argue, are not necessarily crazy ones.

But there's a problem with the argument that the South Sea Bubble inflated through a purely rational process: it contradicts how those who lived through it described their experience. Perhaps the idea that people are either completely in command of their reason or mad is too binary. The economic historian Anne McCants suggests another alternative: humans are social animals. The emotions we feel and communicate to each other mold what we can persuade ourselves are objectively "rational" decisions. It may not require an epidemic of madness to break a financial market; that is: the ordinary flow of human emotion, swept up in the excitement, can be enough of the trick. If Isaac Newton, his time's avatar of reason, couldn't control his passions long enough to think his choices through, how many others would?

ANOTHER EXPLANATION BESIDES mere money mania would make it possible to see reason in the events of 1720. Perhaps good people had simply fallen afoul of bad ones, con men perpetrating a fraud on the innocents abroad on Exchange Alley. In that case, the Newtons and Popes and Portlands could be seen as having made rational decisions based on bad—false—information. If so, they were done in not by folly or greed but by the malice of others. So: Did Company insiders know what they were doing, know that the stock could not be propped up forever, and choose to conceal those facts so they could pocket their fortunes before everything fell into ruins?

There is no clear way to answer that for the decisions made in August, with the Company still in full command of its affairs and jealous of its secrets. But even if the signs remained mostly hidden from view over the summer, there is no doubt that at least some of those running the show at South Sea House were getting nervous—

even while others, at a slightly greater remove from the heart of the matter, remained at ease.

The older Craggs, for one, father to Secretary Craggs who had watched the crowds clog the Alley in July, did not behave like the mastermind of a giant con. He boasted that he'd obliged "incredible numbers" of his friends and connections by making sure they won places in the August sale of new stock and then, as he left London for his summer holiday, told friends that he was wholly content. "I am at last settled," he wrote on August 2. His country stay had begun with "a mighty handsome entertainment"—a celebration that included "a bonfire at night, barrels of ale and illuminations."

This was not a man worried about the walls closing in. The new baronet, Sir John Blunt, was not nearly so sanguine. As the stock neared its top he moved a significant amount of his wealth out of South Sea shares and into the old standbys: cash and land. Blunt's accounts reveal three separate sales in the week of June 10, more in July, and then, in August, enough to make him rich by any measure: 265 shares sold in that one month for roughly £100,000 in cash and "an estate at Wenham." That last was only one of the land purchases he made that summer, until, by August, "it became known he had recently purchased considerable estates in Norfolk." Blunt was also clearly nervous about the summertime South Sea offerings. In August every Company director was expected to put up £3,000 of his own money in the money subscription. Blunt tossed just £500 into the kitty.

He and his inner circle made decisions for the Company that summer that also suggest they were men who knew something others didn't and mustn't be allowed to learn. For example, Blunt himself turned that money into loans for shareholder stock on a heroic scale: "The Cashiers lent upwards of three millions in one day without acquainting the Committee of Treasury with it"—communication that, if not required, was, the writer implied, the norm for such a consequential decision. Blunt could not have made such a move on his own, but the underlying facts remain. Company directors, whose

role it was to oversee financial operations, were not just kept in the dark on that one day; they were actively prevented from learning the details of the loans, as Blunt made sure "there was no Committee of Treasury summoned during the whole month of July."

Such high-handed behavior didn't pass without protest. Members of the committee complained about "the irregularity and confusion in the Treasury," which, they feared, might leave "the Company great sufferers by it." Blunt responded that bafflement was the point. He said (or was said to have said), "The more confusion the better; People must not know what they do, which will make them the more eager to come into our measures." There is no way to avoid the implication: Blunt understood that the Company was on perilous ground and did everything he could to hide that fact from those who most needed to know the truth.

But—and this is key—even this most hostile of narrators acknowledges that Blunt had a motive beyond personal gain behind the summer's moves. The pamphlet quotes him as saying, "The execution of the Scheme is our business." Britain's finances were at stake: "The Eyes of all Europe are upon us," he said. "Both houses of Parliament expect to have it done before their next meeting," and if the South Sea treasury should suffer a little in defense of the stock? "One million or two is nothing, to the speedy execution of the Scheme."

So: there is no doubt at all that Blunt tried to secure his fortune over the summer of 1720. And there was no question that rules and prudence were for little people: the South Sea Company's managers tiptoed along the edge of permissible behavior to keep the party going and crossed the line when the Company secretly bought its own shares on the open market. But it's not clear either that Blunt and the rest fully understood how this all would end or that they didn't care about a general catastrophe as long as they got theirs.

Most strikingly, as late as August, even Blunt, with his eye on his own fortune, did not foresee imminent disaster. He and his allies continued to act as if they were still masters of their moment. The argument that the architects of the South Sea scheme were as much ignorant as intentionally malign turns on the one sure fact of the Bubble: it was the first of its kind. There was no historical memory to help anyone in Europe see how it would end. The mathematical tools of financial analysis were being worked out on the spot—and no one yet knew if such calculations truly described reality.

Of course, people grasped that money could lead their fellow human beings to lie, trick, and cozen their fellows. No one with any knowledge of the Alley had any illusions about Blunt and friends: it was understood that they were after as much wealth as they could grab. But while convenient villains like the Alley's small-time con men did indeed know how to fleece the unwary, the much more sophisticated architects of the South Sea scheme were themselves still novice financial engineers. Blunt and Knight, Craggs and the rest did not fully grasp the terrible vulnerability, growing with each successive attempt to preserve, at all costs, the "execution of the Scheme"—because none among them had lived through such times before.

They would learn.

Part Three
The Fall and Rise of Money

Ye wise philosophers, explain
What magick makes our money rise,
When dropt into the Southern main;
Or do these jugglers cheat our eyes?

—Jonathan Swift,
The South Sea Project

Chapter Seventeen

"the People were now . . . terrified to the last Degree"

By mid-August, the Bubble Act had been on the books for two months. To that point, the new law hadn't proved much of a deterrent to unsanctioned joint-stock offerings. Adam Anderson commented acidly that "for a few days, indeed" after the king had given his assent to the new law, "some check was given by this measure to that frantic traffic." But then, "in the face of all authority, it soon revived, and increased more than ever." It was, according to Anderson, a repeat of earlier scenes, "continual crowds all over Exchange Alley, to choak up the passage through it."

Anderson noted that the bubble companies made sure their subscription terms were both easy and cheap—as little as six pence could buy into an offer. He was a bit of a snob, and he took offense that such bargains might make it too easy for "the lower class of people" to enjoy "luxury and prodigality, as well as their betters." Still, he acknowledged that "persons of quality of both sexes were deeply engaged in many of these bubbles, avarice prevailing at this time over all considerations of either dignity or equity."

Anderson was right. Among those enjoying this latest riot of speculation was the Prince of Wales, the future King George II, who was said to have invested heavily in the Welsh Copper Company and who then agreed in July to serve as its governor. Robert Walpole told him that this was a terrible idea, warning that "he would be prosecuted, mention'd in parliament and cry'd in the alley." If—when—it failed, he was told, it would be called the "prince of Wales' bubble." The prince ignored his counselor, and the younger Craggs reported to the ministers attending the king in Germany that he did so for the simplest of reasons: "He has got 40,000£ by it."

These bubble companies were once again in direct conflict with the Company's interests, more so when South Sea shares slipped slightly over the summer. So it must have seemed an obvious ploy to turn the so-far-unenforced Bubble Act against such pesky competitors. It's impossible to determine who pushed the government to act. It's plausible—barely—that the odds and ends of the government still in London that summer decided on their own to crush the new speculative furor. But according to the Company's Anderson, who was in a position to know, it was "The South Sea Junto," which brought its complaint to the Treasury in August.

In response, on August 17 Treasury officials contacted the attorney general for legal direction on what to do about "the Trade now carryd on in Defiance of the Laws with relation to the buying and selling" of bubble-company stocks. The reply came swiftly: companies engaged in the sale of such stock could be served with a particular legal instrument, a writ of *scire facias* (literally, a writ to make known). Such an action would assert a violation of the Bubble Act, putting each entity thus informed out of business. The next day, writs were approved against four companies trading without an approved charter—among them, the prince's Welsh Copper Company.

The York Buildings Company went first, shedding a third of its value on the news, and, Anderson wrote, "in two days after, neither it nor the three other undertakings expressly named in the *scire facias,* had buyers at any price whatever." Nothing more needed to be done. Anderson again: "The more barefaced bubbles of all kinds immediately shrunk to their original nothing. . . . Their projectors shut up their offices, and suddenly disappeared; and Exchange-Alley, with its coffee-houses are no long crowded with adventurers." Many, he wrote, "now found themselves to be utterly undone; whilst on the other hand, such as had dealt in them to great advantage, became extremely shy of owning their gains"—which is to say that any of the promoters of the fleeting projects who had pocketed their investors' money were making themselves as invisible as possible.

So far, then, a clear win for Blunt and friends. The South Sea men had aimed to eliminate the last distractions that could tempt speculators away from their stock, and their well-aimed writs had done just that. Some who had no particular love for the Company also cheered the move. Daniel Defoe captured one strand of popular opinion in one of his regular newspaper polemics, noting that it was on "just Principles" that the ministry had moved to enforce the Bubble Act—a decision made necessary by "the growing Evil of Projecting, Bubbling, and Jobbing of Bubblers . . . as must in the End have been universally ruinous to Trade, to public Credit and to private Fortunes."

SUCH SUPPORT MISSED the heart of the matter: Blunt's urgency to "execute the scheme" and complete the absorption of Britain's debt into South Sea capital. A healthy market in South Sea shares remained the first priority—and would become acute once the Company's transfer books reopened. That happened on August 22, four days after the little bubble companies had been crushed. At first there didn't seem to be much immediate effect: South Sea shares traded at £750, down more than 10 percent from the beginning of the month. Two days later, with the enthusiasm for the fourth money subscription, the stock bumped upwards, changing hands in the Alley at £820.

Then something new happened. South Sea stock began to slide, with none of the usual upticks to reassure nervous investors. It fell just a little at first, a mere £10 on the twenty-fifth; then £10 more the next day; then another £25 for the next two days, sliding back to £750 by August 30. The retreat unnerved Blunt and his fellow directors, and they tried some old tricks to prime the market. They closed the transfer books again and announced new dividends: £30—in cash, not shares—to be paid at Christmas, and an astounding 50 percent payout per year for the next twelve years. The following day the market responded with what we'd now call a dead cat bounce, boosting South Sea shares back up to £810, bang in the middle of the range they had traced all summer.

It didn't last. It is impossible to say for certain what wrecked investors' confidence at that moment. Economic historians have identified several possible triggers for or accelerants in the collapse. Overseas competition may have played a role, the lure of copycat speculative booms in Amsterdam and Lisbon. Such satellite bubbles fared no better than the ones in London—most burst by the turn of the year. But they diverted money from Exchange Alley for a while, reducing demand at the same moment South Sea sellers were chasing buyers from Jonathan's to Garraway's to the Jerusalem and back again.

But even with such contributory pressures, there was a clear suspect in the collapse: the South Sea Company itself, panicking. Company clerk Anderson believed that it was those "fatal writs of *scire facias*" that had led investors to wonder whether the South Sea Company might itself be a bubble company, no more solid than those it had destroyed the week before. "The court of Directors now saw their mistake, too late," he wrote. The dividend announcement compounded the error, as it focused investors' attention on exactly the kind of financial calculation that few had been prepared to make for the last six months. The question of whether the implied promises of Company returns were too good to be true now came to the fore. Critically, the promised yearly dividend was calculated on the par value of the stock, which meant that the fabulous-sounding 50 percent payout amounted to just £50 per share per year. That was a bare

5 percent return on the £1,000 price of the last two money subscriptions.

Five percent was boring, and worse, easy.

Everyone even tangentially involved in financial life knew that it didn't take a high-flying, high-risk stock to earn that kind of return: 5 percent was the usual interest rate on a private loan. Damnably, it was significantly less than the long-term irredeemables had delivered before being swallowed up in the scheme. Worse, as Archibald Hutcheson had been shouting for months, there was very good reason to doubt that the company would be able to pay even that much. Anderson confirmed that dire thought; at best, he wrote, "the company might possibly have been able to have made so vast a dividend for at least part of the said twelve years . . . but that would have been attended with the certain and grievous future loss of their principal as well as interest." In other words, no combination of interest earned and trading profits could keep up with that promise.

Such numbers pointed to only one conclusion: it was time to cut and run. In the early days of September growing numbers of shareholders did their best to escape with whatever they could rescue. The brief hope of the dividend promise lasted all of two days. On September 3, the stock relapsed to £750. One week later, Company shares could be had for £640. In seven days more, they cost £440, until, on October 1, the stock dropped below £300, never again to climb above that line.

The slide continued throughout the autumn, until in December South Sea shares completed their journey back to almost exactly the same price they'd brought before the first rumors of the scheme leaked around the turn of the year. They would fall a little further in the new year, but all those months of relentless losses merely ratified what was obvious by the end of September.

It was over.

The "damn'd South Sea" had drained, never to rise again.

DANIEL DEFOE KEPT his faith in the South Sea deal almost as long as anyone, as scornful as he had been about the emotional excesses of the Bubble. But he had recognized early that the financial craze involved unfamiliar ideas about money as something more than just cash on the nail. So even before he was fully able to analyze what could go wrong with the South Sea Company itself, he began to worry about what would happen should Alley fever break.

He mused on the danger when news came from abroad during the summer. On August 15 he wrote that "the *French* have, among the rest of their Merchandizes, imported the Plague at *Marseilles.*" Defoe told his readers they should not worry yet; Marseilles was a long way off, and it appeared that the threat was not as severe as some in living memory. But, he asked, "Pray, what would be the Price of Stocks if such a dismal Particular [an epidemic] . . . should be our fate?" The answer was obvious: "I am loth to remind our People of . . . the Time of the last Infection . . . How Grass grew in the streets of London, and on the Exchange." Given the mob thronging every coffeehouse between the Cornhill and Lombard Street, "How many People it may be supposed would be seen in Exchange-Alley every Day? And, how many Transfers of Stocks be made at the *Bank* or *South-Sea* House?" Would the "Bubbles . . . all Vanish and Dye, like Vapours in the Morning?" Could the plague break the London stock market, taking every paper fortune with it?

Defoe was thinking beyond the mere financial impact of such a crash. His journalism, and even his most explicitly partisan polemics, always had a moral frame. To Defoe, the plague was never just a disease; it was a metaphor for other contagions that could shatter society. In 1722 he would publish perhaps his strangest novel, *A Journal of the Plague Year*. It was ostensibly a fictionalized account of London in 1665, suffering through a pestilence so consuming that a burial pit in Aldgate became the gate to hell itself. The South Sea year was a recent memory when he drew that portrait of a city dissolving before his readers. In his tale, he described how the hucksters appeared before the plague fully took hold, eager to prey upon a community desperate for any scrap of good news. "The common people," he wrote, "were now led by their fright to extremes of folly." At the first reports of plague along the river, "they ran to conjurers and witches, and all sorts of deceivers . . . who fed their fears . . . to delude them and pick their pockets."

A
JOURNAL
OF THE
Plague Year:
BEING
Obſervations or Memorials,
Of the moſt Remarkable
OCCURRENCES,
As well
PUBLICK *as* PRIVATE,
Which happened in
LONDON
During the laſt
GREAT VISITATION
In 1665.

Written by a CITIZEN who continued all the while in *London*. Never made publick before

LONDON:
Printed for *E. Nutt* at the *Royal-Exchange*; *J. Roberts* in *Warwick-Lane*; *A. Dodd* without *Temple-Bar*; and *J. Graves* in St. *James's-ſtreet*. 1722.

The title page to the first edition of *Journal of the Plague Year*

That was what the early stages of the Bubble looked like, when mountebanks appeared to feed dreams of riches that would slip away if one didn't buy in fast enough. The "crash" in this telling was the full onslaught of the epidemic, with its brutal demonstration that quack remedies were worthless. Those charged with keeping order couldn't: "it was not in the Power of the Magistrates, or of any human Methods or Policy, to prevent the spreading of the Infection." In response, "People began to give up themselves to their Fears, and to think that all regulations and Methods were in vain, and that there was nothing to be hoped for, but an universal Desolation."

As Defoe's epidemic peaked, the bonds of social life dissolved. "It is hardly credible to what Excesses the Passions of Men carry'd them in this Extremity of the Distemper," Defoe wrote, and "the Aspect of the City itself was frightful." London fractured. "Nothing answered; the Infection rag'd, and the People were now frighted and terrified to the last Degree [and] abandon'd themselves to their Despair."

Defoe's readers could easily see in this fictional metropolis a mirror of what they'd just experienced. By the end of September 1720, it was not hard to see that the spiraling financial catastrophe had more sorrow to inflict. It was as yet unclear what could be done, whether magistrates of money could mitigate the present distemper before all Britain abandoned itself to misery and rage.

Chapter Eighteen

"No Man understood Calculation and Numbers better than he"

Those in London wondering what could come next—what to *do* next as the South Sea Bubble deflated—were lucky, after a fashion. They could look across the Channel to Paris, where a very similar sequence of events had just unfolded, offset from their own calamity by just a few months. There were clear differences between what the French monarchy had just attempted and what the South Sea Company's directors and the British ministry had devised. But there was enough overlap—even some direct imitation—to make the fate of the Parisian adventurers a matter of vital interest to their British counterparts, gamblers and leaders alike.

One difference between the great French speculative bubble and the one bursting that autumn along Exchange Alley came in the figure at the center of the Parisian crisis. That man, a wandering Scottish duelist, gambler, and mathematician, was, unlike John Blunt and his peers, a genuinely distinctive thinker about money. When he reached the pinnacle of power in France and tested his ideas in the real world, he was driven by intellectual passion as much as by any excess of avarice. His had been an unlikely journey, one that almost died in a burst of violence when he was barely out of boyhood.

JOHN LAW KILLED his man on the ninth of April 1694.

His victim, Edward Wilson, was a dandy—in the parlance of the day, a beau. Just before their fatal rendezvous, Law had gone to Wilson's house. The conventions of the duel were honored: they drank together—Law had a "pint of sack in the parlor"—and then he left. Wilson soon followed, taking a coach to Bloomsbury Square.

Law was ready. "Before they came near together, Mr. Wilson drew his sword, and stood upon his guard." Law unsheathed his own weapon. They made one pass. Law's blade bit two inches into his adversary's belly. Wilson died on the spot.

That pool of blood, spreading on the dirt of the square, almost finished John Law along with his opponent. It would not have seemed much of a loss to most who knew him then. As a boyhood friend put it, he was "handsome, tall, with a good address, and . . . a particular talent for pleasing the Ladies"—but also a gamester, a spendthrift, and, if contemporary rumors were true, a blade for hire. Those same whispers also hinted that the "insult" that had led to the duel was no mere accident. Wilson, it seems, had become the wrong person's lover—the king's mistress in one story, a nobleman who would not survive exposure of a same-sex liaison in another. Either way, a powerful person had a very good reason to find an impulsive, broke swordsman willing to provoke a fight to rid him (or her) of such an inconvenience.

If the rumors were true, working for such an influential client may have saved Law's life. After the duel, he was jailed, tried, and condemned to death—and, just to show that history has a sense of humor—was slated to go to the gallows alongside two currency criminals, a counterfeiter and a coin-clipper.

Law's friends, whoever they were, managed to delay his execution long enough to get him moved from the reasonably secure Newgate Jail to the much more porous King's Bench Prison, on the far side of the Thames. From there, it was said that he cut his own leg irons, drugged a guard with opium, climbed a high wall, and jumped to freedom, injuring his leg in the leap, before making his way to the coast and thence to France. This was surely nonsense. Law himself noted that "Romances must be embellished with Resemblances of Truth to make them go down." It appears that whoever had backed him simply walked Law out of the prison.

That was the end of his life as a wastrel. The danger awoke what proved to be one of the most formidable intelligences of his day, and, once he turned his attention to money and credit, ignited a career that made him, briefly, the most powerful man in Britain's ancient antagonist France.

MOST ACCOUNTS OF Law—this one too!—highlight that youthful duel. Partly that's because, in Law's day and ours, blood sells. But even with centuries of hindsight, this youthful eruption of violence is often treated as a kind of metaphor for Law's motivation throughout his life. He was, it seemed, a man so in love with money he was willing to kill to fill his purse.

That picture is false, or, rather, so incomplete as to distract from the larger theme of Law's life. He was not in love with money; he was *fascinated* by it. According to a friend, "No Man understood Calculation and Numbers better than he." At first, after his escape from prison, he used such "superior and very uncommon Skill" purely for personal gain, playing cards with "People of Quality wholly ignorant" of the math behind their games. It's fun to bet when you know the odds and your opponents don't.

Law soon outgrew fleecing easy marks. His travels took him to city after city where experiments on money were under way—new kinds of banking, lotteries, stock speculation, and more. He wrote his first financial essay in 1704 and at about the same time produced the earliest in a series of proposals about money and its management, to be sent to officials in England, Scotland, and last, France. The word *economist* gained its present meaning only in the early 1800s, but even if it's an anachronism, the first two or three years of his exile transformed John Law into one of the earliest who could be seen as a recognizable example of the species.

His thinking matured in 1705, when his next treatise, titled *Money and Trade Reconsidered: with a Proposal for Supplying the Nation with Money,* set out what the economist and central banker François Velde has called "the first . . . theoretical framework" for much of the modern conception of money. There, Law argued that money was, like any commodity, subject to supply and demand; that the amount of money available within a society determined prices within a market; and that this price mechanism shaped a nation's trade, its employment, and the productive capacity of its economy. This was not wholly new thinking—recall that Isaac Newton answering Pollexfen had offered some similar notions, and that others were working on related ideas. But there was more to come: Law's book offered only the first steps in an argument he would work out over the next decade.

As he thought through the problem, Law first analyzed the various things money does: it can store value (money in the bank), it serves as the medium of exchange (buying and selling), and it is the device, as he put it, through which "contracts are made payable"—which is another way of saying that money allows bargains to extend across time, with set, established payments to be made at some agreed-upon later date. Anything performing that function has to hold its value as the months and years pass, but, Law pointed out, it doesn't necessarily have to be traditional hard currency—coins made of silver or gold. Different representations of money can perform many or all of these tasks. So far, he followed similar arguments previously posed by others. What distinguished Law from many of his contemporaries, though, was his eagerness to test out his theories in the real world. His first such attempt was a variation of a project that had already been proposed by others: a land bank—a money-creating institution that would issue banknotes supported not by deposits of precious metal but by the production of the acreage within any given jurisdiction.

Such land banks had been suggested in London in the preceding decade, but Law moved past his predecessors to come up with a much more modern view of what gives money its value, in a series of intellectual steps traced by the historian Antoin E. Murphy. Law argued that because the supply and demand for money set its value, conscious decisions about the amount of money to be made available within a nation could shape economic outcomes for any government willing to create the institution—his land bank—that could play that role.

Equally important, he recognized that the forms of money that could affect such outcomes were not restricted to legal tender, coins, or even the paper currency a land bank could issue. "The stocks of the East India Companies, of the Bank, Irish debentures, etc.," he wrote, "are received in some payments because their value though uncertain what it will be yet at the time it is known and those who think these stocks will rise rather than fall and are willing to run the hazard will prefer them to the same sum in silver money"—or to a banknote.

In one sense, this was a penetrating glimpse of the obvious; certainly the British financial markets already recognized what Law described. But such formalizations of daily practice are often central to constructing a useful model out of the messiness of real life. For Law the goal was to interrogate everything within the world of trade, commerce, production—anything to do with getting and spending—that could illuminate just what money is and does.

A land bank was barely the beginning of Law's intellectual journey. It was a long way from the experiment he would ultimately attempt, his version of what would be attempted by the South Sea Company. But it set him on course for that destination. Law's next step was to push past these initial thoughts. Money made of metal, or even, as in his early land bank proposals, backed by actual fields of dirt, was intimately connected to the material world. Once he realized that shares in trading companies could perform many of the same duties handled by money derived from such solid pieces of the material world, Law came to argue that such financial inventions formed a numerical representation of much of the economy—all the growing and manufacturing and trading and craft that could be parceled out as shares. In that view, the hard reality of a stack of coins becomes a constraint. As the English had in fact experienced in the 1690s, running out of cash hurt trade, raised interest, and generally disrupted daily commercial exchange. Paper money—whether official notes or stock in some going concern—could, Law argued, ease such shortages, both boosting an economy and making it easier for the state (and anyone else) to borrow as needed. That suggested a policy—a designed solution—that could shape the economic life of a nation. What was required was an institution or institutions that could both create and manage the supply of cash, along with the credit, the lending and borrowing that this new money would fund. Properly done, this would form a system—a *système*—that would enable an enlightened minister to ensure the prosperity of a nation's people while bolstering the authority of its monarch.

FOR HALF A century, there had been no greater exemplar in Europe of absolute monarchy than the French Sun King, Louis XIV. Even so, he wasn't omnipotent: there was a limit to the amount of cash he could extract from his subjects using France's traditional methods of raising revenue: direct taxes on individuals, other forms of taxation, the sale of offices, and periodic devaluations of France's metal money. In consequence, by the time Louis died in 1715, the same wars that had driven Britain into its crushing debt left his France similarly close to bankruptcy. By some measures, the situation in Paris was actually worse than that in London. In the year of his death, France owed 3.5 billion livres (very roughly, £250 million, or around six times Britain's long-term national debt in the same period). Interest alone consumed between 70 and 80 percent of state revenues, and, as in Britain, many of those loans carried terrible terms, with high rates of interest due for long or perpetual terms. As King William had discovered, debt on that scale imperiled the monarchy's ability to prosecute the inevitable next war.

Louis was succeeded by his great-grandson, Louis XV, in 1715. The new monarch was all of five years old when he took the throne, and power lay with his cousin, Philippe, the Duke of Orleans, regent to the child-king. Philippe was neither an administrative nor a financial genius. After decades of the Sun King's wars, French coins, its hard currency, were in short supply, to the point that his ministers had faced deepening difficulties in securing new loans over the last years of Louis XIV's reign.

Enter John Law.

Law proposed a solution. He began modestly, persuading Orleans and the minister of finance, the Duke d'Noailles, to authorize a joint-stock bank loosely modeled on the Bank of England. Law's Banque générale would, he promised, at least partly address both the shortage of cash and the monarchy's difficulties in borrowing. Like the Bank of England, it would issue its own banknotes, thus providing ready cash for Paris and France.

John Law in 1720

This was hardly a revolutionary idea, but Law was the first person with the political power to make public policy out of ideas that had been coalescing over decades on either side of the English Channel. His Banque opened for business in 1716. Almost all of the bank's documents have been lost, so it is difficult to reconstruct exactly how it worked, but the record does show that it prospered. Founding subscribers swapped interest-bearing notes forming part of France's floating debt for shares in the new company. As it took in these notes, the Banque became a lender to the government.

French debt was trading at a deep discount—reflecting fears of an impending default. So, echoing the experience of the first South Sea Company transaction in 1711, those who swapped their holdings at face value for shares in Law's banking business got a bargain: in just two years, the bank was reorganized as an arm of the state—by which time its stock price had advanced enough so that those first shareholders had more than doubled their money.

The bank delivered on its promised financial services as well. It issued almost 150 million livres in banknotes that in 1717 became legal tender: "real" money that could be used to pay taxes to the crown. And it became, possibly in violation of its charter, an active lender to private businesses. Perhaps most important for what came next, over the years he operated it as a private business, Law dropped the interest his bank charged for the floating loans it made to France's treasury from the then-prevailing 6 percent rate down to 4 percent, anticipating, and to some extent inspiring, similar moves in London in 1720.

Sufficient money in circulation, accessible loans, and a drop in the cost of state borrowing: this was success—enough to persuade his official masters to let Law take the next step. He'd gotten his bank; now he wanted a trading company, modeled on British examples, especially the East India Company. He proposed a monopoly on trade with France's Louisiana territory, to be capitalized by the same kind of debt-for-equity swap with which the South Sea Company had launched itself in 1711. Investors would trade their holdings of *billets d'etat,* more of France's floating loans, for shares in the new trading company. Just as before, France would pay a reduced rate of interest to this new Compagnie d'Occident—better known in English as the Mississippi Company. That income would fund trading operations that, it was expected, would enrich all its creditors-turned-shareholders—all this an anticipation of the full South Sea scheme launched in 1720. Law got the go-ahead in August of 1717, and within another year enough debt had been taken in to complete the authorized capital funding of 100 million livres (approximately equivalent to £7 million).

From there, Law's ambition rocketed. In early 1718, Law argued that his Banque générale should be nationalized to take full control of the state's finances. In December the state bought out the private bank's shareholders to create the Banque Royale. This new bank continued to issue notes as legal tender, but with this difference: they were no longer convertible, which meant that the bank wasn't obliged to swap new notes for their face value in coins. This was a genuinely radical shift, a step further than almost all contemporary financial thinkers had contemplated. It allowed Law to provide France with a purely paper currency by late 1719—the first time a European power took this step. Gold and silver ceased to be legal tender, which meant that they could not be used to meet tax demands. The Banque Royale thus became a prototype of a central bank. It had total control of the money supply, the amount of cash that could circulate throughout the realm, with the ability to create as much more money as it chose (or to call it in), unconstrained by how much precious metal France contained at any given time.

In those same months, Law maneuvered the Mississippi Company through a gluttonous expansion. In August 1718, he detoured from international trade to swallow the national monopoly in tobacco. Next up, the Senegal Company, the supplier of slaves to the company's Louisiana territory; two more mercantile companies followed, bringing exclusive rights to the Indies and China trades. In July 1719 Law bought the rights to the royal mints. In August he took over one of the two main tax-farming concessions—a business in which private investors bought the right to collect certain taxes for the state, delivering an agreed amount and keeping any surplus they could extract as profit—and he followed that up in October by grabbing the other. Last, he performed the trick that caught Blunt's attention, and Aislabie's: Law offered shares in the Mississippi Company in exchange for the rest of France's remaining debt, while reducing the interest rate on all outstanding state loans.

To pay for this spree, Law employed many of the same devices the South Sea Company's men would imitate a few months later. His Mississippi Company issued new capital stock in a series of increasingly complicated offers. Shares were sold for installment payments, and some speculators in Paris made eye-popping profits trading option derivatives, those rights to buy or sell shares that cost a fraction of the stock itself. Access to new issues was restricted to those who held previously released stock, which (as long as everyone stayed optimistic) buoyed the market for existing shares, as those who wanted in on the action needed to get their hands on older shares to gain tickets to the game. There was, however, one key difference between Law's *système* and what was about to follow in London: with his control of the national bank, Law could pump into the marketplace as much currency as needed to help investors bid up Mississippi Company shares.

Thus, by early 1720, Law ruled virtually all of France's international trade, its official banking system, the national money supply, and the entire tax apparatus. As a finishing touch, he merged the Banque Royale and the Mississippi Company into a single enormous enterprise, his *système,* while persuading his patrons to appoint him minister of finance. This was the climax to one of the most improbable ascents to power anyone has ever accomplished: as far as its economy was concerned, John Law, duelist, adventurer, and Scot, ruled France.

FOR A WHILE, it all seemed to work. It was more complex than the South Sea scheme, as there was no real counterpart to its monetary side in London. No one there proposed a pure paper "fiat" currency instead of metal money, nor could the Bank of England control the money supply. Britain did not farm its taxes, and no one involved in the great mercantile companies saw any value in bringing all those various monopolies under one roof. But as far as the stock market went, the broad outline of both booms was much the same, with Paris becoming, if anything, even more exuberant than London. Mississippi Company shares moved from an initial price at 500 livres to a peak of over 10,000 in the eight months between May and December 1719, about twice the South Sea Company's gain.

For most of that rise, Law's underlying theory appeared to be working: it was possible both to transform France's ability to borrow and to improve the overall health of trade and industry by managing money. As 1719 ended, those of the state's creditors who had swapped into Mississippi Company shares were perfectly happy. Official France regained ready access to the credit market and was able to borrow enough to fund new military adventures in 1718 and 1719. And, with a state bank at his command, Law could take actions Blunt could only dream of to support Mississippi Company share prices. Through the latter half of 1719, the Company would on demand buy back its stock at about 9,000 livres per share—a promise that was in essence an exchange rate peg. That peg worked because the Banque Royale would issue new notes as needed, which meant that Law's system could self-deal in its own shares.

But as the year turned, Law seemed to grow concerned about the amount of newly printed money it took to maintain that fixed exchange rate, which led him to try to void that buy-back guarantee in late February. Mississippi Company stock lost a fifth of its value in a single week—and he caved. Law hastily restored the peg, effectively confirming that company shares could be seen as equivalent to the Banque's notes: never worth less than the fixed exchange price. But maintaining that price level meant that the Banque had to keep on issuing new money—and with that, the *système* began to fracture. France found itself awash in all that newly printed cash, more and more of it chasing bread and wine and anything else that could be bought and sold. Prices for anything that could be bought began to rise, in a phenomenon we now call inflation. Law was sensitive to the danger, and in May 1720 he tried again to bring matters under control. He might as well have tried to stop a maddened horse with reins of string.

On May 21, Law again tackled the pricing for Mississippi Company shares, proposing to lower them from 9,000 livres to 5,000 in a series of monthly steps, while dropping the face value of banknotes by 50 percent over the same period. The hope was to cut prices in the economy as a whole by a similar proportion.

The Paris stock market responded as one would expect: grimly. For a couple of weeks shares declined gently, but from the first of June, Mississippi Company shares cratered. Law was briefly placed under house arrest. He was soon released and restored to power, but nothing he did revived Mississippi stock or restored confidence in a paper currency that, the public now realized, could be printed into worthlessness.

The final bill was staggering. By one accounting, measured by the contemporary exchange rate for specie, the total value of the Royal Bank's notes and Mississippi Company shares in May 1720 was over three billion livres. Half of that evaporated in the first month of the crash, and by December the total value of all of Law's paper stood at just 411 million—an almost eight-fold drop.

John Law, fleeing on an ass, from a Dutch volume on the financial follies of 1720

Law fled Paris on December 18, stuffed into a borrowed coach. He had not committed any crime except failure. But enough people had lost more than enough to put his life in danger. He left too fast to gather sufficient traveling money. He was able to borrow pocket change at an early halt and reached Brussels—beyond the reach of French vengeance—on December 22. On his first night in exile, he went to the opera.

AS DEEPLY AS any contemporary, John Law both recognized and demonstrated that money didn't have to be real. His system expressed the purest form of the logic that flows from that insight. Gold or silver did not possess some unique property that gave them significance—value—beyond what the market said it had. An economy could make similar judgments about much more abstract units of exchange—livres or pounds listed on sheets of paper. Those abstractions enabled a mathematics of credit to build an understanding of money as it evolves over time. Such knowledge, Law had believed, could make national prosperity a matter of reasoned calculation. And yet it had failed to do so in twin debacles in France and then, months later, in Britain.

That Christmas, with Law gone from Paris and the South Sea Company down to roughly one-tenth of its high, financial officials of both nations faced a common task: how to restore their nation's credit in the face of those collapses. Nothing like the events of the preceding months had happened before in either kingdom, which meant that trying to navigate a way out of the disaster meant charting their own paths to safety. As always in the midst of a disaster, there were opportunities that would not come again, to be seized by states—and people.

In London, Robert Walpole, that once banished and newly resurrected man, had been preparing for his chance for a very long time.

Chapter Nineteen

"a great and general Calamity"

How hungry for power was Robert Walpole? So much so that in the spring of 1720 Lady Mary Cowper could write in her diary that he "let the Prince lye with his wife, which both he and the Princess knew" . . . and expect to be believed. Walpole was then deep into the third year of political exile, growing ever more hungry to return. Strong ties to the king's heir were useful, and, as his circle of acquaintance knew, Walpole was no sexual innocent himself. The arrangement was at least plausible.

Next question: How lucky was Robert Walpole? Chance had been kind to him in both his friends and his enemies through much of his early career, but in 1720, both what he didn't do and later what he did would make that year the most fortunate of his life.

Pimping his wife out to the Prince of Wales—if that ever happened—would have been one of his many maneuvers in his reach for power. Beginning in late 1719 he tested each lever of influence. He deepened ties to the Duchess of Kendal, the king's longstanding mistress. He flexed his muscles in the House of Commons, defeating the ministry's chiefs, Stanhope and Sunderland, on the so-called Peerage Bill that would have given their faction total control of the House of Lords. Then he turned around to offer a carrot that would appeal to the king himself—a necessary gambit given how deeply George distrusted Walpole. His hook was the control Parliament had gained over royal finances in the settlement that had placed William and Mary on the throne. By 1720, the account Parliament budgeted to cover the royal household's bills, along with much of the civil service, had fallen £600,000 in the red. Walpole let it be known that he was prepared to use his mastery of the Commons to persuade his fellow MPs to settle that debt. Most important, he did everything he could to persuade the king and his son—who loathed each other—to reconcile. He succeeded at last in the spring of 1720, and if the royal family's public detente was consummated without any sincerity or affection, it at least opened up enough political space for Walpole to once again take his place in the governing coalition.

This flurry of political cut and thrust occurred at the same time the South Sea scheme was coming together, but the two had nothing directly to do with each other. As Lady Cowper, never Walpole's friend, told her diary, the only goal of this campaign was "to procure *Walpole* and *Townshend* the Benefit of selling themselves and their Services at a very dear Rate to the *King*"—the services in question being the use of the parliamentary faction they led in exchange for office and power. At last, in June of the Bubble year, Walpole regained the post of paymaster general of the armed services, the job that had made his fortune back in the midteens. Though it wasn't a senior post, it would do as a stepping-stone.

The timing of his resurrection was Walpole's first big piece of luck during the Bubble year. On the outside during the feeding frenzy when the Company was buying political support in early 1720, he had been one of the few members of Parliament clearly not worth bribing. Counterfactuals are always risky, but it is fair to note that Walpole was no more scrupulous than most of his contemporaries, so it is plausible that had he made his way back to power a few months earlier, his name too would have been found on those lists. But as the year played out he reached the autumn untarnished by any South Sea association simply because it had been his enemies' project.

Fate's kindness didn't end there. Before his political exile, he had been recognized as a master of government finance; in the years to come that reputation would grow. Early histories of the Bubble conflated his real skill in managing Britain's money with the false notion that he had worked out what was going on with the stock market in the first half of 1720, making an actually honest profit on the Company. Walpole had many skills, and he was unequaled as a parliamentary infighter, but there is no evidence that he grasped what his rivals had missed throughout the year. He did avoid losing his shirt, but there is a strong case that he dodged some very expensive mistakes by sheer chance, or rather, still more good fortune: his man of business was cautious enough to save Walpole from himself.

Walpole's stock exchange adventures had begun well enough. In late April a minor boom in a pair of insurance companies doubled a small investment in just sixteen days, netting him over £2,500. At about the same time he bought Bank of England and the Royal African Company stock, both sound investments. But his dalliance with South Sea Company stock itself confirms that his objections during the parliamentary debate in March hadn't been because he saw some fundamental flaw in the deal. Out of power, his primary goal had been to discomfit the ministry. Once the scheme passed he was perfectly willing to play the market in South Sea Company shares. He was just bad at it: he sold too soon and bought again too late. He unloaded the last of his original South Sea portfolio—bought well before the Bubble year—by March 18, 1720, at £194 per share, just weeks before the shares would double and double again. He returned to the market at June's third money subscription, buying back in at £1,000. He was so sure of the Company's prospects later that spring that, according to Lady Cowper, he had even convinced his ally, the Princess of Wales, to become "a Stockjobber in the *South Sea.*"

Similar behavior would wreck many others, but Walpole hadn't risked a great deal in June, and he largely escaped the worst consequences of his folly—through luck and a good choice of agent. He was ready to buy again in August, and he put together a list of family, friends, and himself, all eager to purchase more shares at the fourth money subscription, again, at a price of a thousand pounds. His banker, Robert Jacombe, had a much better sense of timing. Without positively disobeying Walpole's orders, Jacombe managed somehow to dawdle long enough before forwarding to miss the Company's deadline for new orders.

Such tactical bumbling saved Walpole from both significant losses and an obvious conflict of interest when it came time to deal with the wreckage once the Bubble burst, a slide he watched from the safe distance of his Norfolk country house. He set out for London in mid-September, reaching the capital just as the market became terrifying enough to prompt the first significant official action to halt the collapse. On arrival, he was almost the only member of the ministry available whose hands were clean.

THE COMPANY'S FIRST attempts to save the situation on its own had already failed. On September 8 its general court—made up of shareholders and not just directors—met for the first time since spring. This broader group voted to ease payment terms for the shares the Company had sold on the layaway plan. The move failed to boost confidence on the Alley. The stock, at £660 on the day of the meeting, dropped below £600 within four days. Walpole reached town by about September 19. By then, South Sea shares were selling around the £400 mark—half their value at the beginning of the month, and down £40 from the previous day.

Several days later, another general meeting approved a more radical move, transferring stock from the Company's books to its shareholders. They did so by dropping the price from £800 to £400 for those who had swapped debt for Company equity over the summer (thus doubling the number of shares they would receive). Buyers of stock for cash in the final two money subscriptions got the same

£400 price, a 60 percent discount from the original £1,000. No matter. Day after day on Exchange Alley, South Sea kept sliding to and then below that number.

No particular genius was required to diagnose what had happened: a stock buoyed by the belief in its infinite possibilities could not be saved by Company promises once that faith was lost. An obvious conclusion followed: if the Company's own attempts to support its stock failed, it would need help, some kind of bailout: new money, enough to reverse its accelerating plunge.

There was only one company with that kind of capital: the Bank of England, which had weathered its own, less extreme rise and fall in its shares with much less disruption. Its leaders were, of course, the same people that Blunt had scorned in the spring, which made any plea awkward at best. So a few South Sea directors made the approach obliquely, using friends involved in the East India Company as the go-betweens to carry a proposal to the Bank's leaders. The meeting was set for the nineteenth—and Walpole attended for the ministry.

The talks continued through the evening, until sometime after midnight Walpole wrote up a sketch of a rescue plan. The bargain took formal shape over the next several days. In its completed form, the Bank promised to raise £3 million from the public to lend to the Company and would, in addition, invest some of its own capital, buying almost £1 million in South Sea shares at £400. Just the fact of these talks achieved, momentarily, the desired effect: Company shares ticked up, reaching £410 on the twentieth.

Once public, the deal became known as the Bank Contract. But it was a contract in name only, and the two parties soon fell into dispute over whether the Bank had legally committed itself to invest in South Sea equity. The Bank argued that it had not, while South Sea interests contended it had—for the simplest of reasons: almost from the moment it was proposed, the deal was an unmitigated pig for the Bank. The "contract" had been for shares priced at £400. The stock held at that level for all of three days, sliding beneath it on September 23. By November, when the Bank finally killed the proposition, South Sea shares could be bought for just £213. The only meaningful outcome of this first attempt at intervention was yet more misery. For a few days after the abortive deal had been announced, a last few optimists bought more South Sea stock in the hope the Bank would save the day.

Walpole had been just a go-between for this stillborn bargain. The lesson he took from its almost instant collapse was to dodge out of the immediate line of fire. Late in September he hurried back to Norfolk. He would remain there for much of the autumn, weighing possible solutions safely out of public view. In London, meanwhile, the full extent of disaster was becoming clear.

ON SEPTEMBER 23, Alexander Pope wrote to his friend Francis Atterbury, bishop of Rochester, that "the universal deluge of the South Sea, contrary to the old deluge, has drowned all, except a few *unrighteous* men." As with Noah's Flood, he added, "No man prepared for it; no man considered it would come like a thief in the night; exactly as it happens in the case of our death." As at that final knell, so now, Pope wrote, heaven sits in judgment: "Methinks God has punish'd the avaricious as he often punishes sinners—in their own way, in the very sin itself."

Alexander Pope in 1719

He admitted his own role: he told Atterbury that among the offenders one had to count "your humble servant." He recognized the cause of the offense: "The thirst of gain was their crime; that thirst continued became their punishment and ruin." Pope was being a little coy here: it appears that overall he ended up in the black on his trades throughout the Bubble year—just not nearly as much as he (like everyone else) had dreamt of just weeks before. His larger point stood—the sudden collapse of South Sea stock had left London an almost biblical desolation. Mashing up Job and Psalms, Pope told the bishop that in the current state of affairs "*Men shall groan out of the City . . . They have dreamed out their dream, and awaking have found nothing in their hands.*"

Pope gave himself credit for stoicism in the face of disaster—but most losers were not nearly so self-contained. Lady Mary Cowper—Walpole's scourge—was the friend he'd so poorly counseled in August. She had invested not just for herself but for others too. Pursued by one of those whose money she had thus lost, she declared herself "the most unlucky in the world"—and fearing what might come in the ongoing dispute, she wrote that she was "in the utmost terror for the consequences" of her descent into the South Sea flood.

At least Lady Mary was not herself bankrupt. That eager Salt Office man James Windham was in much greater peril. His brother William wrote to a third sibling, Ashe, on September 27 that "poor Jimmy's affairs are most irretrievable"—not that he was unique among William's acquaintances, given that "almost all one knows or sees are upon the very brink of destruction." By January, "Jimmy" finally recognized that he'd lost it all, confessing to his brother Ashe that he was one of Britain's many bankrupts: "I have come into the same state for being a very silly fool." There was nothing for it, he concluded, but the usual escape for a ruined Englishman: "The sea is fittest for an undone man," he wrote, "so I am for that."

Others too tried to float away from ruin. Leading aristocrats were reduced to begging for colonial governorships to rebuild the fortunes swept away in the deluge. "There are very many and considerable families reduced by extravagant bargains," wrote one member of the Board of Trade in November. The Duke of Portland, on the hook for all those forward contracts that were now irretrievably out of money, was forced to plead for a job in Jamaica, "which is not thought proper, but will get a pension, for he is very much worse than nothing."

It was understandably fun to mock such fallen giants, but there were more gruesome tales to be heard. One told of a man who stopped a farmer returning home, asking to borrow a knife, allegedly to cut some string. Instead, "to the Countryman's great Terror, he got off from his Horse, and sitting on the Ground, cut his own throat with it." The farmer leapt to stop him, but "the Gentleman, being enraged that he had not cut his own Throat effectually . . . turn'd the Point of the Knife upwards, and stabb'd it into his Jaw." The letter writer did not know if the man survived but offered the most plausible guess he could muster to explain why he might have done it: here was another lost soul, drowned in the South Sea.

Beyond such horrors, there was no shortage of schadenfreude throughout the fall. In early October, a self-described (and almost certainly fictitious) "plain Country Fellow" published his account of visiting Exchange Alley in October. There he encountered what he took to be "some honest sorrowful Persons come together to some great Burial." He met "a Man, wringing his Hands and crying out, *I am Undone, I am undone.*" The man from Leicester asked, who were all these people? A conveniently lounging chimney-sweep answered: "*There goes a South-Sea-man,* you may see he has *South-Sea Face,* he looks pale, Frighted, Angry and out of his Wits."

In the end, behind such pride-goeth-before-a-fall prating lay a full butcher's bill for the South Sea scheme. As 1720 opened, some thirty thousand people and institutions had owned pieces of the national debt, most of which ended up as Company stock. More thousands—overlapping in part with the debt holders—had bought shares directly from the Company in the four money subscriptions in April, June, and August.

Those buyers included, as the financial historian P. G. M. Dickson has documented, more than half of those in the Commons and a substantial minority of the Lords. Even more damaging, the emergence of options and other exotic bets on the Company left a huge tangle of contracts and commitments to be settled in the months and years to come. There were £10 million in sales, many uncompleted, on record at South Sea House, untold more in unregistered side bargains. All of these deals—private deals, futures contracts, options to buy or sell, loans offered with South Sea shares as collateral, and more—showed how leverage can bite, putting more and more money at risk on what was, ultimately, the same underlying share of stock.

Most threatening, such deals increased the number of people who could be hurt, not by their own actions, but because someone they'd dealt with couldn't settle their side of a bargain. "Jimmy" Windham's brother, William, feared for his own fortune because "they upon whom I have obligations"—who owed him money, for transactions agreed when the market was high—"are bad paymasters." Walpole himself faced a similar risk, having sold stock worth £9,000 to Lord Hillsborough, who, after he lost his market bets, tried to recoup by wagering on the horses running at Newmarket. The inevitable ensued, and his creditors—including Walpole—were left with worthless claims on a vanished fortune.

Such financial contagion made itself felt from the earliest stages of the Bubble's collapse. Several London private banks went bust in the first weeks, as Thomas Brodrick, a member of Parliament, noted on September 2. "A great many goldsmiths are already gone. . . . I question whether one third, nay a fourth can stand it." These bankers had issued loans on the security of South Sea stock, pledged at £500 and £600 a share. Now they couldn't collect from their bankrupt customers, and they were running out of capital to satisfy their own depositors.

That particular danger was among the first to be recognized after South Sea stock began its slide. By then, the accumulating reports of individual catastrophes were recognized as the building blocks of a "great and general Calamity," as Archibald Hutcheson put it in October. "There are many Persons who have been intirely unconcerned in the Stocks, who will be hurt by the Distress which has been brought upon others." As many observers recognized, the threat extended beyond any one speculator's bankruptcy triggering losses for others; if the cycle of losses extended far enough, the finances of the nation as a whole would be at risk as well.

It was this system of credit, the market mechanism that allowed Britain to borrow, that was beginning to fracture. Recall that Defoe (and others too, of course) had already pointed out that the evolving market in debt securities had made it possible for the Treasury to raise the unprecedented sums that allowed Britain to stand toe to toe with France. If buyers and sellers in that market lost the belief that their counterparties would complete the bargains struck on the Alley, then money might flee the Exchange—and the Treasury might starve for cash. In the last months of 1720 it was far from certain that Defoe's earlier bravado about the clockwork reliability of the British credit system would hold. "Will Paper Credit revive any more?" asked one popular weekly sheet. "Can Parliament do any Thing to restore it as it was before?" To that thought, this correspondent added perhaps the key question: "And if they should, Will it be any better than before?"

It was the lack of a good answer that had driven Walpole away from London at the end of September. The disaster was financial, but if Britain were to become unable to wage war at times and places of its choosing, the crisis would be political: a government that cannot raise and maintain its armed forces is a vulnerable one. If the current ministry couldn't fix matters, then the Whig hold on power could easily be broken. At worst, the Protestant monarchy itself was at risk. Given how recently the Hanoverian king George had taken the throne, the Catholic Stuart dynasty still pressing its claims could conceivably use the crisis to its advantage, especially if it finally persuaded European Catholic monarchs to provide help in force. Walpole, like many of his fellow subjects, had seen one crowned head driven from the throne. It was easy to imagine that it could happen again.

SUCH FEARS WERE more than a mere excess of caution. One of the illusions that historical narrative can produce is that people at the time can see how the plot is going as clearly as safely removed readers do. We know how it comes out, but King George could not; Walpole did not; no one living alongside them knew how their story would end. There was plenty of good reason for them to believe their situation to be not just difficult but on the brink.

Throughout the years of the financial revolution, this sense of precariousness centered on the deposed Stuart heirs and the threat, lurking within their Catholic and Scottish dynasty, of renewed religious and national conflict within the still very new union of Scotland and England. The Glorious Revolution of 1688 had been an overwhelming rejection of the last reigning member of the family, James II. But his cloddish misrule didn't mean that all the members of the British elite immediately and sincerely transferred their allegiance to Dutch William or later to that German prince, George I. It didn't produce unanimous allegiance to the Church of England over that of Rome; perhaps most important, it didn't call a halt to the struggle for power between the various grandees and factions aspiring to dominance in Parliament and at court.

For decades after William and Mary seized the throne, political tension erupted at times into outright rebellion, usually centered on Scotland. Ongoing resistance to the Acts of Union that had united the two countries in 1707 and the emotional connection and potential practical advantages of a Scottish royal family combined to create fertile ground for insurrection. The first rising in the Jacobite cause came in 1689 as an immediate reaction to the loss of the throne the year before. That conflict blended into the Irish "War of Two Kings" (1688–91), in which Protestant king Billy's army ultimately defeated King James's forces, including a significant French contingent. That connection raised the stakes for—or at least the fear evoked by—every fight between France and Britain for the next century. The French pursuit of British disunity reappeared in an attempted invasion in 1708, when the United Kingdom of Great Britain was just a year old; the landing was foiled by bad weather and fast action by the navy, but the point had been made.

To be clear: no Jacobite threat to the Union or to the Protestant succession of the monarchy came close to succeeding—but again, those facing the seemingly endless repeat of the same alarms could not know that. In the run-up to the Bubble year itself, the rising of 1715 represented the single most potent Jacobite military expedition to that date. Scottish leaders raised an army of twenty thousand, and James I's son, James Francis Edward Stuart, the "Old Pretender," landed at Peterhead. That rebellion had some initial success, reflected in the scare on Exchange Alley—driving home the vulnerability of the new Hanoverian and Whig state. A renewed rising in 1719 was more swiftly crushed, but the fact that this one was supported by Spanish troops was a reminder that Britain remained a target for foreign enemies who could combine with domestic dissidents.

The South Sea scheme took shape in the immediate aftermath of that latest danger facing the newly singular kingdom of Great Britain, then barely over a decade old. Taming the debt was more than an administrative necessity. The evident failure to do so in 1720 turned the South Sea collapse into a test of a still novel experiment in British governance, in which Parliament had taken charge of the national purse. Could it meet and master a clearly new kind of crisis—one born of its own decisions?

To many, the answer was obvious: a foreign king and his ministers had failed at the most basic job of government. In 1721, just two years after the last rising had been crushed, men with connections to the center of British politics began to plot yet another Jacobite revolt, to take place in England as well as Scotland. That conspiracy, led by Pope's confidante Bishop Francis Atterbury, collapsed in 1722, but not before its members had again sought French support in the overthrow of their king and his ministers. That was what was waiting for Robert Walpole, timing his return from the wilds of Norfolk. As Hutcheson put it: "The Evil complained of, is plainly seen and felt." Fixing it, he wrote, would require nothing less than "the consummate Wisdom and Goodness of the *British* Legislature."

Thus Walpole's task in the last weeks of autumn of 1720: fix it. Failure to do so could plausibly end very badly—for him and for the emerging shape of the British state.

Chapter Twenty

"without a breach of parliamentary faith"

For most of the fall Britain's moneyed men—and the mostly absent ministry—could do little but digest the ever-grimmer news from Exchange Alley. Parliament wasn't due to return from its summer recess until November, and in the meantime Walpole's seclusion had the curious effect of keeping him very much in the mind of political London. People remembered that he had been the ministry's expert on national finance the last time he held power. Could he rescue Britain in this latest hour of need?

In the years to come, after his ascent to the very top of government, and still later, in historical memory, it seemed that he could and had. Walpole has been mythologized as a political-economic genius—the one man among the king's ministers who had figured out how the realm could recover from the debacle. This much is true: Britain did rebound while he led its government, and Exchange Alley and the institutions around it did too—imperfectly but well enough, accumulating into a uniquely potent system of national finance.

But it's not correct to say that Walpole was the mastermind as well as the political maestro who orchestrated this victory. His actions at the point of crisis and through the recovery reveal a more plausible story, one in which an ultimately successful course of action emerged as people muddled through events they did not fully understand. Walpole's contributions were important—vital, even—but he was no visionary. Rather, he had the right instincts about whose advice to seek and heed, and he was a genuinely great politician whose mastery of the House of Commons gave him a source of power none of his rivals could match.

During Walpole's sojourn in the north, his banker and man of business Robert Jacombe served as his eyes and ears in London, acting as a kind of outboard brain as well. He wrote to Walpole on October 11 with an idea about how to replace the failed Bank contract. He started from the simplest fact: having just absorbed most of the national debt, the Company was by far the largest joint-stock enterprise in Britain (and the world, given the collapse of Law's *système*). To deal with such a giant, the first step to bringing the crisis under control, Jacombe argued, would be to break up the Company, dividing its holdings into three parts. The move would "engraft" some of its capital, all those government obligations, onto the Bank of England's books; transfer more to the East India Company; and leave the rest in what would be a rump South Sea operation, thus reduced to a manageable size. In October, this remained a sketch of a proposal, but, Jacombe told Walpole, a mutual friend (presumably someone with influence) had been "pleased with the thought" and "he commanded me to consider how it might be practicable." So, Jacombe wrote, he would work on the plan, "against your return to London."

That last hint became a shout as Jacombe pushed Walpole to take charge of events over the next weeks. Two days after mentioning the engraftment scheme, he wrote again: "Everybody longs for you in town," and then again, on November 1, "They all cry out for you to help them," though, as Jacombe acknowledged, the ruin of so many had advanced so far by then that "you will be prodigiously importuned by all the sufferers to doe more than any man can doe." But still, he repeated, Walpole was coming to be seen as the one essential man who could save the day—and perhaps more significantly, Walpole himself seems to have grasped that this was a moment when power was in play. To be absent was to abdicate as master of events.

WALPOLE ARRIVED IN London on or just before November 8. He had thought about Jacombe's notion and appears to have reviewed other proposals as well. The rest of the ministry, especially its leaders, Sunderland and Stanhope, still saw him as a threat. So he negotiated more or less on his own with the Bank's representatives for three weeks. Then, having brought them on board (leaving the East India Company's participation for a later conversation), he wrote up the plan for the king. The moment was still incredibly perilous—both to the nation and to anyone who could be tainted by political failure in the face of the crisis. Walpole himself was aware of the hazard, writing in his message to the king, "It was with great reluctance, and in obedience only to your majesty's commands, that I was prevailed upon to undertake any thing relating to the South Sea affairs." Fools rush in, and all that: "I am too sensible of the many difficulties that will attend any scheme formed to regulate the perplexed and unfortunate state of the South Sea company, to hope that satisfaction can be given, to the infinite number of sufferers." But for all such diffidence, something had to be done, and Walpole was willing to do it.

The plan had three parts. First, the proposal ratified the revisions the Company had announced in September, so that those who had exchanged government securities for stock over the summer would get double the amount of stock originally promised, with some significant adjustments. Second, Walpole wanted to aid those who'd bought South Sea stock for cash. In this new plan, the first money subscription, the April 14 sale at £300, was allowed to stand—those who'd bought then would have to complete their purchases. But the other three subscriptions were to be canceled; no further demands would be made for the payments due over time. Instead, the subscribers would receive shares priced at £400, in return for whatever amount each investor had already paid in.

Four hundred pounds was still a high price to pay for what at that moment cost less than two hundred on Exchange Alley. To cushion that blow, Walpole suggested handing over all the extra stock the company had gained when the public swapped its official debt for shares priced above par. Shareholders would receive this extra stock in proportion to their existing holdings. Some of that stock had been sold in the four money subscriptions, but the Company still owned plenty. Now, in a move that effectively transferred that gain from the Company's books to its shareholders, that stock would increase the amount of shares every former debt-holder would own and thus would boost the value of their holdings at whatever price Exchange Alley offered for South Sea equity. For almost everyone who had played the market during the Bubble, this would reduce their losses rather than carry them into the black, but it was better than nothing.

Finally, along the lines Jacombe had proposed, Walpole aimed to restructure the Company. Over the course of the year, it had taken in £37 million in government debt. As Hutcheson had spent the spring pointing out, there were no realistic South Sea trading prospects that could produce enough profits to make Company stock a good investment. In theory, there was a possible solution: for a smaller Company, that same trading operation would deliver its revenue to a smaller investment base, delivering more income per share. This could be done, Walpole suggested, by another stab at "engraftment"—which meant turning over £18 million of the Company's debt holdings to the Bank of England and the East India Company in exchange for stock in both. The Company's capital would shrink by over one-third—which meant that the trading operation would become a larger proportion of that smaller enterprise—and, through Company ownership of stock in its rivals, its surviving shareholders would gain a stake in those profits as well.

There were problems with the proposed deal, one of them obvious from the start. While those currently stuck with overvalued South Sea stock would clearly gain from engraftment, there was no obvious reason for the other two companies—and their shareholders—to suffer from their rival's mistakes. The three parties formally agreed to consider the notion, but neither the Bank nor the East India Company took it seriously. The possibility—really just the forlorn hope of South Sea investors—lingered until 1722, when it was formally laid to rest. Walpole's and Jacombe's first two ideas survived, however, and formed the centerpiece of the settlement that Parliament finally approved on August 10, 1721, almost a year after the start of the crash. At the same time, to spread the pain, the government gave up its claim on most of the £7 million fee the Company had offered in the bidding war with the Bank.

THROUGH ALL SUCH private negotiations and public maneuvering, Walpole refused to entertain one obvious move, and the final bill before Parliament ratified that omission. The original South Sea Act was framed to transform Britain's debt into shares in a private company. Walpole's measure preserved that core aim: no South Sea shareholder could reverse the decision to surrender his or her (mostly) secure government securities for what had turned out to be wildly overpriced South Sea stock. Calls for such a do-over began almost as soon as the market began to collapse. Over the next several months different groups of losers in South Sea transactions would complain in print, petition, threaten lawsuits, lobby Walpole himself, and, in the extremity, tiptoe to the edge of riot in the lobby of the House of Commons, to the point that the House sent for the justices of the peace and constables of Westminster to disperse them.

It was not just those who had lost by the scheme who sought a do-over. Some in Parliament saw the matter as one of both fairness and hard-nosed, practical politics. Those tens of thousands who owned government securities before the scheme launched were, in this view, the backbone of the nation—some aristocrats, of course, all those improvident dukes and the rest, but also the county elite, landed gentlemen, or the better sort of townsmen, merchants and the like. These were the people from whom the Commons drew its members; they formed the social network through which power flowed within the nation. They had been undone, their spokesmen argued, by "the Plunderers of their Country," who used the trickery of the new to cheat honest subjects of King George out of their property: "Diabolical Artifices . . . such as 'till now were never known or heard of in Britain." Thus, again, Archibald Hutcheson wrote in October 1720, in the hopes that he could persuade Parliament "to detach the present Sufferers from the Workers of Iniquity."

Hutcheson argued that no mere adjustment in the terms of the swaps or the money sales would do, for even at £400 instead of £800 or £1,000, such a fix "would not repair a Moiety [a part] of the Loss . . . sustained by the detestable Execution of the scheme." To Hutcheson and his allies in the House of Commons, and especially all those disappointed (or beggared) by their South Sea trades, the only true resolution was one that treated the entire affair as a mere nightmare from which it should be possible to wake.

It was true, Hutcheson conceded, that "Dealers in this Stock have acted freely and without Compulsion." But given all the tricks the Company and its allies had come up with to get the deal going in the first place, and then to prop up the stock, the right thing to do would be to declare "that all Bargains since the First of January last, relating to all kinds of Stocks, ought to be esteemed of no more Force or Validity, than the Bargains of Children, Lunaticks and Madmen." It was, Hutcheson concluded, a matter of "Justice and Equity, to vacate those which have been actually executed, and oblige the Sellers to refund."

Infants and idiots; that's an unkind portrait of South Sea dreamers—though the Bubble's losers would swallow the insult if it could persuade the government to make them whole.

It did not. Hutcheson's eloquence had no effect, nor did the threats and pleas of the former debt holders. Hutcheson acknowledged the legal obstacle to his demand: no one had been compelled to strike a lousy deal, which meant that to undo the Bubble, Parliament would have to annul valid contracts. That would directly undermine the claim that Britain was a nation subject to the rule of law (and hence governed by the Parliament that made the laws), as opposed to one in which the king wielded arbitrary power.

Hutcheson responded to this by arguing that the South Sea experience had been so far out of the norm that the law of contracts should bow to circumstance—that market trickery was reason enough to unravel the sanctity of a bargain. That uncertain claim of principle did not move Walpole. His true insight, his genius at the point of crisis, was that shareholders' ruined hopes were the price of a much greater gain. He was pragmatic enough to try to ease the pain of South Sea collapse as much as possible, but he was clear on this one point. The idea of a contract, and much more, the clear interests of the state, meant that the actual deal at its heart could not—*must* not—be made to disappear.

So, through the winter and into the new year, Walpole kept his focus on the original goal of the whole affair: the still-vital reform of national finance. "The South-Sea act was made," Walpole told the Commons on December 20, 1720, to relieve the nation from "its encumbrances and public debts, and putting them in a method of being paid off in a few years." This could not have been done, he told the House, "unless a way had been found to make the Annuities for long terms redeemable." It was this that had been "happily affected by the South-Sea Scheme"—and, crucially, "without a breach of parliamentary faith." The operation was successful, that is, even if some patients died.

Walpole was right. All the sound and fury and final catastrophe along Exchange Alley had ended in a financial triumph for the nation: the motley and expensive pile of loans and terms the Treasury faced in March had by the end of August become a single, simple obligation: a pool of capital, salable as shares in the stock of a single company, on which the government owed an agreed and modest interest rate, and which it could repay at any time. This, Walpole reminded his listeners, was what would be lost: "If they should now unravel what had been done, they should not only ruin the South-Sea Company, but instead of alleviating, aggravate the present misfortunes." After he spoke, the Commons voted on the resolution that "all Subscriptions of public Debts and Incumbrances, Money Subscriptions, and other Contracts made with the South Sea Company, by virtue of an Act of the last Session of Parliament, remain in the present State."

The South Sea scheme had worked. Walpole reminded Parliament of that fact. The collateral damage was real but secondary. To preserve that success, every bad bargain must stand.

Two hundred and thirty-two members voted in favor; only 88 voted no.

WALPOLE'S ARGUMENT HAD the virtue of being true. Politically, it was trickier to defend. He was still waging a mostly silent battle with Stanhope and Sunderland, the two Whig earls who had earlier banished Walpole and Townshend from the cabinet. That faction retained more of the most important offices in the ministry. Coexistence between the rival leaders was an ugly necessity of the crisis but was likely temporary; whoever best survived the next several months would most likely force their counterparts from power, just as Stanhope and Sunderland had done to the losing pair in 1717.

But in the maelstrom of late 1720, the rival sides shared a common interest. Those destroyed in the crash wanted revenge. If they couldn't get their fortunes back, then those responsible for their ruin should damn well suffer too. Everyone knew that the Company couldn't have launched its scheme without allies within the ministry and in Parliament—Stanhope and Sunderland among them. And yet, Walpole rose to their defense. The bribes paid in February and March of 1720 had been too broadly distributed to be secret, and it was common knowledge that much of the leadership, as well as many senior officials and dozens or hundreds within both houses of Parliament, had received their sweeteners. While Walpole would probably not have been too upset if the scandal could surgically remove his chief opponents, the danger was that the carnage might spread to the entire Whig administration, taking him and his friends down with the rest. Similarly, his enemies would have been happy to watch Walpole fail in his handling of the recovery, if only his botch wouldn't also bring them down—*and* again bring the nation's finances to ruin. The two sides had to hang together; otherwise, they'd surely dangle one by one.

The attempt to dodge public anger, though, began poorly and got worse. Walpole had tried to avoid serious scrutiny of the history of collusion between the Company and Parliament, but on January 4, 1721, he and the ministry were defeated as the Commons approved the creation of a committee to "enquire into all the Proceedings relating to the Execution of the Act" that had launched the South Sea scheme. "All the Proceedings" definitely included tracing the bribery trail: who had paid how much to whom for exactly what services received in return. Five days later the House chose who would serve on the committee. It was a key vote, given how many within or allied to the government were vulnerable to a real investigation. It couldn't have gone worse for the ministry, especially the Stanhope-Sunderland wing. Most of their candidates lost, while several leading parliamentary critics, notably Hutcheson, won places on the panel.

This Committee of Secrecy met immediately, accepting a punishing schedule: "sitting from nine in the forenoon till eleven at night daily, Sunday's excepted." The South Sea Company's cashier, Robert Knight—Blunt's closest collaborator—was one of the first to testify before Parliament, appearing on January 12 and then again on the sixteenth and seventeenth. It was clear what his inquisitors wanted from him: culprits, those who had been bribed with offers of stock that had not been recorded in the official books—everyone Knight had listed in his secret ledger. Knight knew what would follow if those records become public. Later in the week Sir Theodore

Janssen—both a South Sea director and a member of Parliament—told him to come clean, "to make a Discovery of whatsoever he knew relating to the whole Proceedings." Not a chance, Knight replied. If he were to spill his secrets, "it would open such a Scene as the World would be surprised by it."

On Saturday, January 21, he returned for another round of questioning. Earlier in the month he had prepared a new document for the committee that purported to account for stock transactions from the previous spring—actually bribes used to push the South Sea Act. The document omitted any actually useful information: whom he had paid and the numbers that would reveal who got what. Asked why, he replied that he "did not think it proper to enter the Names of the Members of Parliament who had any Part of this Stock, in the Cashbook"—confirmation, if any were needed, that the corruption had been widespread. The committee demanded his original records, "his Paper and Letters," and told him to repair his memory forthwith. Knight then said that he had to testify before the House of Lords and begged leave to deliver the documents on the coming Monday.

That night Knight wrote to his South Sea colleagues, saying that he found "the weight of Inquiry too heavy for me." He had no doubt about how the case would end for him. As one member of Parliament wrote, Knight ran "the risk of losing his own great share of the gains that had been made by this undertaking"—and quite likely faced a stay in the Tower of London as well. He also had to think about the dangerously powerful men and women he could implicate. He had told Janssen that the stock had gone to "Persons whose Names were not proper to be known to a great many." These men were, he told another director, "Forty or Fifty of the Company's best Friends"—the kind of people for whom "He would go through Thick and Thin" rather than reveal.

Knight was hardly alone in fearing the consequences of his testimony. Everyone he'd paid understood their peril—members of Parliament, peers, ministers, and even grandees and their mistresses at the court of King George. Those who wanted a full reckoning knew how vulnerable witnesses would be to pressure from such people and tried to ensure that Parliament would nonetheless get the full story. A bill to prevent Company directors and officers from leaving Britain had been in the works for some weeks and passed the Commons on the nineteenth. The Lords took it up immediately, but it wasn't until January 24 that the final act received royal assent and thus became the law of the land.

Knight's letter to his colleagues reached the Commons before Monday, January 23. The House responded immediately, asking the king to close the ports and to issue a reward for "discovering, apprehending, and detaining Mr. *Robert Knight.*" His Majesty issued the necessary orders—but too late. Knight had fled Westminster two days before, immediately after his appearance before the Committee of Secrecy on Saturday the twenty-first. He fetched his son, and by early the next morning the two were afloat on the Thames. They reached Dover on Monday and took passage for the Netherlands, just before the closure of the ports. A few days later, they reached the Duchy of Brabant, which conveniently refused to extradite the fugitives back to Great Britain.

Knight's escape was memorialized in "Lucipher's new Row-Barge," the cover illustration for the *Weekly Journal or British Gazetteer* on May 21, 1721.

The stench of rat was overwhelming. Arthur Onslow, then a member and later speaker of the House of Commons, recalled that it was common knowledge that Knight "had been in the depth of the project from the beginning, and in all the secrets of it afterwards; so that whatever corrupt dealings had been with people in power, he was thought to know the most of that work." As Onslow implied, the corruption extended throughout the networks of influence at the center of British life. It was also known, he wrote, "from undoubted authority" that as a reward for shielding Knight "and for not delivering him up" Brabant's ruler "had out of the King's Civil List . . . £50,000." That was money from the account usually used to cover the monarch's household expenses and many of the nonmilitary expenses of his government, paid out to ensure that the most damaging witness never faced Parliament's inquisitors. This fee for service, Onslow wrote, had been arranged by "my Lord Sunderland"—still one of the two chief ministers ruling the kingdom of Great Britain.

IT'S HARD TO imagine a more transparent ploy. Sunderland and the ministry's hope seems to have been that without Knight's testimony and the unequivocally damning evidence of his books there wouldn't be enough to convict anyone in the cabinet or its allies beyond it. That fantasy died fast. Knight had taken some crucial records with him, but not all, and he left behind plenty of people who had worked (or conspired) with him. These witnesses now faced the full fury stoked by his escape. It didn't take long for one of them to break—a man as well placed as the absent treasurer: the Company's co-founder and secretary, Sir John Blunt himself.

Impelled by Blunt's testimony, the Hutcheson-led Committee of Secrecy's investigation would become something of a blood sport as its members chased those at fault. But while Hutcheson himself sought some way to reverse the events—and the bargains—of the previous year, and while his committee assembled its account of an astonishingly widespread pattern of corrupt misbehavior, the inquiry itself took place after the real battle was done.

That is: the South Sea scheme had completed a transformation akin to that achieved in the scientific revolution. It had taken all the different species of debts and obligations Britain had accumulated over the preceding three decades and abstracted them into a single form, South Sea stock. This was an imperfect echo of a similar process of abstraction that Britain's natural philosophers had performed in the preceding century when they turned the travels of the stars and the rise and fall of the tide into numbers. The hope of a true Newtonian revolution in matters of money was unfulfilled and still is. But by the 1720s the drive to encompass experience in ever more general forms had made the leap from nature to social life. Stocks and shares were measured in the simplest of numbers, a price. That number could be probed by calculation; it could form the input to models analyzing the variety of human possibility years into the future. Mathematical approaches could create derivatives of that single number to create new ways to trade and new concepts of what could be tradeable—the risk involved in an investment, for example, and not simply the investment itself.

On the streets of London, this new way of thinking would evolve in the decades to come. The national debt as of 1720 was now almost entirely in the form of South Sea stock. But the question going forward was the same as the one France faced once Law's *système* collapsed: How would the British state attempt to finance itself going forward? How would it borrow? Would the Treasury authorities revert to their old pattern, launching a lottery or a long-term annuity at the point of need, paying whatever the market demanded as a new set of awkward and expensive obligations once again consumed the national budget? Or was there another way? Had the Bubble opened up new possibilities that a clever or fortunate leadership could exploit?

Before Walpole could answer that question, he had to survive the immediate political crisis. The decision to preserve the South Sea transactions was the first step of many; the settled outcome that would arrive by the middle of the century was not inevitable in 1721. The parliamentary investigation, with Blunt's account at its center, still posed a threat. If enough cabinet members and other officials were shown to have been corrupt, then the whole Whig interest could fall. Walpole had decisions to make in the face of the Commons' indictments of the ministry: whom to defend and whom to abandon; how to respond to the rage that specific revelations evoked in the Commons or on the street. At stake: the chance to set the course of Britain's credit, and hence the nation's ability to project power around the globe.

Chapter Twenty-One

"Mercy may be Cruelty"

The fourth of February was a busy day at the Palace of Westminster. That Saturday, the House of Lords gathered in the Queen's Chamber, on the south side of the palace, expecting to hear about the South Sea Company from Sir John Blunt. Knight had run, but Blunt could not; his family was far less movable than his erstwhile partner's, and all those estates he'd bought in the glory days of the previous spring and summer weren't portable either. So he took a different approach: fully cooperating with the Commons' Committee of Secrecy while ignoring the Lords, where the two leaders of the government had every reason to discredit him. Earl Stanhope and Earl Sunderland—First Lord of the Treasury and the senior secretary of state—feared, or knew, that if he chose, Blunt could implicate most of the ministry. They wanted to know what he had already told the Commons committee and, if possible, to discredit the South Sea man before he could ruin them. There was just one difficulty.

The witness refused to answer.

Sir John Blunt was many things—ferociously ambitious, unscrupulous, fortune-hungry, and desperate, for starters—but no one had ever called him stupid, certainly not to the point of doing favors for his enemies. Too many in the House of Lords—not just Sunderland

and Stanhope—had excellent reason to undermine his testimony about the Company campaign of bribery that everyone already knew had taken place. So he stood silent. He left the chamber. He was called back. He repeated his refusal. He was released again, then called back, then excused once more, and then brought again before the assembled peers. He would not take the oath required of witnesses. He would not speak.

At last, Sunderland and Stanhope gave up. Blunt left for a final time. The floor was opened to the ministry's enemies. The Duke of Wharton seized the opening. Wharton, profligate and a drunkard, had been both an early opponent of the South Sea scheme and a heavy loser in the Bubble (he was said to have held a funeral for the Company, complete with musicians and a hearse). He savaged Stanhope in an elaborate allusion to ancient Rome that would have been perfectly clear to his classically educated fellow peers. Stanhope, the duke said, was like that "favorite minister, by name Sejanus," who persuaded the emperor Tiberius to cloister himself in his palace on the island of Capri. Thus his servant, Sejanus, was able to take charge. "And so," Wharton said, "Rome was ruined." The Duke's meaning was clear: for Sejanus, read Stanhope; for the empire's ruin, see the wreck of Britain under the earl's rule. Stanhope rose, consumed in rage, to hurl a Roman insult of his own. Wharton was so lost to decency, he said, that like Brutus before his treacherous sons, the duke's own father would have watched his execution without pity. Suddenly, in full cry, Stanhope stumbled, then collapsed. Semiconscious, he was carried out of the House and borne to his house. Within a day, he was dead, probably of a stroke.

WHARTON HAD BEEN unfair. Stanhope himself had not accepted bribes from the Company in the previous year, nor had he claimed to be expert or even particularly adept at financial administration. His rage at the false charge likely reflected a genuine sense of personal affront. Though he was broadly seen as honest, if ill-informed, his death stripped away the protection his personal integrity might have given to his less worthy colleagues. With his passing, the slaughter, some of it literal, could proceed. James Craggs the Younger—as one of the secretaries of state, a powerful member of the cabinet—had pushed the South Sea scheme in Parliament, and he was one of those Knight had fled to protect. On February 6, the day after Stanhope died, he complained of pain in his back and head. It swiftly resolved into smallpox, and he died a few days later. Despite his role in the scandal, he was beloved by his friends and so was buried in Westminster Abbey. In the spirit of saying nothing ill of the dead, Alexander Pope's epitaph for him began, "Statesman, yet Friend to Truth!"

No one said the same of his father, James Craggs the Elder. The postmaster had been one of the key beneficiaries of the scheme, profiting earlier and more than most from corrupt dealings in Company stock. Once he buried his son, he did not choose to discuss the matter further. On March 16, just before he was to be tried for corruption by the full House of Commons, he was found dead, "not without very strong suspicions of having used violence to himself." Unsurprisingly, this was seen as a presumptive confession.

Craggs was not the only man the Bubble killed. Charles Blunt had worked in the family shoemaking business before his cousin John led him into the world of money. As a stockbroker, then a South Sea director, he was deeply enmeshed in the Company's backroom operations. He recognized the true extent of the disaster as early as the third week of September in 1720. He decided not to watch the inevitable play itself out. On the twentieth of the month, he slit his own throat.

Those deaths were just the most public of those that followed in the wake of the Bubble. Most others would have passed unremarked, and some of what may be reckoned homicide-by-Bubble proceeded more subtly, as in the fate of Henry Bentinck, Duke of Portland. He owed far more to various creditors than he could possibly repay. The utterly-beneath-him post of governor of Jamaica offered an escape: he would rebuild his fortune through the ordinary habits of colonial corruption, while, as the direct representative of the crown, he could not be imprisoned as a debtor. Against those advantages, he had to survive in his fever-ridden post long enough to resume his rightful place in London. As with so many of his other wagers, this one was a loser. In 1726, forty-four years old, still living in Spanish Town, he died of the black vomit—yellow fever. He was, almost certainly, the wealthiest casualty of the South Sea year.

SUCH LOSSES WERE exclamation points to Parliament's ongoing investigation. Tracing who was at fault for the Bubble continued through the first half of 1721. The Commons' Committee of Secrecy did most of the work, issuing seven reports between February and June, though with Stanhope gone the House of Lords also managed to pursue some leads. Blunt and other witnesses reconstructed the Company's campaign of bribery, making concrete the rumors of the more than a million pounds laid out to beat the Bank and shape the act that launched the Bubble.

In that testimony, the committee learned that Blunt and a small inner group of directors handled the distribution of stock to those who could help pass the South Sea bill. Blunt admitted that the first set of accounts laid before the committee was false, and known to be so by Company insiders. Next he revealed exactly how the bribes made their way from hand to hand, using the example of Postmaster Craggs to show how the trick worked. At the "Latter end of February or the Beginning of March last," Blunt testified, Craggs told Knight to reserve £80,000 of stock at the current price, £50,000 for Sunderland to dispose of and the balance for himself and his friends. This was the general pattern throughout that season of persuasion: "This Stock was set down as sold at several Days, and at several Prices." It was then recorded as held for "the Benefit of the pretended Purchasers, altho' no mutual Agreement was then made for the Delivery . . . of the Stock." Then there was the step that transformed such deals from insider trading into plain bribery: "No Money was paid down, nor any Deposits, or other Security . . . so that if the Price of the Stock had fallen, as might be expected, if the Scheme had miscarry'd, no Loss could have been sustained by them; but if the Price of Stock should advance . . . the Difference by the advance Price was to be made good to the pretended Purchasers"—all as Knight cleaned up the paper trail.

In repeated rounds of questioning, other directors and staff produced ever more detailed and damning information. Knight's assistant Robert Surman proved especially helpful in fleshing out the techniques used to bribe the great and the good, while a number of the Company's directors told stories of systematic, wholesale corruption reaching through the ministry, much of Parliament, and the court. Ever more prominent figures were implicated, Sunderland along with Craggs and Aislabie, each wholly entangled in incriminating detail. Some names had been recorded in code, but the committee soon penetrated the ruse. That was how Charles Stanhope—Earl Stanhope's nephew—appeared, barely concealed behind the pseudonym Mr. Stangape, Esq.

The amount of money involved was staggering. As more testimony rolled in, the committee recorded a total of £1.7 million in bribes, alongside almost £1.8 million in loans from the Company, going to almost two hundred members of both houses. Even worse for the ministry itself, an appendix gave a precise tally of what John Aislabie had done. Aislabie, chancellor of the Exchequer and, with Postmaster Craggs, one of the two principal government architects of the South Sea Deal, had accomplished an epic amount of self-dealing in the days when Parliament weighed the competing bids from the South Sea Company and the Bank: eighteen separate transactions between January 1720 and the first of March.

The committee's second report followed up that raw table of numbers with a narrative entirely devoted to Aislabie's misdeeds. The third added to Aislabie's bill of particulars and moved on to the elder Craggs's. And so it went throughout the spring, updates delivered in grim, deadened, bureaucratic prose. A sample: "The said Mr. Surman

[Knight's assistant] being examined says, There was £2000 Stock charted to Lord *Blundel,* read to the Examinant out of the Tree Book, as he has said in a former Examination."

Such blandness accumulated into thudding certainty. The litany of names spun out in endless columns numbering the shares and money that went to ministers, to members of the Commons and the Lords, to the rich, the royal, the well connected—page after page of typeset venality. The moral of that story? Someone, many, all across Britain's ruling elite had sinned.

Walpole had hoped to avoid exactly these kinds of revelations; he might have been safely unbribed, but if too many of his colleagues fell, he too would be ruined in that general disaster. Once the Committee of Secrecy proved impossible to control, he switched tactics. With its revelations pouring out, there was no way for the ministry to avoid taking some damage. So Walpole's became a rearguard fight, sacrificing some to protect the rest.

The first to go fell in mid-January: six South Sea Company directors who also held government posts. That didn't do the ministry much harm, but the losses stung, especially as the House of Commons then expelled four members who were also directors, costing Walpole their potentially vital votes. Next, the Commons chased the money. Company directors and officers were ordered to catalog their estates and submit those inventories to Parliament. Then, in February, Parliament directed that the wealth of the directors should be seized for the benefit of South Sea investors.

Public rage would allow no less. As one polemicist writing under the name Cato argued, "Tis true, it is both prudent and religious in private Persons, to stifle the Notions of Revenge. . . . But Jealousy and Revenge, in a whole People, when they are abused, are laudable and politick Virtues," adding that "when the Dignity, or Interest of a Nation is at Stake, Mercy may be Cruelty." To that end, Cato concluded, "You may, at present, load every Gallows in *England* with Directors and Stock-Jobbers without . . . so much as a Sigh from an Old Woman." It would have been futile—and costly—to get between such wronged souls and their revenge. Walpole, no one's fool, stood aside.

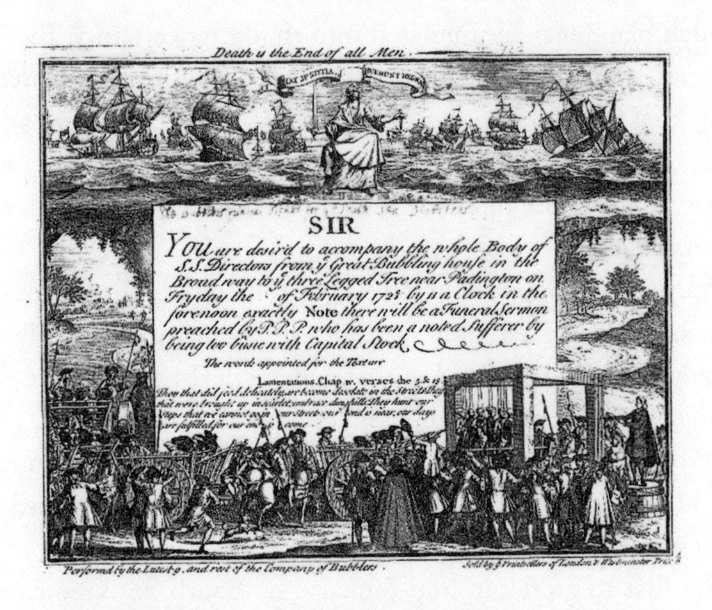

"Death is the End of all Men," a not wholly in jest suggestion for the proper fate of the South Sea Company's directors

His calculus shifted once the investigation turned to his colleagues in the ministry. As the Committee of Secrecy labored on, the Commons voted to try the leading malefactors for "corrupt, infamous, and dangerous practices." Charles Stanhope, the dead earl's cousin, went first, and, to protect the Whig interest as a whole, Walpole led the defense.

What followed was a perfect example of political justice. On the facts, Stanhope was guilty. The committee's digging had produced a clear narrative of a powerful man using his position to profit at the expense of just about everyone else. Five witnesses, led by Blunt, put that story before the House. Walpole followed with just enough legal distraction to muddy the prosecution's tale. In his account, Stanhope claimed he had actually paid for the stock credited in his name, while his bankers were persuaded to assert that they had spent his money without his knowledge. Both stories were implausible; each was incompatible with the other. But the bafflement served, just barely: Stanhope was acquitted by a margin of just three votes.

There was to be no rescue for the next defendant. The Committee of Secrecy's Sir Thomas Brodrick wrote to a colleague that Stanhope's escape had "putt the towne in a flame, to such a degree as you cannot easily imagine." Aislabie, heroically corrupt as chancellor of the Exchequer, was an easy sop for that rage. The House heard the evidence the Committee of Secrecy had accumulated of Aislabie's self-dealing and enrichment in a prosecution even more overwhelming than that implicating Stanhope. Crucially, as Brodrick observed, "Mr. Walpole's corner sat mute as fishes." That signal was clear. It took barely a day for the members to convict Aislabie. He was taken immediately to the Tower of London, and as word spread from Westminster celebratory bonfires flared in the streets.

On the next day, March 9, Walpole allowed the Commons the pleasure of condemning another, more minor figure. But then the House took up the most important case of them all: that of the surviving leader of the ministry, Earl Sunderland. Though Sunderland was the last serious obstacle between Walpole and the top job, a successful attack on such a powerful man would threaten the existing political alignment in ways Walpole could not control. Accordingly, he again intervened, with tactics that Brodrick admitted were "entirely different from what I expected." Blunt laid out Knight's dealings with Sunderland. Rather than denying that any such corrupt transactions had occurred, Walpole set out to impeach Blunt as a witness. He created a fog of doubt by persuading the witnesses that though Blunt had told them of dealings with Sunderland, Blunt hadn't been there at the time or, if he had, had been out of earshot. To Brodrick, "Such trifling stuff never surely was insisted upon in any other case." No matter. Walpole wanted to sow doubt—or rather, to create enough of a distraction to provide cover for a vote in Sunderland's favor. The members understood what was at stake: not simply just deserts for one bought-and-paid-for politician but, as one of

Walpole's allies told the House: "If you come into this vote against lord Sunderland, the ministry are blown up, and must, and necessarily will bee succeeded by a tory one." Even for Walpole's factional enemies, that was not to be contemplated, and the Whig majority asserted itself. For all the evidence against him, Sunderland was easily acquitted, with a margin of fifty votes.

WITH THAT, THE major political battles over the South Sea debacle came to end. For Walpole and his allies, it could have gone much worse. The Committee of Secrecy had given the opposition plenty to work with. But the ministry's enemies were themselves divided: financial losers in the crash—county gentry and city adventurers alike—jostled uneasily with High Church Tories and Jacobite supporters of the house of Stuart. After Sunderland's trial, the Whigs were able to brush aside that uneasy coalition to maintain their hold on power.

But if the ministry survived mostly intact, its center of gravity had moved. Walpole's rivals had accumulated major political debts; now he called them in. In April, he took over Aislabie's office as chancellor of the Exchequer and then added Earl Stanhope's post, First Lord of the Treasury, to his portfolio. That combined both leading financial posts in his hands alone, giving him power over both the broad direction of national finances and the management of day-to-day fiscal matters. He wasn't yet the undisputed head of government. He still needed his partner Townshend, and though Sunderland had resigned after his trial, he retained King George's appreciation, which gave him considerable influence. Still, for the first time in his two decades in Parliament, Walpole stood on the highest rung of government. Sunderland obligingly died in 1722, and none of his followers managed to take his place. With the field clear, Walpole would hold his place as the leading man in British politics for more than two decades.

IN APRIL 1721, in the first days of his ascendancy, his ministry proposed a measure described as "Relief of the unhappy Sufferers in the South Sea Company." This was the bill that confiscated the accumulated wealth of the Company's directors, along with anything to be found in Knight's accounts, along with those of his talkative deputy, Surman. The bill was soon amended to add Aislabie's and Postmaster Craggs's fortunes as well. All that money was to be shared with those who bought South Sea shares in the money subscriptions or acquired them in the debt swaps. This was the same session in which the Treasury forgave most of the £7 million the Company owed for the right to do the deal. Walpole managed the debates through which Parliament offered relief but also imposed what the preeminent historian of the period, P. G. M. Dickson, called "the harsh cautery of reality"—in which the shareholders "were forced to reconcile themselves to drastic losses of income and capital."

In the end, those losses turned out to be somewhat less brutal than many feared. It took several years to unwind the Company's affairs and dispose of the directors' property. When the final numbers came in, those who had bought shares for cash and those who had given up their holdings of the redeemable, floating debt faced the loss of about half their capital; Dickson calculated that buyers in any of the last three money subscriptions would by the middle of 1723 own stock worth £51 18s., 8d. for every £100 they'd originally spent. Those who swapped long-term annuities received between 65 and 70 percent of the income they'd have received if they had avoided the deal, while those who surrendered shorter-term paper found themselves much worse off, with only one-third of their prior return. Few suffered the full ten-to-one slide suggested by the full collapse from the top of the market in June to its bottom in December (though those in the Duke of Portland's predicament, people who bet more than they owned through leveraged or forward commitments, were still comprehensively ruined).

The South Sea Company itself had more work to do to resolve its remaining financial difficulties. But on August 10 the king approved the measures Walpole had guided through the legislature, bringing the immediate crisis to an end. Britain as a nation maintained its credit. The Treasury could still borrow. A national government that since the Glorious Revolution had been extending its influence across Europe and beyond would continue to play on the world stage.

THERE WAS ONE more bit of business the new Walpole-led ministry faced before it could return its attention to the ordinary business of government. Parliament and public opinion agreed: the Company's directors had to suffer. Their estates were forfeit, but one question remained. How much should transgressors be allowed to keep as the bare minimum to feed their families? Walpole preserved significant sums for many of the prominent targets, especially those who were connected to the Whig interest. Theodore Janssen, for example, held on to £50,000 out of his reported £250,000 fortune. Some, mostly the passive directors who had played no part in the key decisions, were allowed half their assets or even a little better. Others were more thoroughly sheared. The historian John Carswell described the case of the pepper trader Samuel Read, who told the Commons he owned a total of £117,000, with only about half deriving from his South Sea holdings. No matter: Parliament seized all but £10,000. And so it went: the closer they had been to the center of Company affairs, the less malefactors could hope to keep. That meant there was one man who could expect no mercy at all. Not only had Sir John Blunt been the public face of the South Sea scheme throughout the Bubble year, he had had the gall to commit the one truly unforgivable sin: he had given joy to Walpole's foes by naming names. Blunt tallied a net worth of £183,000. Walpole left him with just £1,000. Invested in government securities, that sum would yield £50 per year: bread and cheese for supper every weekday, with the hope of a chop on Sunday.

Blunt's allowance was ultimately raised to £5,000. He left London, abandoned finance, and spent the last fourteen years of his life in Bath. His family survived the blow, ultimately achieving enough prominence to suggest that however much Walpole managed to take, Blunt had successfully squirreled something away. The baronetcy he had gained in his glory days still survives, now on its twelfth incumbent (Sir David Richard Reginald Harvey Blunt), and in between, Sir John's descendants and connections include at least one general, a bishop, and the current Earl of Cromartie.

But Parliament's message was clear: Blunt had become a symbol of the worst that too much cleverness about money could produce. In its place, Walpole aimed to make government finance as monotonous as possible. He would spend his two decades in power doing so, using the remains of the South Sea Company as a key tool.

TO THAT END, Walpole launched Britain into a different kind of arms race, a contest that turned on techniques of governing rather than tactics on the battlefield. Britain's financial crises had come in response to its expanding ambition, both in Europe, where it took part in Continental wars of succession, and beyond, as the British constructed an empire. Those conflicts continued after the twin financial crises of 1720 in Britain and France. But after John Law's system crashed in Paris and London's South Sea drained, the decisions taken in the two capitals diverged, with the French choosing an approach to rebuilding their national credit very different from the one Walpole and his ministry pursued. Those different choices reverberated through the global war France and Britain waged over the next century. The outcome of that fight would settle which nation had gotten it right.

Chapter Twenty-two
"ringing their bells"

Taken together, the South Sea and Mississippi bubbles offer that rarity in history: a genuine control on an experiment in statecraft. In it, two nations confronted similar debt crises and responded with matching, partly imitative attempted solutions.

It is true that there were significant differences between the Mississippi and South Sea bubbles, most notably in Law's use of a "fiat" paper currency, one unconnected to any material good like silver or to a stream of revenue (the kind of thing the English did when they tied a bond issue to a tax on malt). But the two schemes were still similar enough to make it possible to compare how each nation responded once each of their gambits failed. Walpole in London and Law's successors in France followed divergent paths to rebuild national credit going forward. Those decisions would be tested as each nation paid for their long eighteenth century of violent confrontation—the wars the French and the British would fight as each strove for global dominance.

In Paris, the first reaction to the Mississippi affair matched London's. There would be no bailout, no do-over. As in London, Law's scheme worked, in that it tamed France's government debt. The crash in company shares and banknotes was a feature, not a bug, on the official side of the ledger. The losers, those who exchanged valuable assets for the *système*'s shares or banknotes, would have to live with their disastrous choices—just like their South Sea counterparts. Otherwise, from the purely national point of view, what was the point?

Almost immediately, though, the two nations diverged. Beginning in 1721, France's new financial administrators, the brothers Antoine and Claude Paris, oversaw one of the most spectacular feats of bureaucracy in the history of the famously bureaucratic French state. Their clerks reviewed 1.4 million separate transactions involving over half a million people. From those records, the clerks constructed a hierarchy of losers. Those who had exchanged debt for equity—who had in essence been swept up in the scheme—were seen as more deserving than mere speculators. That ranking set the table of compensation to be paid out of whatever could be recovered from Law's various institutions. No one would be made whole, or even close: the average claimant recouped only 17 percent of the face value of his or her banknotes. The one consolation was that whatever compensation claimants did receive came in the form of either life or perpetual annuities to be paid by the French treasury—exactly the same kinds of long or everlasting irredeemable obligations that both the South Sea scheme and Law's innovations had sought to abolish.

From the state's point of view, this was a model of conservative monetary practice, literally a return to prior methods. The French treasury had been the clear winner in the crisis, as all the money lost by Mississippi Company shareholders was debt the state no longer had to repay. The Paris brothers' reorganization took that as a one-off victory. With the state's debt crisis thus solved for the time being, they decided to return to more or less the system of public credit that had been in place prior to Law's radical ideas. Experiments with paper money were at an end; metal coins returned as the fundamental tokens of exchange, legal tender, and the true measure of wealth; and France returned to the old system of selling the right to collect taxes—an approach that made it difficult to increase tax revenue in times of need.

It's not just people; nations too can be neurotic. Repeating the same actions in the hope of a different outcome is rarely a good bet. It took just five years before revenue lagged behind expenses enough to bring on post-Law France's first partial default. In 1726 the French treasury refused to make payments on some of the annuities it had just created to clean up the Mississippi mess. Similar cycles of ballooning debt and some form of default would follow over the next six decades. Compounding such difficulties, the use of annuities tied to individual lives marked a return to the bad old practice of issuing irredeemable debt, exactly what Law had been determined to avoid. Perhaps worst of all, clever men figured out how to game the system. It was possible to buy an annuity on a life other than one's own. Over the eighteenth century, syndicates emerged to buy annuities on the lives of groups of people—women and girls from well-off families by preference, chosen for their on-average long life spans. Some of those enlisted were as young as four. These packaged investments, most famously known as the "Genevan heads," after the groups of Genevan girls that assembled as annuitants, were an ingenious, even modern response to the revival of France's old and brittle approach to credit. They responded to the possibility that any one girl within the group would die expensively young by spreading the risk across a big enough group who would, statistically, enjoy long and perhaps happy lives. As low-risk, high-return investments, these bundles of annuities became vehicles in which those with money to spare all over Europe could park their cash.*

The Genevan heads and similar arrangements did have one limitation: they weren't easily bought and sold on a secondary market. That meant no analogue to Exchange Alley emerged in Paris that could readily absorb each new attempt by the state to borrow. Ulti-

* These buckets of insurance policies on the lives of young women are the distant ancestors to modern asset-backed securities that are now in use and that were deeply implicated in the Great Recession that began in 2007–8—about which, more below.

mately, as the economic historian Larry Neal points out, this created a system in which the French government couldn't raise money easily or on good terms at precisely the times it was most needed: in the midst of war against its most persistent antagonist, Great Britain.

AT FIRST GLANCE, any fight between eighteenth-century Britain and France would have seemed a mismatch. If national wealth alone were the measure of power, France should have been able to brush aside its upstart antagonist on the edge of Europe. From beginning to end of their long struggle, France was at least twice as rich as its adversary, and French subjects outnumbered Britons by as many as three to one. Yet there was never a moment when France overwhelmed its poorer, seemingly weaker enemy by the sheer weight of men or treasure. Somehow, when it came to the tooth end of national power, the ships and guns and soldiers arguing the point of empire, His Majesty was able to stay level or ahead of whatever *le Roi* could put into the field. It came down to the bloodless tally of national accounting. In times of war the British were able to spend as much as four times the amount of money per subject as the French state did, which meant that Britain was able to match France pound for pound, livre for livre, throughout their eighteenth-century wars.

A common, easy explanation for France's lackluster fiscal outcome is that a feckless French court stumbled from one financial disaster to another as a sequence of overmatched rulers tumbled inexorably to their doom. We know how the story ends, after all, in revolution, regicide, and terror. But French rulers were not all financial illiterates, nor was France itself remotely a failed state. Its officials attempted various expedients, which along with the occasional bump in tax revenue brought the French budget almost into balance as late as 1774—while the national debt had been managed well enough to that point that as a proportion of state revenue it cost just half as much as it had at its peak under Louis XIV.

Meanwhile, France itself remained rich. Private lending was as common as it was in the other great commercial capitals, Amsterdam and London—which meant that the economy was still bustling. Businessmen and entrepreneurs raised funds through networks and sources separate from those available to the state. From the 1760s forward, such commercial borrowing significantly outpaced public credit—which helps account for Paris's reputation as the most luxurious city in Europe. Whatever might have gone awry in public finance, private enterprise still thrived. Last: imperfect doesn't mean incapable. Throughout the century France remained unmistakably a great power, able to project military force around the globe and to sustain its side of its battle with Britain through to a decisive victory in the revolution that gave birth to the United States.

Still, the fact remains: smaller, poorer, less populous Britain was able to fight France on a level footing at every turn. How did it do it? Much of the answer lies with a series of choices made in London in the early 1720s by Britain's new master of affairs.

THE OFFICE, OR, rather, the role, of chief minister to the crown is an old one, dating to the earliest days of the English monarchy—naturally enough: even the most energetic or megalomaniac of divinely anointed rulers needs someone to talk to, and a vessel through whom royal wishes become deeds. The title and job of "prime minister," though, is a relatively new one. For most of the history of the English crown, much of the power monarchs deputized to their leading servants was divided into specific offices. As parliamentary government matured in the seventeenth century, and especially after the Glorious Revolution, faction and party politics tended toward a more hierarchical organization of power within the cabinet. In the wake of the partial vacuum created by the South Sea debacle, this culminated in Walpole's ascent to primacy within his coalition.

Walpole himself hated the suggestion that he held any extraordinary role or post, declaring in the House of Commons after twenty years in power, "I unequivocally deny that I am sole and Prime Minister." His colleagues knew better. By convention, Britain's first prime minister took office in 1721, when Walpole combined the crucial jobs of First Lord of the Treasury, chancellor of the Exchequer, and leader of the House of Commons. It took him a couple of years to consolidate his position, but by no later than 1723 Walpole had clearly exceeded the authority of his recent predecessors. Most important, his authority derived from what is recognizably an ancestor of modern prime-ministerial governance: he was at once the king's top deputy and the leader of the party that controlled the Commons—the elected body that controlled the nation's purse. He could thus serve as a bridge between often-opposed forces in British governance—and, just as important, possessed his own base of power, largely independent of royal whim, through his command of Parliament.

All this meant that as Britain faced the question of how to pay for its national affairs after the debacle of 1720, Walpole and his closest partners could act with unprecedented official authority to settle the matter. As the Paris brothers had demonstrated across the Channel, it was perfectly possible to take advantage of a financial crisis, consolidating the debt by turning the public's losses into the state's gain, but then to go right back to the same kinds of borrowing that had led both nations toward their respective bubbles in the first place.

This Walpole refused to do, in a series of decisions made over his two decades in office. That is: he did not have a fully formed plan ready to go in 1721. But he allowed himself and his ministry to be guided by ideas developed over the preceding half century as it partly created and partly responded to the rise of the first recognizably modern market in money. This process, in turn, underpinned Britain's drive toward global might in ways that none of its competitors, France first among them, would match.

That transformation began with the South Sea Company itself. Even after the crash it remained a behemoth, swollen by all the debt it had absorbed. At the Bubble's end, the Company's capital totaled

£38 million. That made it by far the largest joint-stock company traded on Exchange Alley. It remained an awkward hybrid, both a financial company and a trading venture. Those two very different businesses demanded distinct expertise while facing divergent types and amounts of risk. Before systematic reform could take place, it had to be rationalized. The first step was to shrink it, bringing it more in line with the rest of the financial system. To that end, in the summer of 1722, the Company transferred £4 million of its capital to the Bank of England in a sale that cleared some of its outstanding liabilities.

With its balance sheet thus improved, the next step was to break up the monster into its component businesses. Midway through 1723, the "old" South Sea stock was split into two. For each original share they owned, investors received a new asset called a South Sea annuity, actually a bond that paid a steady stream of income derived from what the Treasury owed on all the different debts taken in during the Bubble. They also got a "new" South Sea share in the trading side of the Company, now severed from the purely financial operation. This maneuver eliminated the confusion that Hutcheson had tried so hard to explain in 1720. Now there was a stable, liquid financial asset that had a publicly known and low-risk return—and another business playing in the high-risk, occasionally high-reward arena of transoceanic commerce.*

It's one of the oddities of the Bubble year that this step actually revitalized the venturing side of the South Sea experiment. The persistent disappointments of the trading ventures of the 1710s were part of what had goaded Blunt and his comrades toward bigger and bigger financial gambles. Now, with this confusion of aims at an end, the Company's monopoly privileges grew relatively more interesting.

* Despite the name *annuity* these differed from the French version (and earlier British annuities) in that they were not tied to a specific life or series of lives—which would have made them illiquid, not readily transferable to someone else. These were simply bonds that paid interest indefinitely and that could be redeemed at any time.

With peace once again achieved with Spain, the annual trade voyage to Spanish American ports resumed. These journeys continued until the Company abandoned the effort in 1732, giving up its last pretensions of becoming a great trading monopoly. After the final voyage, South Sea clerks summarized the post-Bubble trading account. There was a net profit—but at just £32,261 over eleven years, the venture simply wasn't worth the effort.

A second seafaring wager proved a mistake. In 1725 the Company lurched into an experiment in whaling in the seas off Greenland, acquiring a dozen ships, then adding more over the next several years to create a flotilla of twenty-five vessels. The whaling fleet was a money pit, primarily because those involved hadn't mastered the technical demands of the specialized and complex business of whaling. After seven years of pouring money into the ocean—almost £200,000 by 1732—they gave up and sent no more ships to the Arctic.

There was one more line of business open to the Company: buying and selling human beings. The Company's early forays into slaving in the previous decade had begun poorly. But after a two-year halt in slaving voyages in 1720 and 1721, South Sea ships returned to the African depots. Over the next seventeen years, until another Spanish war brought the trade to an end, South Sea vessels would land at least thirty thousand African men and women in the New World. That number is certainly lower than the true tally of stolen lives, as the best estimates suggest that around one in seven captives perished on the journey.

But even though slaving may have pushed the Company's mercantile efforts into profit, overall the results never came close to creating a transatlantic twin to the East India behemoth. After the Bubble, the Company demonstrated that it was indeed possible to make money by sending ships across the Atlantic. But it was hard, complicated work, and it produced generally meager returns. In 1733, a shareholder revolt forced Parliament to convert two-thirds of the Company's remaining trading capital—the "new stock" created ten years before—into more of the other kind of South Sea security to emerge from the Bubble, those bonds, backed by all the faith and credit of the king and Parliament.

That cost the nation, as it increased the total national debt. But it also acknowledged the obvious: the South Sea Company would never become an ocean-spanning wealth machine. There was to be no further attempt at trade in any form. All that was left of what had been, briefly, the largest private business in the world was a financial back office in which clerks recorded a steady flow of official interest payments from the government and an equal and opposite flow of money back out to its bondholders.

That was, of course, precisely the fate Blunt had dreaded back in the dog days of the 1710s—the tedious, repetitive, and not particularly lucrative job of scratching numbers on paper that passed from desk to desk. It was, though, a triumph of the Walpole-led effort to create a distinctly British way of paying for empire. The Company's annuities remade the national debt into a new form, a single pool of liquid, easily bought-and-sold bonds. That achievement set the stage for the final task: the British government had to prove that it could pay what it owed when it owed it, reliably enough so that whenever it became necessary to borrow again, the king's subjects would be willing to forgive the Bubble and trust their leaders with their funds.

Creating this trust would be one of Walpole's central preoccupations throughout his long tenure as prime minister. He began by reviving the device he'd first proposed in 1717, the "Sinking Fund"—so called because it was designed to use any spare government revenue to pay back, or sink, the total of what the state owed. Walpole's fund got its money from whatever was left over after the payments had been made on loans tied to specific revenue streams (like the excise tax on malt tied to the Malt Lottery of 1697). After the original South Sea Act cut the Treasury's interest rates, the Sinking Fund began to swell, until by 1727 it was spinning off more than £1 million a year.

Under Walpole's direction, the Treasury used part of that hoard to retire the residue of redeemable obligations that their owners had (wisely) refused to turn into South Sea stock during the Bubble year. It continued to cut the nation's total debt by buying back some of the Company's paper—thus reducing the Treasury's future interest burden. Walpole continued to play the role of good steward into the 1730s, pushing any available funds toward the most expensive remaining loans: Bank of England holdings, old lottery deals, and others. There was truth to the reputation he had for such fiscal prudence, and some fiction as well. Critics noted that on occasion Walpole was willing to turn the Sinking Fund away from debt repayment, using it to cover current expenses. Even so, during his time in office the Walpole ministry repaid much more than the Treasury borrowed—a net drop in the national debt of over £6 million.

Such fiscal rigor drove a virtuous cycle. Within two decades, Walpole's management helped push the price the British government paid for its borrowing down to just 3 percent per year. Less principal combined with lower interest cut the annual cost of the nation's debt by almost a quarter in the twenty years after the Bubble. Even if Walpole was not without sin when it came to spending the Sinking Fund, the financial system he shaped provided clear evidence—testimony that became more persuasive each passing year—that Britain had crossed a threshold. It could manage its debt prudently. Those who entrusted their capital to the state would receive what they'd been promised—which meant that when events created a need for the Treasury to come up with funds it did not possess, the ministry could be virtually certain that the money would show up regardless. Given why the British government borrowed, this was a victory in the purest meaning of the term, one that would play out on battlefields in every corner of the globe.

HERE IS A story that produces horrified joy in every classroom hearing it for the first time:

On April 9, 1731, *La Isabela,* Juan de Léon Fandiño commanding, was cruising near Cuba in search of smugglers. A lookout on *La*

Isabela sighted a brig in the waters off Havana: the *Rebecca,* bound for London from Jamaica with a cargo of sugar. Spain and Britain were at peace, but by treaty the Spanish navy could stop and inspect British vessels for contraband. *La Isabela* fired several shots to compel *Rebecca* to heave to. The brig launched a boat, and the first mate carried over the vessel's papers, "expecting that would give sufficient Satisfaction."

It did not. Fandiño seized the sailors and used their boat to carry an armed boarding party the other way. On board the *Rebecca,* the Spaniards turned into brigands (according to the British crew). "They broke open all her Hatches, Lockers and Chests" in search of plunder, and when they found nothing to satisfy them, "their Lieutenant order'd [the captain's] hands to be ty'd, as also his mates, and seiz'd them to the Fore-Mast." They beat and cut "a mulatto boy"—the captain's servant. Next they tied a rope around the captain's neck and hoisted him till his feet dangled just above the deck. They let him down, asked him where he'd hid the money, and when he told them again that there was none, they hauled him up again, then once more "til he was quite strangled, and then let him fall down the Fore-Hatch." The Spanish sailors rifled his pockets, stole the silver buckles on his shoes, and dragged him—still by the neck—along the deck. Pressed again, the captain maintained that there was no more money aboard than the "four British Guineas, one Pistole, and four Double Doubloons" he'd already revealed. This so enraged the Spanish officer in charge of the boarders that he raised his cutlass and slashed at his left ear. Another Spanish sailor then ripped the remains of the ear free, and "gave him a Piece of his Ear again, bidding him carry it back to his Majesty King George."

This was, of course, the ultimate insult, and the London papers took up the cause of Captain Robert Jenkins's ear as soon as he reached home waters. Gory tales of the Spanish beastliness and the honest *Rebecca*'s sufferings filled news columns all that June. For the moment, the incident served more to entertain than inflame. Among Robert Walpole's virtues was that he was never eager for war.

Though relations with Spain had been prickly for years, his ministry managed to avoid any immediate rush to battle. But Jenkins did not disappear, nor did the tale of his severed ear, preserved against the day when Britons might need to be reminded of unfinished business.

That day came in 1738. In the retelling, Jenkins appeared before the House of Commons early that year, brandishing his mutilated body part—different accounts hold that it was preserved in alcohol or that Jenkins had kept it dried and folded into a wad of cloth. There's no direct confirmation of either story, but even mention of the shredded ear was enough to embolden Walpole's parliamentary opposition to rebuke him for weakness in the face of ongoing Spanish arrogance and insolence.

Walpole swooning at the sight of Jenkins's ear—an imagined scene drawn in the mid-eighteenth century

All this excited the eighteenth-century equivalent of the tabloids, but for the war party in Parliament the insult was merely cover for their real aim: expanding the reach of Britain's trade. The Spanish grip on so much of American and Asian wealth was seen as both an offense and an opportunity, which meant that seven years after the worst day of his life, Captain Jenkins had become useful to those trying to force the issue with Madrid.

They succeeded. British warships attacked La Guaira, Venezuela, on October 22, 1739—thus sparking what would come to be fixed in British classroom memory as the War of Jenkins' Ear. It would drag on for nine years, becoming a sideshow within yet another European dynastic struggle known as the War of Austrian Succession, fought between 1740 and 1748. That larger conflict became a contest between coalitions, with Britain and the Austrian Hapsburgs at the center of one alliance, Spain, Prussia, and France on the other.

FOR ALL OF Walpole's fiscal care, it still wasn't easy for Britain to cover the extraordinary costs of such a global struggle. National revenue hovered around £7 million annually, while public spending demanded by the new war climbed to more than £13 million a year. The ability to borrow easily and cheaply was thus vital throughout a conflict that would add £20 million of debt. The war was the test of the post-Bubble reconstruction of Britain's credit.

The nation met the challenge. As late as 1745, the Treasury could still borrow at 3 percent, but in 1746 the military's needs pushed the price for new loans up to 4 percent, where it stayed till the end of the conflict. That was still well below the worst of the cost of money in the decades leading up to 1720—and crucially, the terms of this new wave of borrowing were the same as those set in the South Sea cleanup. Each of these new loans took the form of redeemable perpetual annuities, coupons that would pay their stated interest rates forever, or until the government chose to pay them off. This new government borrowing behaved exactly like the South Sea annuities that had been in circulation over the previous two decades—and so were more of the same things that Exchange Alley had already learned how to trade and value: uniform, abstract expressions of money traveling through time.

Such financial consistency, the ordinary daily business of a government bond market, was Britain's secret weapon—the key to the Treasury's ability to borrow, at reasonable rates, even in an emergency. The South Sea Company, once the engine of national disaster, had become instead a machine processing the credit that allowed British leaders to spend as needed whenever needed to protect what they saw as the national interest.

That success did not come without its costs. Robert Walpole had his faults—among them a willingness to self-enrich, no great sense of marital constancy, and a gift for holding grudges. But he was never reckless with other men's lives. As long as he remained head of government, he had done his best to deflect and defuse any sudden surges of war fever. He had tried to avoid any fight on the pretext of Jenkins's missing appendage—and when forced to surrender to public passions, "They are ringing their bells, soon they will be wringing their hands."

His inability to keep Britain out of that war was an indication of his weakening hold on power. By 1740, his government was more divided than it had been since the early 1720s. His mastery of the House of Commons was eroding. A new election in 1741 left him with a majority of only nineteen pro-ministry votes out of 558 members—a margin that shrank to three in January 1742 and then to an outright loss in a crucial vote days later.

That was the end. On February 6, 1742, King George II transformed plain Mr. Walpole into the Earl of Orford, Viscount Walpole, and Baron Walpole of Houghton. Thus elevated to the House of Lords, he surrendered his part in the daily workings of the ministry. He was succeeded by a ministry effectively headed by an old rival, John Carteret, later Earl of Granville—who soon learned that the old man still retained some bite. Carteret soon fell to one of Walpole's protégés the following year, when Henry Pelham took the prime ministership in 1744, serving until his death eleven years later.

Pelham hadn't been regarded as a likely candidate to reach the top of the greasy pole, but his twenty years in fairly obscure posts had turned him into the ideal keeper of the financial house Walpole had built. He took over as the war shifted from what had been a distant series of skirmishes in the New World into a Europe-wide affair. That increase in scale meant the Treasury would have to borrow still more, which meant that Pelham confronted an increasingly skittish market for new debt. In 1746, it appeared as if Britain might follow France in reverting to old bad habits, floating a new lottery loan at an interest rate of 9 percent. Worse, as news of setbacks in the war hit Exchange Alley, financial stocks began to fall, which had a knock-on impact on the interest rates the government would be forced to pay. What happened at Jonathan's and Garraway's set the parameters for every new loan his government might need. (For example: the £4 that a bond issued at 4 percent on a par value of £100 would become a 5 percent return if a jobber would sell you a similar bond for just £80.) So Pelham entered office with twin problems: he needed to find a way to drive down the interest rate at which the Treasury was borrowing, and he had to forestall any further lapses back into the bad old days of emergency loans secured on terrible terms.

THE FIRST TASK was to make money cheaper. For that, peace was the answer. It took some careful maneuvering, but with the end of the War of Austrian Succession in 1748, Pelham was able to push a broad cut in the interest rate charged on public loans, moving in two steps from 4 percent to 3. Each of the major monied companies, the Bank, the East India Company, and the South Sea Company, had to be persuaded to take less on the official bonds they held, and it took some time for their shareholders to accept the blunt reality that the old days were indeed gone: the Bubble had washed the irredeemable loans out of the system, and any bond that could be redeemed would be bought out whenever the interest on new money was lower than the rate paid on older loans. The last major institution to agree to its haircut was the one that was still the nation's largest creditor: the South Sea Company, with its almost £26 million in redeemable stock.

The larger goal was to create a market in government credit so effective that Britain would retain a permanent advantage over its enemies. In a hangover from the pre-Bubble era, a tangle of loans and other obligations—many tied to one specific source of revenue or another—remained on the books. These were the remnants of the desperate moves the Treasury had been forced into from the 1690s forward. Pelham spent the early 1750s doing what Walpole could not: persuading Parliament to combine most of that remaining tangle of government borrowing into a single instrument paying the newly standard 3 percent per year. By 1752, the job was done. The new financial device became known as consolidated bonds—consols for short. Though the consol would not be the only tool that the British government used to raise cash in coming years, it represented the last key advance in the process of financial invention that had begun with the first Parliament-backed loan back in 1693.

In the years to come, the power to raise capital simply with a uniform and trusted instrument paid for the advance of empire. By 1750, Britain owed close to £70 million. The next conflict with France, the Seven Years' War that ran from 1756 to 1763, pushed the total amount of official bonds in circulation higher still. That sum rose again over the course of Britain's losing fight against its American colonies—and then skyrocketed during the climactic fight sparked by the French Revolution in 1789.

At each time of need, the Treasury was able to place each new issue of debt, millions of pounds—billions in current money—that would permit the nation to press the fight. That is: with the capstone accomplishment of the consolidated bond, in combination with the rest of the maneuvers that smoothed the lingering traces of the South Sea Bubble, Britain had not merely adjusted the way it did its official business. Instead, step by step, not necessarily consciously, but guided by a consistent logic, London's leading financial movers and shakers managed to exploit the idea that money is not a thing—not, necessarily, a sliver of material reality. South Sea annuities, consols, and all the rest were instead functions that operated over time. In practical terms they represented the future wealth of the nation that could be deployed to serve national ends in the here and now.

IT'S EASY TO believe that ideas are never nearly as lethal as a cannon's roar or the cut of a saber. Something as banal as a bond might seem far removed indeed from the battlefield. But Britain's mastery of money extended the capacity of its armed forces to deal violence and grab territory from Plassey to the Plains of Abraham. By the end of Britain's North American wars, the South Sea Company was a mere shell, a clearinghouse for a part of the national debt and a name inscribed on slips of paper that entitled their bearers to their constant 3 percent. Such an unexciting end was hardly what exuberant crowds had dreamt of back in 1720. But over the next century, what came out of that debacle—the unprecedented financial agility of the British state—outstripped even the most extravagant hopes of the Bubble year.

By how much would become clear at a single moment in 1815, when the very different decisions taken in Paris and London were tested, on what is remembered as the last day of the long and blood-soaked eighteenth century.

Chapter Twenty-Three

"it cannot be cured"

At about eight P.M. on Sunday, June 18, 1815—still sunlit on that midsummer evening—the Duke of Wellington stood in his stirrups over his famously cantankerous chestnut, Copenhagen. He waved his hat above his head. His army, men he had once called, with respect, "the scum of the earth," poured over the ridge, chasing the remnants of Napoleon's Imperial Guard back down the slope. At the battle-ruined farmhouse called La Haye Sainte, three battalions of France's elite soldiery stiffened to a last stand. Facing a combined charge of infantry and cavalry, they broke, and the rout was on, with fleeing French soldiers shouting (in legend), "The Guard retreats! Save yourselves if you can."

About an hour later, Wellington crossed the battleground to greet Field Marshal Gebhard Leberecht von Blücher, the commander of the Prussian army that had crushed Napoleon's reserve at a decisive turn in the fight. There would be a few more skirmishes in the coming days, but this was the end: the long century of conflict between Britain and France was done at the moment those two commanders met, victorious, on the field of Waterloo.

Wellington and Blücher meeting on the field of Waterloo, in a late-nineteenth-century idealization

TO THE LAST hours of that seemingly endless war, nothing about the British triumph was inevitable. Battles are decided by those who do the actual fighting, of course. But it took more than simple courage, or even the mastery of a Wellington, to lift Great Britain past its larger rival. The Napoleonic Wars were national conflicts. The forces the two sides deployed were vastly larger than any either had previously deployed. The fleets that met at Trafalgar represented huge capital investments in the most advanced machines of their day. Ships of the line, red coats, muskets, beer and beef, gunpowder, shot, and everything else had its price. Somehow King George's government managed to keep on paying for a war that, at times, pitted Britain virtually alone against the wealth of Napoleon's conquered Europe.

It's too simple to say that what happened in the banks and markets of Paris and London after 1720 determined the outcome at Waterloo. But it remains true that Britain and France had pursued two very different responses to very similar financial disasters—and that over the next decades, those choices made material differences in Britain's ability to punch above its weight class.

So Waterloo might very well have been lost had Wellington been absent. But it was highly improbable that Britain could have maintained the fight long enough for the Iron Duke's army to reach that ridge if Britain, like France, had reverted to old habits of public finance after the South Sea crisis. Over the course of the eighteenth century, the raw size of the nations at war came to matter less than the ability to mobilize the resources that each state's economy could produce.

This was where the financial system centered on London excelled. By the Napoleonic era, the scale of British borrowing was staggering, dwarfing previous experience. In 1814, the last full year of the war, public spending climbed to over almost £111 million. Almost two-thirds of that went to meet the costs of the military. In that one year, the Treasury raised £36 million in the capital market centered on Exchange Alley—almost exactly the same as the accumulated debt of a quarter century that the South Sea Company had taken in through the Bubble deal. Somehow, the Treasury managed to extract vast amounts of credit from the British economy—not just for those twelve months, but for every year for the more than two decades required to secure the final victory.

They were able to do so as a direct result of what Walpole and his successors had achieved in the decades after the Bubble. It wasn't easy; the system could have failed. The war-driven demand for funds in the early years of the fight in the 1790s drove up the cost of money for the government. The price for consols and other bonds fell on Exchange Alley, which forced interest rates on new government borrowing to over 6 percent by 1797—double the rate of the last years of peace. To help pay those extraordinary bills, in 1798 Parliament imposed the first income tax in British history, which, with a brief exception during the momentary peace of 1802, would be collected throughout the war. Over the same years, hard currency—coins!—had to be exported to help would-be allies in Europe. Those payments left Britain perilously short of gold, to the point that in 1797 the Bank of England stopped converting its banknotes into coins on demand—turning them into a paper currency unsupported by underlying assets. After what Walpole and Pelham had labored to do to stabilize British official finance decades before, the situation seemed dire, an echo of that of the 1690s, with its debasement of the coinage and the first unsustainable, expensive explosion of national borrowing.

But then, by 1800, the credits markets turned, decisively. Freed from the need to maintain gold reserves, the Bank was able to issue enough notes to keep cash in circulation in Britain, a distant homage to John Law's monetary ideas. Once the income tax and other measures strengthened the Treasury's ability to make its debt payments, the market for government securities rebounded, and interest on consols dropped—with occasional hiccups—to between just under 4 and 5 percent for the rest of the war. Once the cash-and-credit crunch of the late 1790s eased, Britain in the Napoleonic era never approached the precipice that King William had faced in the 1690s—that cliff-edge past which the nation might simply run out of money. All that followed—the navy's victories from St. Vincent to Trafalgar; the army's work in the Peninsular War and beyond; that little side project known in the United States as the War of 1812—every skirmish and grand conflagration rode on the steady flow of borrowed money from the heart of London to the tip of the spear.

PUT ANOTHER WAY: The United Kingdom's secret weapon in the fight against Napoleon was the successful implementation of new financial technology. The institutions and techniques that emerged in the three decades after the Bubble made it possible and then desirable for private actors to put their cash at the state's disposal, in exchange for a share of taxes yet to be collected from the economic activity of generations yet unborn. There was an exchange in which such bonds were easy to trade, supported by the Treasury's demonstration over the years that government-issued credit was a safe investment (this in contrast to the French experience of defaults at intervals throughout the eighteenth century). That combination of a market and an increasingly plausible official guarantee meant that the ultimate legacy of the Bubble for Britain was the emergence of an elastic, expandable pool of credit available for any national purpose.

This was what Defoe had glimpsed long before. Ready access to debt, Defoe noted, was his country's decisive advantage in the never-ending struggle for power: "Foreigners had been heard to say," he wrote, "that there was no getting the better of England by Battle." It was not that the inhabitants of that green and sceptered isle possessed greater bravery or martial skill. Rather, Defoe wrote, potential enemies understood "that while we had thus an inexhaustible Storehouse of Money, no superiority in the Field could be a match for this superiority of Treasure."

Defoe wrote that at the height of the South Sea spring. This doesn't mean Defoe understood and anticipated the evolution of Britain's financial system in the decades to come—he clearly did not. Still, he was a sharp enough observer to recognize great change as it unfolded, and he grasped that Britain's financial system had experienced a revolution. We can see what Defoe didn't: that the financial revolution was as transformative as the other upheavals in ideas that convulsed Europe from the late 1600s forward. The seventeenth-century scientific revolution had not been solely or even mostly a British invention, for all that Isaac Newton has come to symbolize its triumph. The various breakthroughs behind the new forms of money and credit that Britain exploited for its ends weren't either. But the cultural changes that flowed from the eruption of natural philosophers' habits of mind—the way many people came to incorporate the values of empiricism, of experiment and the importance of measurement and calculation—had a profound effect on British civic life. What animated people and events in London, that is, was the eagerness, almost urgency, to apply emerging ways of thinking to everyday human experience.

The South Sea scheme itself was no deeply reasoned application of mathematical insight. It was, rather, born of specific historical circumstances, the immediate pressures of governance and the urgency of war, power, and strife. But it emerged within an intellectual and political world in which the apparatus of calculation and the willingness to accept abstractions from material reality—replacing jangling coins with the elusive and elastic notion of "credit," for example—fed a public culture in which a vast experiment in the manipulation of money could seem plausible, even the obvious thing to try.

Yes, that experiment failed. But the longer view captures companion truths: the nation's debt was indeed transformed, and in place of its prior tangle of unmanageable obligations Britain gained the ability to conjure up money more or less at will out of little more than trust in the future. Though it would be foolish to say that London's bankers and its exchange secured the final victory in the wars of the long eighteenth century, the fact remains: a war against a French Empire that could, at times, command the resources of most of Europe could not have been fought without them.

YET FOR ALL that the market in public credit gave it a critical advantage in tests of national power, it is also true that Britain, like many first movers, failed to take the fullest advantage of its invention. The financial methods employed in its early version of a modern securities exchange could have been turned to all kinds of ends, a dizzying range of bets on the future. Instead, during the 1700s and well into the next century the British government limited who could use London's capital market. By law, only a handful of joint-stock companies could play in that sandbox, almost exclusively familiar names: the Bank of England, the East India Company, a handful of insurance companies, and a few others. For the most part, the dominant institution that had ready access to the market for debt was the Treasury itself.

This was due in part to the fevered politics at the peak of the South Sea season. The "Bubble Act," which impeded the formation of new joint-stock companies, remained in effect until 1825. While plenty of clever people managed to find other ways, often partnerships, to organize business ventures, it still blocked the majority from the most obvious source of capital for private enterprise.

Britain's banking system was similarly hobbled. Except for the Bank of England itself, with its joint-stock structure and a broad base of shareholders, every other English and Welsh bank was restricted to the partnership form. (Scottish banks operated under different rules.) Every partner in such an operation faced unlimited liability, which meant they could lose all they had should their bank fail. That clearly restricted both the amount of capital and the appetite for risk any private bank could deploy. It's important not to exaggerate the implications of such constraints; plenty of major new businesses formed under these seemingly hostile conditions. For example, Josiah Wedgwood launched his to-be-iconic company in 1759 and raised much of the cash he needed to get started the old-fashioned way: his family owned the building he used to get going; he married well—a cousin who carried with her a handsome dowry; and he was himself a canny businessman, with an enduring knack for invention that led him to new ceramics, glazes, and designs.

But while clever and well-connected entrepreneurs could and did flourish, Britain's eighteenth-century financial infrastructure was not particularly welcoming to private business. The new social technology of markets in shares and credit was too valuable to the state to share. With most of the money the Treasury raised through its domination of the apparatus of finance going to pay for war, the role of the London Exchange through its first century was clear: it was a tool of empire, not the economic life of the nation as a whole.

How important for private business was it to break that state grip on access to capital? Very, as another natural experiment demonstrated. Until Britain's North American colonies won their independence, they were subject to much of the same legal framework that home-country enterprises faced, along with additional restrictions all their own. But when the fledgling United States of America slipped those fetters in 1783, the resulting financial emancipation had an instant impact. Banks sprouted across the American landscape until by the early nineteenth century there were over three hundred across the nation, many organized as corporations, not partnerships. With that business structure, these new institutions could raise operating capital on any of the several stock exchanges in the major cities, while limiting the liability of their managers. That might seem a mundane distinction, a boring bit of business law, but it meant that US private banks in this period were prepared to take on the risk of creating credit on margin—letting out more money as loans than they held in deposits. The US in the 1820s had only half the population of Great Britain, but its bankers were able to place four times the loans into the US economy as their British counterparts provided their fellow subjects of the crown.

Similar stories played out across the financial landscape. American securities markets traded in shares that represented at least as much capital as, and quite possibly more than, that which flowed through the London Exchange, despite the difference in the population of the two nations. As a result, more or less anyone who wanted to try some new enterprise could. Data from the historian Richard Sylla show that between 1800 and 1830 the ten northeastern states alone added 3,500 such new businesses.

Most simply: American finance, more democratic in its reach than anything experienced in Britain, funded an explosion of commercial enterprise and wealth that was without precedent. In 1790, the first full year of George Washington's presidency, estimates put US GDP at $189 million (about $4.5 billion in twenty-first-century terms). Britain at the same moment could boast five times as much. By 1825 the US GDP topped $822 million, while the UK's had climbed to roughly $1.6 billion—still double that of its former colonies, obviously, but that lead was closing at a furious rate.

IT WOULD NOT be true to say that the sudden burst of economic growth experienced by the US after the revolution was simply, wholly, or even primarily driven by the financial institutions the new nation embraced, or that it was due to the different regulatory regimes operating on each side of the Atlantic. Great Britain clearly suffered some consequences from its self-imposed inability to fully exploit the ideas about money and credit pioneered in and after the South Sea Bubble, but there are more prosaic reasons why its economy may not have grown at the same speed as its former colonies did. For one thing, it's easier to expand from the smaller economic base from which the new United States began. For another: even though Britain's financial revolution allowed the state to amass huge debts to wage its wars, there was still a cost beyond the obvious one of the interest to be paid. The economic historians Peter Temin and Hans-Joachim Voth have shown that Britain's public borrowing mostly monopolized the public exchange and reduced the amount of credit available to the private customers of London's goldsmith banks. Temin and Voth note that it's hard to determine just how much such crowding out occurred. But it is a fact that even with the debts left over from the War of Independence, the US government did not need to borrow heavily in its early decades—except during the war of 1812—while its mother country continued to fight and pay for a world war. The resulting drag on the British economy partly accounts for its slower economic growth at the end of the eighteenth century, relative to its breakaway possessions.

Still, it is also true that after independence the US embraced the full range of possibilities created by the inventions pioneered across the water. US credit markets supplied funds that didn't have to pay for regiments and fleets and could instead fund canals, or railroads, or the cotton mills of Massachusetts, and on through the crucial opening moves of the industrial revolution.

This isn't just hindsight speaking. Contemporaries saw the US's economic explosion as extraordinary, a phenomenon to be explained. The most famous of these early observers, Alexis de Tocqueville, marveled at the new nation's preternatural economic energy. He came to the US in 1831 and published the results of his nine months of travel there in *Democracy in America.* He began by emphasizing how ridiculously young was this society lodged on the ocean's western margin: "The Americans arrived but as yesterday on the territory which they inhabit, and they have already changed the whole order of nature for their own advantage."

In particular, he emphasized, "No people in the world has made such rapid progress in trade and manufactures as the Americans." His tally of the accomplishments he observed reads like a draft of a Woody Guthrie song: "They have joined the Hudson to the Mississippi, and made the Atlantic Ocean communicate with the Gulf of Mexico, across a continent of more than five hundred leagues in extent which separates the two seas. The longest railroads which have been constructed up to the present time are in America." Even more than the mere size of such projects, de Tocqueville was struck by the quantity of ambition in the new nation. "But what most astonishes me in the United States, is not so much the marvelous grandeur of some undertakings, as the innumerable multitude of small ones. . . . The Americans make immense progress in productive industry, because they all devote themselves to it at once."

That judgment fit with de Tocqueville's larger theme: in his account, the United States was animated by its passionate embrace of the democratic impulse, an urge that extended, he argued, from politics to every corner of daily life. Given that conclusion, he didn't see the need to ask what, beyond national culture and habits, propelled all those eager entrepreneurs. For his purposes, the mechanisms of economic life were secondary at best. He was, after all, an observer of people, not a theorist of capital.

This is where hindsight does help: the numbers reveal how the American idea of liberty extended to the belief that finance should be as accessible to citizens as it was to the state. And those same numbers tell the rest of the story: it worked. As de Tocqueville himself attested, US economic performance by several measures was in the process of matching or bettering its European competitors within the first fifty years of the new nation's existence.

AT THE SAME time, even though he was no financial analyst, de Tocqueville was an exceptionally keen observer, and he put his finger on a crucial vulnerability within the new republic: the striking degree of interdependence within an economy bound together by promises that come due in an uncertain future. "As they are all engaged in commerce," he wrote of the inhabitants of the new country, "their commercial affairs are affected by such various and complex causes that it is impossible to foresee what difficulties may arise. As they are all more or less engaged in productive industry, at the least shock given to business all private fortunes are put in jeopardy at the same time, and the State is shaken."

This was, de Tocqueville reasoned, a necessary property of an open economy, one in which the financial health of any individual enterprise is deeply engaged with that of every other with which it does business. That led him to a warning. By the time of his travels in America, stock market crashes and other financial disruptions had occurred often enough for him to see them as inherent in the economic life of the United States and any imitators. "I believe that the return of these commercial panics is an endemic disease of the democratic nations of our age," he concluded. "It may be rendered less dangerous, but it cannot be cured; because it does not originate in accidental circumstances, but in the temperament of these nations."

Temperament is a loose term, but to put de Tocqueville's warning into a more modern form, this is an argument that the more connected an economy becomes, the more vulnerable it becomes to shocks that reverberate through the entire web of its relationships. From de Tocqueville's day to our own, such ties are woven by finance, by credit, by money in forms that embody expectations about the future—expectations that can fail.

Here's one example, beautifully documented by William Cronon in his classic work *Nature's Metropolis*. Cronon was fascinated by Chicago, which for nineteenth-century Americans and foreign visitors alike was the exemplar of a new kind of commercial city. The most famous of its new institutions was the Board of Trade, which opened for business on April 3, 1848, barely fifteen years removed from the time when the city had been a mere village with perhaps two hundred inhabitants, mostly French or Potawatomi. Initially more of a booster organization, by the mid-1850s the Board of Trade had become a market in futures, a specific kind of derivative. Here the abstractions of money move yet further away from any material chunk of reality. Investors could buy grain (and later, other commodities) that hadn't even yet been planted.

The Chicago Board of Trade moved to its current location in 1885, into what was then the tallest building in the city.

Such commodities futures weren't loans. They were payments for goods that did not yet exist, embodying judgments on a harvest to come. Chicago's futures market turned vast tracts of land, their produce, and the labor required to cultivate them into simple sets of numbers: prices agreed for a bushel of red winter wheat, say, to be delivered at some defined date in the months to come. A vast real-world system existed to handle the physical commodity represented on these slips of paper: railroads to carry grain from the field to Chicago and on to end users, steam-powered grain elevators to classify and store the crop, and so on. But the system that could support farmers between harvests, the elevators as they held and shipped grain and so on, turned on the ability to represent all that work as a simple, anonymous, and completely interchangeable security, a marketable piece of paper that enabled people to both fund and bet on events to come. That is: futures can certainly be used to gamble, to lay a wager on a good crop or a bad. But they are first and fundamentally a way to capture the value of labor and resources in play over time in a way that allows the present to keep going long enough for human goals to materialize.

WHEN FUTURES AND the rest of the modern forms of finance do that work, they allow humankind to express both risk and expectation in increasingly flexible and potent ways. In doing so, they support human endeavor by marshaling the resources and ambitions of the future in our service—all of us living in the present moment. Most important, credit in all its forms is not a one-to-one transfer of wealth from a distant moment to today; our future selves and heirs don't necessarily lose when we ask them to pay for a desire in the here and now. Rather, there's a kind of multiplication that takes place: "I will gladly pay you Tuesday for a hamburger today" gets me dinner tonight, while my promise to repay the loan (with interest!) creates a note that has a value (a price) of its own that can be sold to someone else (for enough to cover a milkshake, perhaps).

Thus, in cartoon form (literally: thanks, Wimpy!), a partial explanation for the haste and energy and above all the explosion of wealth de Tocqueville recognized on his travels. The financial liberty the Americans claimed for themselves and the eagerness they displayed enabled them to launch new projects at an incredible rate. Over time, the impact of the role of finance in the economic emergence of the US extended well beyond its borders. Great Britain shed many of its restrictive financial rules in the 1820s and by some measures soon possessed a financial system that was more effective at delivering capital to business than the American one, especially after Andrew Jackson led his populist charge against the Bank of the United States.

But even with such local retreats, the fundamental ideas of large-scale financial capitalism spread, with dramatic effect. The emergence of the institutions, laws, and intellectual apparatus needed to calculate the value of a possible future in this present—and then to securely, legally, trade on those representations of human actions over time—paid for the nearly continuous industrial revolution whose third or fourth wave bears us forward today. Steam power, transport, communication, machinery, a world now webbed with networks of airplanes and of bytes, the chemical knowledge that feeds billions, and all the rest of the multiple industrial and intellectual revolutions have by now allowed an unprecedented proportion of humanity to live at a level of well-being undreamt of for most of history. All our experience, everything we possess in the world, has been advanced by the creation of formal tools to trade promises, to bet on the success of plans that may not deliver (if they ever do) for years to come. We literally bank on the future. The concepts that allow us to do so, in turn, all derive from the core idea behind the financial revolution—the use of numbers in combination with analysis to allow abstract reasoning about the diversity of actual stuff in the world.

The catalog of financial tools is still evolving, of course—some of the time to good effect. For just one example: by the 1990s, it became possible to trade on the future value of the song "Rebel Rebel." That particular bit of financial engineering became known as the Bowie Bond, in which for the first time musically inclined (or celebrity-dazzled) punters could buy and sell shares of a pop star's future earnings. Selling shares in a bundle of royalties to come from his songs brought a tidy $25 million to the man behind Ziggy Stardust. That was exactly the result financial innovation is supposed to deliver: Bowie got cash in hand to use as he saw fit, while buyers of his bonds got a stream of income extending into the future, and, perhaps, an extra, unquantifiable happy jolt every time Major Tom spoke on their radio, floating in his most peculiar way. Isaac Newton and Robert Walpole might have been amazed by the reality of David Bowie, but they would have had no problem with the thinking behind selling a stream of future earnings. Turning the value of a song over time into cash is, after all, exactly the same transformation as the one that converted a lottery ticket or an annuity into a share that could be sold up and down Exchange Alley.

Still, Pangloss is no more persuasive as an economic analyst than as an observer of social life. It is absolutely true that increasingly complex and mathematically intricate financial innovation has, for example, made it cheaper to buy a house or drive off in a new car. But as de Tocqueville feared almost two centuries ago, the modern, utterly interconnected and increasingly powerful financial system is not always the best of all possible worlds. Here again, the South Sea year has some lessons to teach about the risks of financial innovation. The Bubble may be past, that is, but its implications are not, as the world would be reminded, brutally, in the reverberations of a truly wretched Monday morning in September 2008.

EPILOGUE
"an endemic disease"

September 18, 2008, dawned clear over Manhattan. The thermometer touched eighty degrees by afternoon, and the evening was still warm enough to enjoy a drink outside at a sidewalk cafe. For most of those out and about in the city, it was a fine end-of-summer day.

At 745 Seventh Avenue, no such pleasure pierced the misery. In its 150 years in business, Lehman Brothers had evolved from a dry goods store that brokered raw cotton on the side into an investment bank with global reach. By the 1990s, Lehman was in the first rank of world financial institutions. All that history shattered in the space of a single day.

LEHMAN COLLAPSED BECAUSE of a lineal descendant of the financial maneuvers of the South Sea Bubble. The furor of 1720 followed the decision to turn a bunch of individual loans, lotteries, annuities, and other obligations into shares in the South Sea Company. That menagerie of debts became a single type of security, ultimately a bond; in modern parlance, a passel of distinct obligations had been securitized. The advantages Britain later gained from the Bubble turned on this action—that transformation of a mountain of illiquid assets into a financial instrument that could be traded on an exchange—which then gave the British government access to all the wealth its subjects chose to put to work in London's money markets. The same securitization trick could be used on many other types of financial instruments and, in the late twentieth century, Lehman and other major institutions did just that: assembling thousands of loans of different types—on cars, or credit card balances, or, most important for this story, home mortgages—into bundles, in which shares could be bought and sold.

Such modern securitization produced the same benefits that flowed from remaking Britain's national debt into bonds. Before mortgages were routinely bundled into these enormous bond-like investments, for example, the local bank that would lend you money to buy a house could offer mortgages only up to a ceiling dictated by its own reserves. But combining those loans into one big pile that could be parceled into shares and sold far beyond city limits opened up the US home market to a literally worldwide pool of capital. As Britain had experienced with its consols, such a vastly enlarged supply of resources made renting money—borrowing to buy a house—easier and cheaper. This was, again, the miracle of finance: a win for those looking for (seemingly) safe returns to put their money to work, and equally a victory for borrowers, now paying less to get their hands on sums large enough to fund major purchases.

And yet . . . as the Bubble reminds us, such seemingly effortless wealth is not truly risk-free. It wasn't obvious at the start of this latest wave of securitization (in the housing market, the new instruments were called mortgage-backed securities, or MBSs), but the mathematical complexity of the new vehicles obscured their vulnerabilities. When those flaws revealed themselves, the result would make the South Sea year seem a mere trip to the beach. The weaknesses of this latest example of securitization have been extensively dissected in the aftermath of the 2008 crisis, but collectively they all derived from one or both of two major pathologies. For one, MBSs claimed to be diverse enough to eliminate much of the risk of what would happen if and when US housing prices stopped climbing. They were not, in particular because if the housing market stalled (in a clear echo of the South Sea Bubble's behavior in September 1720) home buyers would be unable to flip themselves—sell their houses—out of mortgages they could not sustain. Worse, MBSs spawned a range of other, related financial instruments that created an almost inconceivable amount of leverage in global financial markets. These were devices that enabled the same kind of speculation that brought down the Duke of Portland: derivatives, ways of placing larger and larger bets with less and less money up front, which vastly multiplied the stakes if the market ever refused to behave as expected.

By the twenty-first century, the simple derivatives available in 1720—put and call (sell and buy) options, futures contracts, and a few others—had blossomed into enormously more complicated forms. There are derivatives based on interest rates. There are derivatives that dissect MBSs themselves into securities that contain just the interest payments from the underlying loans, or just the principal portion, or another of the seemingly infinite ways to divide rivers of money as they run from one place to another. It's now possible to trade on the volatility of an individual instrument or a market as a whole—making bets on how rapidly and how much the price of an underlying asset may move. There are trades on the volatility of such volatility, which are bets on how rapidly the curve of changing prices alters its slope.

That's a third- or fourth-order derivative—and it was here that the securitization wave of the 2000s turned itself into the Bubble on steroids. The luckless Portland doomed himself through leverage, seeking control of much more South Sea stock than he could afford. In the growing enthusiasm for the elaborate new ways to place secondary and tertiary bets on the risk and return of fragments of the flow of money associated with housing, leverage in the first years of our century multiplied far beyond anything experienced in 1720.

LEHMAN *LOVED* LEVERAGE. By 2008 the bank owned a lot of the same MBSs it had sold to others, and it had acquired its portfolio by borrowing to build a portfolio of $680 billion—thirty times Lehman's actual capital reserve of just $23 billion. That meant that Lehman had put up as little as three cents of its own money to control each dollar's worth of the MBSs it had piled onto its books. The reward for such boldness was clear: a three-cent gain on Lehman's notional dollar stake in an MBS would be just a minor advance for the underlying security but would reward Lehman with a 100 percent return on its true investment.

But what if the market price of that MBS slid just a little? A 10 percent drop in the value of its portfolio would wipe them out and leave them short another $55 billion to settle with their counterparties. Worse: if any of the exotic and mathematically complicated flavors of MBSs Lehman possessed didn't drop but instead failed completely, falling to zero, then the bank would be on the hook for the full disaster, a thirty-fold loss on its original investment. Just as in 1720, leverage is your best friend when the market is kind—and a killer when it is not.

We all know what happened next. The American housing market began to waver in the mid-2000s, then dropped during 2008. Lehman tried to wiggle out of its trap. A multibillion-dollar loss in the second quarter was followed by an announcement on Wednesday, September 10, that another $3.9 billion was gone. Lehman scrabbled to find a partner, but rumors of a takeover fell through, and Lehman stock, already down by over 75 percent since the start of the year, plunged yet further. On Friday, Tim Geithner, later President Obama's Treasury secretary, then the president of the New York Federal Reserve Bank, convened a meeting of every major Wall Street bank to try to craft a rescue for a bank facing debts hundreds of times greater than its assets.

Those bankers stayed in the room for the weekend. Echoing the South Sea Company's desperation moves in September 1720, Lehman tried to sell itself to a presumably more solvent partner, but the Bank of America backed out, and then London-based Barclay's walked away when British financial authorities forbade it to climb on board a trainwreck in progress. By Sunday afternoon, the inevitable became fact. The lawyers started to prepare the paperwork.

Lehman's filed for bankruptcy the next morning, before US financial markets opened.

The crash that followed the news wiped out almost 5 percent of the value on the New York Stock Exchange in a single day. Unlike prior market scares over the preceding few years, there was no rebound. The Great Recession—the worst global financial downturn since 1929—had just announced itself.

Lehman's death did not cause the Great Recession. An extensive scholarly literature has examined the different choices and mistakes that produced a financial system so interconnected—as de Tocqueville had warned—and so overleveraged (the missed message of the Bubble) that the collapse of a single bank could evoke such wider ruin. But a common thread runs through the analysis of the disaster. The elaborate and complex web of financial abstractions produced by securitization and its derivatives had created a mathematically opaque tower of numbers and interdependent promises that obscured just how much risk supposedly prudent investors had accepted.*

AS IN THE South Sea Bubble, there was nothing abstract about the Great Recession. It caused terrible pain to untold numbers of individuals and families, lives wounded by lost homes and drained savings. It slashed the wealth of nations: the US economy fell by over 4 percent in the immediate aftermath of Lehman's failure and didn't

* This increasing complexity also helped fuel a decades-long trend toward greater income inequality in the United States and other advanced nations, as those in a handful of sectors, including finance, managed to capture more and more of the profits of whole economies.

get back to precrisis levels of employment and production for years. Millions lost jobs, and, as always, such widespread misery cost lives: suicide rates rose in parallel with the unemployment rate in the US and through much of Europe.

The term *financial engineering* has become something of a cliché. The phrase does illuminate an important historical thread, though, allowing us to understand what began in London between the 1690s and 1750s as the birth of a type of technology. As this book has argued, such ideas have played a profoundly important role in furthering human ambition, from their first deployment that underpinned Britain's eighteenth-century pursuit of national power, and subsequently within the economic life of every form of human endeavor.

But as the Great Recession and all its exacted treasure and blood remind us, the Bubble's history is not simply past. The sound and fury of the South Sea year echo still. The financial inventions that failed in the 2000s are enormously more complicated than those London confronted in 1720. But crucial similarities remain—with one great difference. Then, the idea that money and its elaborations are mathematical objects was brand-new and barely understood outside a tiny circle. Now that same thought is a commonplace.

De Tocqueville may have been right in his diagnosis: the temperament of nations, really, of humanity as a species, may make it inevitable that "the return of these commercial panics is an endemic disease of the democratic nations of our age." Still, since that earlier crash, we now know that all significant financial innovation carries with it the near certainty that ignorance and confusion about any new investment trick, combined with the fact that human emotion drives decisions about money, create conditions in which the behavior that leads to disaster recurs in broadly predictable sequences.

It's true that it's difficult to determine in advance the precise trigger in each case, and harder still to fix the timing of the eventual wreck. But the fact that there is a pattern in financial failure means that, even if we cannot anticipate exactly what will come, seeing the signs of such events should make it possible to render the broadly predictable consequences of human nature less dangerous.

To put it plainly: this is why we need to regulate financial markets, especially to reduce the risks that expanding leverage imposes—and, for those who wish to gamble to that extent, to fence them in so that, if someone goes all in on black at the roulette table, the pain they feel when the ball lands on red stays where it belongs.

In the infancy of British financial capitalism under Walpole and his successors any such constraints were simple and were generally blunt-force affairs: limiting the number of joint-stock companies and restricting any private market in bonds. But as financial markets evolved, embracing more and more of human economic activity, the need to establish rules has become more urgent, even as it has become more complicated to do so. So there are indeed difficult arguments to be had about *how* to reduce the risk inherent in modern financial engineering; how to make the system less brittle, less fragile, less likely to produce the kind of global pain imposed by the failures of the 2000s.

But *hard* does not mean *impossible.* We do know the broad outlines of what needs to be done. At a minimum, that includes demanding that those who enter into such markets do so with a clear and discoverable relationship between the capital they possess and the amount of money they seek to put at risk. There's much more to require, of course, but an extended technical investigation of modern financial regulation doesn't belong at the end of a book about events that roiled a corner of London so long ago. The point is that we now know that a probabilistic risk of destructive crashes follows from each significant new wave of financial engineering. Such dangers have been amplified in the last several decades by the increase in the scale of modern finance, in the speed with which, thanks to computers and digital communication, money can now flow around the world—all that combined with the mathematical complexity through which the kind of leverage that sank Lehman works its way throughout the system. We can't predict the day or week or even the year when the ax will fall. But still, know that the blow will come.

It is that knowledge that makes what's happening as I write these last lines so frightening. In the immediate wake of the Great Recession, a handful of very modest measures were adopted, requiring certain caution in banks, among other restrictions. The scale of reform was almost negligible compared with what followed the crash of '29 and the Great Depression. But after the US election of 2016, many of even these weak safeguards have been dismantled, while leverage in financial markets is once again on the rise.

As ever, unsustainable financial adventures are much more easily recognized in hindsight than in advance, and it is impossible (for me, certainly) to say when the next crisis will occur. But diagnosing vulnerability is much easier—and the makings of the next disaster are with us as I write these lines.

Thus the last lesson of the South Sea Bubble: at some point, ignorance cannot be a defense. Each error does not need to be made again and again. In 1720, what happened was a true experiment, something unfamiliar, as yet unexperienced. Newton and his peers may be forgiven for failing to anticipate a species of catastrophe that was brand-new.

We cannot claim that excuse.

ACKNOWLEDGMENTS

This work spent a long time in gestation, which means that its arrival has been attended by the kindness and smarts of a great number of people.

Topping the list are the editors who have given so much attention and care to making this book as sharp and elegant as it could be. At Random House, Sam Nicholson acquired and shepherded the project to its first full draft. Molly Turpin provided a generous, rigorous, and acute edit to that first draft and each succeeding one. Both Sam and Molly are editors whose intelligence and taste made the work better. Thanks as well to my copy editor, Elisabeth Magnus, who repeatedly saved me from myself, and to the book's designer, Edwin Vazquez, whose fine work you hold in your hands.

This book and, more broadly, the way I think about the task of writing about ideas grounded in moments and places have benefited from a conversation that's gone on for a decade now with my editor at Head of Zeus, Neil Belton. Every writer should have a mentor-friend-guide like Neil in their corner.

My agent, Eric Lupfer, has been the other constant. He shaped a raw idea into a very persuasive proposal (I wanted to read the imagined book after he got done with it!) and has offered just the right balance of sharp critique and unstinting support all the way down the long road that led to here.

Several scholars and friends helped provoke this book. Cambridge University historian of science S. J. Schaffer gave me early encouragement, as did UC Berkeley's Brad DeLong. Ann Harris, the editor everyone should wish they had, and mine for *Einstein in Berlin,* loved the idea that became *Money for Nothing,* and even if she had the infinite gall to retire before she could work with me on this one, she gave me all the encouragement one could seek. Peter Galison has been an ear and adviser to my projects for many years, and I greatly appreciate his help on this one (and those to come . . .).

The John Simon Guggenheim Foundation's fellowship award in 2016 was invaluable, the gift of time that allowed me to produce much of the first draft of the manuscript. My thanks to everyone at the foundation, and to my friends and colleagues David George Haskell and David Kaiser, who helped make it happen. I received additional support from my home institution of MIT, both in the leave granted to make full use of the fellowship and in a Levitan Prize to fund my research. My thanks to MIT's provost, Martin Schmidt, and to the dean of the School of Humanities, Arts, and Social Sciences, Melissa Nobles, for that support. Thanks as well to Harvard's Department of the History of Science and to its then-chair, Janet Browne, for making me welcome on my writing leave from MIT.

Money for Nothing is lucky enough to be aided by three excellent research assistants, two of them former students. Christine Couch dug up source after source, helping dive deep into the glorious world of print in the late seventeenth and early eighteenth centuries. Cara Giaimo got me going on a fact-and-sense check of the edited manuscript, and Katherine Sypher magnificently finished that job.

Next, in the chronology of the project, come those who read the manuscript in whole or in part, at various stages of composition. Science writer extraordinaire David Bodanis went first, and gave me great advice. My MIT colleague William Deringer both shared his wonderful work on exactly this period as it was in process and offered a critical reality check on the sections where our interests overlapped—truly a model of scholarly generosity. Ian Preston of the University College London Department of Economics gave me a careful read with a view to intercepting economic howlers before they reached print. Later in the rewriting process, my MIT colleagues Ellen Harris, who knows more about Handel's foray into London's stock exchange than anyone else, and economic historian Anne E. C. McCants both gave the book close and careful reads, raising valuable questions and, again, saving it from a number of errors. Physicist Sean Carroll helped me work through some of Newton's mathematical innovations. Historian of science Thony Christie chimed in with a last-minute read and catch in a critical passage. As always—because it's always true—every mistake that remains to be revealed is mine and mine alone.

As the years of writing played out, so many of my writing and science colleagues listened to me talk about the book that really, truly was going to emerge someday, and helped me stay the course. In no particular order, I thank Susan Faludi, Russ Rymer, Jennifer Ouellette, Carl Zimmer, Annalee Newitz, Kelly Roney, Martin Finucane, Kate Wooderd, Jon and Bonny Eckstein, Michael Kosowsky, Jennifer Drawbridge, Brett Oberman, Steve Silberman, Veronique Greenwood, Maryn McKenna, Lisa Randall, Jason Pontin, Robert, Toni, and Matt Strassler, John Timmer, Rebecca Saxe, Allan Adams, Melody Meozzi, John Rubin, and Nancy Kanwisher. My MIT home roost of the Comparative Media Studies/Writing program has been a great nest within which to work; I thank all my colleagues there, noting in particular the support offered by Ed Schiappa, my department head during most of the writing of the manuscript. Thanks as well to my students, both for their explicit encouragement and what I've learned about writing from engaging with their work. And a special shout-out and deep thanks to the great writers, teachers, and staff with whom I work most closely in MIT's Graduate Program in Science Writing: Marcia Bartusiak, Deborah Blum, Tom De Chant,

Alan Lightman, Seth Mnookin, and Shannon Larkin, all of whom, at vital points (read: sloughs of despond), reached in and picked me up.

Almost finally: No thanks are sufficient to pay back the love and support I've received from my extended family over the years it took to complete this book. The English cohort put me up on research trip after trip *and* listened to each new snippet of discovery with (or so it seemed to me) undiminished interest. Lucinda and Kate Sebag-Montefiore, Robert Dye and Adam Brett, Adam and Caroline Raphael, Geoffrey Gestetner, and Simon Sebag-Montefiore: my thanks and my love—with extra gratitude to my aunt, Juliet Sebag-Montefiore, whose absence is such a loss. My in-laws, John and Kricket Seidman, Judy Seidman, Gay Seidman and Heinz Krug, and Neva Seidman and Zeph Makgetla were all equally generous throughout the journey. Y'all are great.

And really last: My siblings are pillars in my life, and they and their families have seen me through this as they have through so much. Richard, Irene and Leo, and Rebecca and Jan—not to mention Joe, Max, Emily, and Eva—words can't express my gratitude. As for my wife, Katha Seidman, and my son, Henry: I wake each day astounded at the good fortune that brought you both into my timeline. You not only made this (and all my work) possible; you give richness and savor to the days we spend together.

Notes

Introduction

xii WRITTEN IN THE VOICE OF "A JOBBER" Daniel Defoe, *The Anatomy of Exchange-Alley; or, A System of Stock-Jobbing* (London: E. Smith, 1719), p. 3.

xii "THE GETTING OF TREASURES" Proverbs 21:6. The quoted verse comes from the King James Version, which is to say the one familiar to Defoe and his contemporaries.

xiii READY, AS OCCASION OFFERS Defoe, *Anatomy of Exchange-Alley,* p. 39.

xiii TREACHERY'S NEAREST COUSIN Defoe, *Anatomy of Exchange-Alley,* p. 28.

Part I

1 NATURAL PHILOSOPHY CONSISTS Isaac Newton, "Scheme" (undated), quoted in David Brewster, *Memoirs of the Life, Writings and Discoveries of Sir Isaac Newton,* vol. 1 (Edinburgh: Thomas Constable, 1855), p. 102.

Chapter 1

3 HE HAD WALKED FOR THREE DAYS Richard Westfall consulted Newton's personal accounts to reconstruct that three-day walk to

Cambridge. See Richard Westfall, *Never at Rest* (Cambridge: Cambridge University Press, 1980), p. 66.

3 CAMBRIDGE HAD BEEN EMPTYING Evelyn Lord's work with the Bills of Mortality for Cambridge yields a population estimate for Cambridge of about 7,500. See Evelyn Lord, *The Great Plague: A People's History* (New Haven, CT: Yale University Press, 2014), p. 128.

4 PEPYS FIRST TOOK NOTE OF THE DANGER Samuel Pepys, *Diary,* October 19, 1663, http://www.pepysdiary.com/diary/1663/10/19/, and November 26, 1663, http://www.pepysdiary.com/diary/1663/11/26/.

4 A FEW CASES WERE REPORTED A. Lloyd Moote and Dorothy C. Moote, *The Great Plague: The Story of London's Most Deadly Year* (Baltimore: Johns Hopkins University Press, 2004), p. 19. This story follows their account.

4 TWO MORE PLAGUE DEATHS John Bell, *London's Remembrancer; or, A True Accompt of Every Particular Weeks Christnings and Mortality in All the Years of Pestilence Within the Cognizance of the Bills of Mortality, Being XVII Years* (London: E Cotes, 1665).

4 BY YEAR'S END Tally in Moote and Moote, *Great Plague,* p. 295.

5 HE SET OUT ACROSS THE CITY TO CRIPPLEGATE Samuel Pepys, *Diary,* June 21, 1665, http://www.pepysdiary.com/diary/1665/06/21/, and June 22, 1665, http://www.pepysdiary.com/diary/1665/06/22/. The incident at Cripplegate—and Pepys's dalliance with the wife of the tavern keeper at the Cross Keys—is an oft-published anecdote; I first encountered it in Evelyn Lord's *The Great Plague,* p. 50.

5 SOME TOWNS BARRED THEIR GATES Lord, *Great Plague,* p. 51.

5 IN CAMBRIDGE, THE BLOW FELL ON JULY 25 Lord, *Great Plague,* p. 1.

6 "THE PRIME OF MY AGE" Westfall, *Never at Rest,* p. 143.

6 HE HAD TO WORK OUT FUNDAMENTAL CONCEPTS See, e.g., Frank Wilczek, "Whence the Force of F = ma?" *Physics Today* 57, no. 10 (October 2004): 11, https://doi.org/10.1063/1.1825251.

6 THE PROGRAM FOR THE WORK TO COME Westfall, *Never at Rest,* p. 114.

7 AN EXTRAORDINARY SERIES OF FORTY-FIVE QUERIES Westfall, *Never at Rest,* pp. 88–90.

7 HE WAS "CONSISTENTLY CONCERNED" Westfall, *Never at Rest,* p. 109.

7 ONE OF THE FIRST PROBLEMS TO CATCH HIS EYE The account of the evolution of Newton's mathematical ideas that follow draws on three principal sources: D. T. Whiteside's magnum opus, *The Mathematical Papers of Isaac Newton, vol. 1, 1664–1666* (Cambridge: Cambridge University Press, 2008); Richard Westfall's *Never at Rest,* especially chapter 4; and James Gleick's *Isaac Newton* (New York: Random House, 2003), chapter 3.

8 "THE EQUATION IS MORE BASIC THAN THE CURVE" Westfall, *Never at Rest,* p. 107.

10 THIS RESULT ALLOWED NEWTON David Goss, "The Ongoing Binomial Revolution," arXiv.org, May 18, 2011, arXiv:1105.3513v1.

11 THE GREEK PHILOSOPHER ZENO See Max Black, "Achilles and the Tortoise," *Analysis* 11 no. 5 (March 51): 91–101, reprinted in *Zeno's Paradoxes,* ed. Wesley C. Salmon (Indianapolis: Hackett, 1970), pp. 67–81, https://books.google.com/books?id=0AzP9WLLJLcC&pg=PA67&dq=zeno+achilles&hl=en&sa=X&ved=0ahUKEwjFtcvXnurhAhVIwlkKHXQ3CocQ6AEINjAC#v=onepage&q=zeno%20achilles&f=false.

11 BUT NEITHER PHILOSOPHICAL RIGOR NOR COMMON SENSE See Gleick, *Isaac Newton,* p. 42.

11 GALILEO KNEW THAT THERE WAS SOMETHING VITAL Quoted in Gleick, *Isaac Newton,* p. 41.

11 NEWTON HIMSELF, IN HIS FIRST MONTHS AT WOOLSTHORPE Westfall, *Never at Rest,* p. 131.

11 "THE CROOKEDNESS IN LINES" Isaac Newton, "Newton's Waste Book," http://www.newtonproject.ox.ac.uk/view/texts/normalized/NATP00221.

12 BUT NEWTON'S THINKING IN THE LAST MONTHS OF 1665 The in-

terplay between mechanical constructions of curves and Newton's analytic approach is a gloss on Richard Westfall's more technical exposition in *Never at Rest,* pp. 126–34.

12 THE TRICKS THAT SCHOOLMASTERS USED Westfall, *Never at Rest,* p. 132.

12 THE "GENERATION OF FIGURES BY MOTION" Isaac Newton, in a document written as part of the conflict with Gottfried Leibniz on who first invented calculus, quoted in Westfall, *Never at Rest,* 55n to chapter 4.

13 THE "INFINITELY LITTLE LINES" Isaac Newton, *Mathematical Papers of Isaac Newton, vol. 1, 1664–1666,* D. T. Whiteside, ed. (Cambridge: Cambridge University Press, 2008), p. 382 et seq.

14 AS HE TOLD THE STORY SIXTY YEARS LATER William Stukeley, *Memoirs of Sir Isaac Newton's Life* (1752), MS/142, Royal Society Library, London, p. 15r.

14 SUDDENLY, AN APPLE FELL No anecdotes about an apple survive from near the time Newton claims to have been inspired by its fall. Newton did tell the story to a few friends in the last year of his life. John Conduitt, his nephew by marriage, described the incident after a conversation with Newton in 1726. William Stukeley, another longtime acquaintance, described a similar conversation from the same year, among a handful of other accounts. See D. McKie and G. R. de Beer, "Newton's Apple," *Notes and Records of the Royal Society of London* 9, no. 1 (October 1951): 46–54.

14 IN THE MOMENT, SO THE STORY GOES Stukeley, *Memoirs of Sir Isaac Newton's Life,* p. 15r with an insert from p. 14v.

14 THIS MUCH IS TRUE See, for example, this account of one of the Woolsthorpe apple tree's descendants at York University: Richard Keesing, "A Brief History of Isaac Newton's Apple Tree," University of York Department of Physics, n.d., http://www.york.ac.uk/physics/about/newtonsappletree/. My home institution, MIT, also possesses a clone. Liz Karagianis, "Newton's Apple Tree Bears Fruit at MIT," *MIT News,* October 4, 2006, http://news.mit.edu/2006/newtons-apple-tree-bears-fruit-mit.

14 BUT EVEN IF NEWTON WATCHED THE APPLE FALL The story of the birth of Newton's gravitational theory is one of the most often told tales in the history of science. For this brief treatment of it, I depend heavily on Westfall, *Never at Rest,* pp. 148–55, to which I refer readers for a more detailed account of the specific steps in Newton's readings. I also consulted James Gleick's treatment of the same material in *Isaac Newton,* pp. 54–59. My *Newton and the Counterfeiter* (New York: Houghton Mifflin Harcourt, 2009) also discusses these events on pp. 15–20.

15 HE "FOUND THEM ANSWER PRETTY NEARLY" Isaac Newton, draft of a letter to Pierre des Maiseaux, probably in the summer of 1718, quoted in I. Bernard Cohen, *Introduction to Newton's Principia* (Cambridge, MA: Harvard University Press, 1971). Newton produced several drafts of this letter, many of which are cited in the notes to the version in *The Correspondence of Isaac Newton, vol. 6, 1713–1718* (Cambridge: Cambridge University Press, 1975), document 1295 pp. 454–562. See also D. T. Whiteside, "The Prehistory of the *Principia* from 1664 to 1686," *Royal Society Journal of the History of Science* 45, no. 1 (January 1991): 14–15, for a discussion of Newton's analysis of the motion of a pendulum.

16 SITTING THERE, AN OBJECT WOULD FALL FOREVER This is not quite accurate; objects under the influence of gravity travel around the center of mass of the total system, not just of the more massive body, as Newton in fact understood. More deeply, at this time Newton did not have a clear understanding of the concept of inertia, nor yet his first law of motion, which holds that objects at rest or in linear motion tend to stay in motion or at rest unless acted upon by an exterior force. Without this fundamental idea, first suggested to Newton clearly by Robert Hooke in a letter in 1679, his conception of gravity remained imprecise. See Westfall, *Never at Rest,* pp. 382–88, for a discussion both of the insight and the conflict between Hooke and Newton that followed. See also S. Chandrasekhar's summary of the sequence of development of Newton's thinking in the first chapter of his *Newton's Principia for the Com-*

mon Reader (Oxford: Clarendon Press, 1995), pp. 1–14. (However, please note: I. Bernard Cohen, one of the great Newton scholars of the twentieth century and the translator of the best currently available English version of the *Principia,* does not think highly of Chandrasekhar's historical skills, and it is true, as Cohen says, that the "common reader" of Chandrasekhar's title had better know a lot of mathematics to make it through most of the argument. Nonetheless, Chandrasekhar, a Nobel laureate physicist, does offer a good introductory summary of the basic concepts in the first section of his book, and it is worth a look.) Another good account of the development of Newton's thoughts on gravity through this period comes in A. Rupert Hall's highly readable biography *Isaac Newton: Adventurer in Thought* (Cambridge: Cambridge University Press, 1996), pp. 58–63.

17 TO FIND OUT, HE TURNED TO THE NEAREST EXPERIMENTAL SUBJECT Newton's notebook, reproduced in Westfall, *Never at Rest,* p. 95.

18 WHAT HE CALLED THE *EXPERIMENTUM CRUCIS* For a detailed reconstruction of the steps Newton took to arrive at his theory of color, see Westfall, *Never at Rest,* pp. 156–72. As Westfall documents, Newton's journey to the crucial experiment was more involved than the legend has it (and Blake's famous painting depicts)—and the demonstration that a spectrum can be recombined into white light passed through several different tests at Newton's hand. In other words: the "experimentum cruces" were actually several experiments.

18 A CAREFUL, LOGICALLY COHERENT APPROACH See Stephen Gaukroger, "Empiricism as a Development of Experimental Natural Philosophy," in *Newton and Empiricism,* Zvi Biener and Eric Schliesser, eds. (Oxford: Oxford University Press, 2014), p. 15. Gaukroger makes a powerful case for the systematic way in which Newton developed a commitment to empiricism, and much of the argument that follows is influenced by his argument.

18 A "SYSTEM OF THE WORLD" *The System of the World* is the third book of Newton's masterwork, the *Principia.* In it, he shows how the

three laws of motion and universal gravitation could account for the motion of all the known bodies of the solar system, the flight of comets, and the mechanism of the tides—a complete cosmology based on a handful of the mathematically expressed axioms.

18 NEWTON RETURNED TO HIS ROOMS Westfall, *Never at Rest,* p. 142.

18 ON WEDNESDAY, JUNE 6 Lord, *Great Plague,* pp. 88–89.

19 LONDON'S PARISH RECORDS "General Bill for This Present Year, Ending the 19 of December 1665 [. . .]," reproduced in Moote and Moote, *Great Plague,* p. 260.

19 THE NEW ST. PAUL'S Wren managed to construct one of the largest domes built to that time by using a new trick. The dome seen on the skyline is not structural; it is instead held up by a masonry cone, within which stands a decorative interior domed ceiling.

19 "A VERY ODD MONSTROUS CALF" Robert Boyle, "An Account of a Very Odd Monstrous Calf," *Philosophical Transactions* 1, no. 1 (May 30, 1665): 10; Silas Taylor, "Of the Way of Killing Rattle-Snakes," *Philosophical Transactions* 1, no. 3 (May 30, 1665): 43; Robert Hooke, "A Spot in One of the Belts of Jupiter," *Philosophical Transactions* 1, no. 1 (May 30, 1665): 3; Robert Boyle, "General Heads for a Natural History of a Countrey, Great or Small" *Philosophical Transactions* 1, no. 11 (May 30, 1667): 186–89.

20 HE PURSUED LINES OF INQUIRY THAT HE ACTIVELY WANTED TO KEEP SECRET Newton's alchemical work was long an embarrassment to at least some Newtonians. There was a lot of it: about one million words out of the ten million that survive in his own handwriting are devoted to alchemical topics; much of that is his copies of alchemical treatises with his notes or commentary. Many of these had been labeled "not fit to be printed" at his death. (The Chymistry of Isaac Newton, "About Isaac Newton and Alchemy," n.d., https://webapp1.dlib.indiana.edu/newton/project/about.do). In 1872 the Cambridge University library rejected a proposed donation of much of his alchemical archive as "of very little interest in themselves"; these papers were among those. (The Newton Project, "The Portsmouth Papers," n.d., http://www.newtonproject.ox.ac.uk/history-of-newtons-papers/portsmouth-papers). They would

become more publicly available after the economist John Maynard Keynes bought them at an auction in 1936. Still, it wasn't until the 1970s that significant scholarly interest revealed the critical role alchemy played in much of Newton's thinking, including his ideas about gravitation and light. The person who had the greatest role in doing so was the historian of science Betty Jo Teeter Dobbs. Her *The Janus Faces of Genius* (Cambridge: Cambridge University Press, 1991) is a good synoptic view of her work; a briefer introduction to her ideas can be read in her "Newton's Alchemy and His Theory of Matter," *Isis* 73, no. 4 (December 1982): 511–28. Two good introductory articles on Newton's alchemical thinking and its connections to different strands in his thinking—"The Background to Newton's Chymistry" by William Newman and "Newton's Alchemy" by Karen Figala—can be found in I. Bernard Cohen and George E. Smith, eds., *Cambridge Companion to Newton* (Cambridge: Cambridge University Press, 2002). Newton's alchemical papers themselves and a significant amount of explanatory material are online at The Chymistry of Isaac Newton (https://webapp1.dlib.indiana.edu/newton/), part of the international collaboration to digitize and study Newton's papers.

20 A DECADE LATER, HE TURNED HIS QUANTITATIVE VIRTUOSITY Westfall, *Never at Rest,* pp. 103 and 118–19.

CHAPTER 2

22 WILLIAM PETTY LIVED THE CLICHÉ I am indebted to Ted McCormick's *William Petty and the Ambitions of Political Arithmetic* (Oxford: Oxford University Press, 2009) for the biographical sketch that follows. McCormick's is the first significant intellectual biography of Petty to appear in the scholarly literature that reaches beyond examinations of specific aspects of his amazingly varied life and intellectual pursuits; anyone seeking a more comprehensive account of an unjustly neglected life should begin with it.

23 HE LEFT HIS LOCAL SCHOOL AT THIRTEEN John Aubrey, *"Brief Lives," Chiefly of Contemporaries, Set Down by John Aubrey, Between*

the Years 1669 & 1696, vol. 2, edited by Andrew Clark (Oxford: Clarendon Press, 1898), p. 140.

23 THE JESUITS OF THE UNIVERSITY OF CAEN TOOK HIM IN See Ted McCormick's discussion of Jesuit curricular approaches in this transitional era in European intellectual life.

23 BEFORE HIS TWENTY-FIRST BIRTHDAY McCormick, *William Petty*, p. 28. The underlying testimony from Petty comes from his papers and John Aubrey's *"Brief Lives."*

23 AN EXAMPLE OF A NEW KIND OF MAN McCormick, *William Petty*, p. 40.

24 BOTH THREAT AND PRIZE There is, of course, an enormous literature on the English conquest(s) of Ireland. For this project, with its emphasis on the interplay of ideas from the scientific revolution with social and political life, I drew most heavily on William J. Smythe's exceptional book *Map-making, Landscapes and Memory: A Geography of Colonial and Early Modern Ireland, c. 1530–1750* (Notre Dame, IN: University of Notre Dame Press, 2006). His account of the reduction of Ireland to measure and number provides a fascinating way into the broader history of conquest, settlement, and the assertion of power over the land and its people. His treatment of the Confederation period and of Cromwell's campaign, its consequences, and William Petty's role can be found in part 1, especially chapters 4 and 5, pp. 103–97.

24 HE LEFT BEHIND AN ARMY Smythe, *Map-making, Landscapes and Memory*, p. 167.

25 THE PROMISE OF . . . IRELAND New settlers were to be placed on eleven million acres of Ireland, more than half the total area of the island. Smythe, *Map-making, Landscapes and Memory*, p. 165.

25 A FIRST ATTEMPT TO CATALOG THE PROPERTY This was the "Gross Survey," which with the Civil Survey was one of two attempts to catalogue Ireland before Petty's work. See Smythe, *Map-making, Landscapes and Memory*, pp. 165 and 170–72.

25 ALMOST TWO YEARS INTO THE PROCESS McCormick, *William Petty*, p. 94.

25 ENTER WILLIAM PETTY Robert Boyle to Frederick Clod, May 15, 1654; Samuel Hartlib to Robert Boyle, May 15, 1654, quoted in McCormick, *William Petty,* p. 87.

26 PETTY OFFERED TO MEASURE EVERYTHING Petty's proposal is described in William Smythe, *Map-making, Landscapes and Memory,* pp. 172–73. Petty's quotes, reproduced here from that account, are from Petty's *The History of the Survey of Ireland, Commonly Called the Down Survey,* edited by Thomas A. Larcom (Dublin: Irish Archaeological Survey, 1851), pp. xiv and 7–9.

26 AND HE WOULD DO ALL THIS FOR A FEE Petty received both a cash payment of almost £19,000, out of which he had to pay survey costs, and land grants that ultimately totaled approximately thirty thousand acres of confiscated Irish land. For a tortuous account of Petty's payment, see Petty, *History of the Survey,* especially chapters 12, 13, and 15.

26 HE BEGAN IMMEDIATELY Petty's work on the Down Survey is covered in Smythe, *Map-making, Landscapes and Memory,* pp. 175–81.

26 HE SENT TO LONDON FOR NEW SURVEYING TOOLS See McCormick, *William Petty,* p. 98. In his account he also detected anticipatory echoes of Adam Smith; it leaps out of the original source, Petty's *History of the Survey,* p. xiv.

27 HE RELIED ON "DRUNKEN SURVEYORS" Petty, *History of the Survey,* p. 50, quoted in William Smythe, *Map-making, Landscapes and Memory,* p. 177.

27 HE INCLUDED AN UNPRECEDENTED AMOUNT OF INFORMATION These come from Down Survey maps published on Trinity College Dublin's The Down Survey Project website: "Unprofitable mountain [. . .]," County Kerry, Barony of Corkagwinny, Clogh Parish, http://downsurvey.tcd.ie/down-survey-maps.php#bm=Corkagwinny&c=Kerry; "Capenaheny's Timber Trees [. . .]," County Limerick, Barony of Abby Othenboy, Abbeyouthneybeg Parish, http://downsurvey.tcd.ie/down-survey-maps.php#bm=Abby+Othenboy&c=Limerick&p=Abbeyouthneybeg; and the neighbors, Edward Dungan Jr. pap [Catholic] and Nenabbey Protestant in the Barony and

Parish of Kilkullen, County Kildare http://downsurvey.tcd.ie/down-survey-maps.php#bm=Kilcullen&c=Kildare.

28 A DECADE-LONG DEMOGRAPHIC CATASTROPHE The demographic disaster in Ireland in the 1640s and early 1650s was transformative, proportionally greater than the impact of the combined death toll and population loss through emigration during the Great Famine that would devastate Ireland two centuries later. The death toll from war, famine, and an epidemic outbreak of bubonic plague has been estimated to be between 600,000 and 800,000 of a total population of at most 2.1 million people. See Smythe, *Map-making, Landscapes and Memory,* pp. 158–61.

28 THERE'S NO EVIDENCE THAT PETTY HIMSELF DELIGHTED McCormick, *William Petty,* p. 189. The calculator and the quote reproduced here are from Petty's *The Political Anatomy of Ireland* [. . .] (London: D. Brown and W. Rogers, 1691), pp. 17–20, http://quod.lib.umich.edu/e/eebo/A54620.0001.001/1:8?rgn=div1;view=fulltext.

28 IRELAND ANATOMIZED ON PAPER This image of anatomizing as Petty's approach to Ireland as a whole as much as to some of its fauna comes from Smythe, *Map-making, Landscapes and Memory,* p. 174.

29 SCORNFUL WITNESSES TO SUCH GATHERINGS John Starkey, *Character of Coffee and Coffee-Houses* (1674), quoted in Markman Ellis, *The Coffee House: A Cultural History* (London: Weidenfeld & Nicholson, 2004), loc. 1376 of 7491, Kindle.

29 THERE WERE ABOUT A DOZEN MEN PRESENT H. G. (Henry George) Lyons, *The Royal Society, 1660–1940: A History of Its Administration Under Its Charters* (New York, Greenwood, 1968), p. 21.

30 CHARLES II GRANTED THE GROUP A CHARTER See Lyons, *Royal Society,* especially chapter 2.

30 HE GAVE HIS PROGRAM A NAME William Petty, *Political Arithmetick; or, A Discourse Concerning, the Extent and Value of Lands* [. . .], 3rd ed. (London: Robert Clavel and Hen. Mortlock, 1690), p. 21.

31 In the 1672 proposal McCormick, *William Petty,* p. 188.

31 "If an exchange was made" Petty, *Political Anatomy,* pp. 29–30; also quoted in McCormick, *William Petty,* p. 194.

32 "To make nearer approaches" Petty, *Political Anatomy,* pp. 63–64.

33 Petty would turn his political arithmetic to other challenges For a full discussion of the extension of political arithmetic, see McCormick, *William Petty,* chapter 6, pp. 209–58.

33 Graunt reduced life and death to data See John Graunt, *Natural and Political Observations, Mentioned in a Following Index, and Made upon the Bills of Mortality,* 2nd ed. (London: Tho. Roycroft for John Martin, James Allestry, and Tho. Dicas, 1662).

34 the "Art of Reasoning, by Figures" Charles Davenant, *Discourses on the Publick Revenues, and on the Trade of England* [. . .], p. 2, quoted in McCormick, *William Petty,* p. 296.

34 King wrote on births and deaths Richard Stone, "The Accounts of Society" (Nobel Prize Lecture, December 8, 1984), p. 120, https://www.nobelprize.org/uploads/2018/06/stone-lecture.pdf.

34 "Mrs Kings fine Calico Gown" McCormick, *William Petty,* p. 296.

34 King offered careful advice Gregory King, *Natural and Political Observations* [. . .] (1696), p. 31, quoted in McCormick, *William Petty,* pp. 294–95.

Chapter 3

36 He was an inventor Alan Cook, *Edmond Halley: Charting the Heavens and the Seas* (Oxford: Clarendon Press, 1998), p. 239.

36 He was a courageous explorer Cook, *Edmond Halley,* p. 256.

36 He mastered Arabic G. A. Russell, *The "Araibick" Interest of the Natural Philosophers in Seventeenth-Century England* (Leiden, Netherlands: E. J. Brill, 1994), p. 154. See the Arabic manuscript of Apollonius's *Conics,* with Halley's marginal notes in Latin: https://genius.bodleian.ox.ac.uk/exhibits/browse/conics/.

36 He generated a "Solution of a Problem" Edmond Halley, "A

Discourse Concerning Gravity [. . .] Together with the Solution of a Problem of Great Use in Gunnery," *Philosophical Transactions* 16 (January 1687): 3–21, https://royalsocietypublishing.org/doi/10.1098/rstl.1686.0002.

36 THE CAUSE OF NOAH'S FLOOD Edmond Halley, "Some Considerations About the Cause of the Universal Deluge, Laid Before the Royal Society, on the 12th of December 1694," *Philosophical Transactions* 33 (1724): 118–23.

38 THAT ACCOUNT OF "A VERY ODD MONSTROUS CALF," Robert Boyle, "An Account of a Very Odd Monstrous Calf," *Philosophical Transactions* 1, no. 1 (May 1665): 10.

38 AN INVESTIGATION OF WATER PRESSURE AT DEPTH "Some Experiments and Observations Made of the Force of the Pressure of the Water in Great Depths, Made and Communicated to the Royal Society, by a Person of Honour," *Philosophical Transactions* 17, no. 193 (December 30, 1693): 504–6.

38 "WHETHER A *FIERCE* DOG" Robert Boyle "Trials proposed by Mr. Boyle to Dr. Lower, to be made by him, for the Improvement of Transfusing blood out of one live Animal into another," *Philosophical Transactions* 1, no. 22 (May 30, 1667) 385; Jean baptiste de la Quintinie, "Extract of a letter of M. dela Quintinie, giving some further directions and observations about Melons," *Philosophical Transactions* 4, no. 46 (January 1, 1669): 923–24. (Note: The dates refer to time of publication, not the date of the Royal Society meeting at which the papers were read.)

38 EDMOND HALLEY HAD A RUMMAGE THROUGH HIS PAPERS Edmond Halley, "An Estimate of the Degrees of the Mortality of Mankind [. . .]," *Philosophical Transactions* 17, no. 196 (December 3, 1693): 597. Some accounts suggest that Halley found Neumann's papers after Justel's death in 1693. A problem with that account is that Justel's death is listed as occurring in September (though there is some uncertainty about that date, and Robert Hooke noted in his diary that Halley spoke to the Society on Neumann's data the previous March). Given that, even with the vague-

ness on the timing of Justel's death, I take Halley at his word in his paper that he encountered Neumann's work through Justel's public communication of it to the Society.

38 HALLEY FIRST EXTRACTED THE SIMPLEST FACTS All of the information in the description of Halley's work comes from Halley, "Estimate of the Degrees of the Mortality of Mankind," pp. 596–610.

39 TO DISSECT THE LIVES OF STRANGERS HUNDREDS OF MILES DISTANT In this summary of Halley's insights, I follow the breakdown of his paper in James Ciecka, "Edmond Halley's Life Table and its Uses," *Journal of Legal Economics* 15, no. 1 (2008): 68. See also David Bellhouse, "A New Look at Halley's Life Table," *Journal of the Royal Statistical Society* 174, no. 3 (July 2011): 823–32.

42 THE FIRST LIFE INSURANCE SCHEMES Geoffrey Clark, *Betting on Lives: The Culture of Life Insurance in England, 1695–1775* (Manchester, UK: Manchester University Press, 1999), pp. 73–74.

42 THIS ANALYSIS, HALLEY WROTE Halley, "Estimate of the Degrees of Mortality," p. 602.

43 HIS BRESLAU PAPER WASN'T THE FIRST William Deringer, "The Present Value of the Distant Future in the Early-Modern Past," Financial History Seminar, Stern School of Business, New York University, April 12, 2013.

45 "HOW UNJUSTLY WE REPINE" Edmond Halley, "Some Further Considerations on the Breslau Bills of Mortality," *Philosophical Transactions* 17, no. 198 (March 1693): 655. For a statistician's evaluation of Halley's work, see Bellhouse, "New Look at Halley's Life Table."

45 THE BRITISH GOVERNMENT DID NOT USE COMPOUNDING CALCULATIONS Ian Hacking, *The Emergence of Probability* (Cambridge: Cambridge University Press, 2006), p. 112.

46 HALLEY HAD COME TO TRINITY COLLEGE For details of the visit, see Richard Westfall, *Never at Rest* (Cambridge: Cambridge University Press, 1980), p. 401. Newton's answer applied to the specific case Halley put to him. The inverse square relationship can produce different paths—a hyperbola, for example—given the right parameters.

47 THE BEHAVIOR OF THE ENTIRE KNOWN COSMOS The third section of *Principia* was published in an English translation under the title *A Treatise of the System of the World* in 1728, the year after Newton's death.

48 PEPYS WROTE A BRIEF SERIES OF LETTERS Six letters between Pepys and Newton, beginning on November 22, 1693, contained in Isaac Newton, *The Correspondence of Isaac Newton, vol. 3, 1688–1694,* ed. H. W. Turnbull (Cambridge: Cambridge University Press, 1959), pp. 293–303.

49 HE GOT HIS SUMS RIGHT Stephen M. Stigler, "Isaac Newton as Probabilist," *Statistical Science* 21, no. 3 (2006): 400–401. As Stigler details, though Newton got the calculation right, his logical argument misstates the problem that Pepys proposed, and does not adequately underpin the quantitative solution he did in fact achieve. Even geniuses can err (and in the autumn of 1693, Newton, recovering from what appears to have been a mental breakdown in June, was hardly at his best).

49 "BABY NEEDS A NEW PAIR OF SHOES!" Neither Pepys nor anyone he knew would ever have said that, of course. It's American slang, dating back only to the early twentieth century. But it's one of my favorites . . .

49 THIS WASN'T NEWTON'S FIRST EXCURSION INTO QUESTIONS OF PROBABILITY See Jed Z. Buchwald and Mordecai Feingold, *Newton and the Origin of Civilization* (Princeton, NJ: Princeton University Press, 2013), chapter 2, especially pp. 90–106.

50 THAT ATTEMPT TO QUANTIFY AND CONSTRAIN ERROR Buchwald and Feingold, *Newton and the Origin of Civilization,* p. 93.

50 "*VERY PROBABLE CONJECTURES*" Charles Davenant, *Some Reflections on a Pamphlet Entitled, England and East-India Inconsistent in Their Manufactures,* quoted in William Deringer, "Finding the Money: Public Accounting, Political Arithmetic, and Probability in the 1690s," *Journal of British Studies* 52 (2013): 665.

50 JOHN LOCKE'S PROBABILITY-CENTERED THEORY OF KNOWLEDGE Carl Wennerlind, *Casualties of Credit: The English Financial Revolution, 1620–1720* (Cambridge, MA: Harvard University Press,

2011), p. 91. Wennerlind's discussion of probabilistic knowledge can be found on pages 85–92.

50 But by the 1690s the notions of risk There is an extensive literature on the emergence of probabilistic thinking in the scientific revolution and in the larger society in which that intellectual transformation took place. As laid out in the notes above, Geoffrey Clark's *Betting on Lives* (Manchester, UK: Manchester University Press, 1999) is an excellent guide to the critical steps in the emergence of life insurance in British financial and social thinking. I am also indebted to David Bellhouse's *Leases for Lives: Life Contingent Contracts and the Emergence of Actuarial Science in Eighteenth-Century England* (Cambridge: Cambridge University Press, 2017), especially the early chapters. A strong general guide to the history of probability is Lorraine Daston's *Classical Probability in the Enlightenment* (Princeton, NJ: Princeton University Press, 1988). There, chapters 1 and 2 shaped my understanding of the historical context in which the quantitative and empirical impulses of the second half of the seventeenth century emerged, and chapter 3 was helpful in situating the mathematical advances against social practice.

51 the Treasury sought outside help The inquiry was headed by William Lowndes, the leading Treasury civil servant. See his *A Report Containing an Essay for the Amendment of the Silver Coins* (London: C. Bill, 1695).

Chapter 4

52 Isaac Newton had first made his way Elements of this chapter are drawn from my earlier work, *Newton and the Counterfeiter* (New York: Houghton Mifflin Harcourt, 2008), especially chapters 10, 11, and 12.

52 "to make Mr. Newton Warden of the Mint" The details of Newton's departure from Cambridge can be found in Richard Westfall, *Never at Rest* (Cambridge: Cambridge University Press, 1980), pp. 549–57.

52 A PASSING STUDENT SAID Quoted in Westfall, *Never at Rest*, p. 468. Westfall cites a manuscript in the Keynes collection at King's College, Cambridge (MSS 130.6, book 2, 130.5, sheet 1), and notes that the anecdote comes from Newton's niece's husband, John Conduitt, who heard it from another friend of Newton's.

53 INITIALLY NEWTON'S NEW LIFE WAS LARGELY CONFINED Westfall, *Never at Rest*, p. 556.

53 HE WOULD, HE PROMISED, FACE "NOT TOO MUCH BUS'NESSE" Isaac Newton, *The Correspondence of Isaac Newton, vol. 4, 1694–1709*, ed. J. F. Scott (Cambridge: Cambridge University Press, 1959), document 545, p. 195.

54 KING WILLIAM HAD ADVERTED TO WHAT WAS ABOUT TO HAPPEN "26 November 1695," in *Journal of the House of Commons, vol. 11, 1693–1697* (London: His Majesty's Stationery Office, 1803), pp. 338–40.

57 "GREAT MASSES WERE MELTED DOWN" Lord Macaulay, *The History of England*, vol. 5 (Philadelphia: Porter & Coates,1890), p. 256.

57 A CRISIS OF SMALL CHANGE See Thomas J. Sargent and François R. Velde, *The Big Problem of Small Change* (Princeton, NJ: Princeton University Press, 2002), p. 87.

57 A LABORER EARNED ABOUT JUST OVER A SHILLING A DAY The price and wage numbers come from the data series collected by Gregory Clark as part of the research underpinning several publications. See especially Gregory Clark, "The Long March of History: Farm Wages, Population and Economic Growth, England 1209–1869," *Economic History Review* 60, no. 1 (February 2007): 97–135; and Gregory Clark, "The Condition of the Working-Class in England, 1209–2004," *Journal of Political Economy* 113, no. 6 (December 2005): 1307–40. Clark's database, updated most recently on April 10, 2006, can be found on the International Institute of Social History's website: http://iisg.nl/hpw/data.php#united. The exchange rate calculation for gold guineas to silver coins comes from Ming-hsun Li, *The Great Recoinage of 1696–9* (London: Weidenfield & Nicholson, 1963), p. 75.

58 MEANWHILE A POUND OF STEAK AT SPITALFIELDS See the spreadsheet under the heading "English prices and wages 1209–1914" in the International Institute of Social History's "List of Datafiles": http://iisg.nl/hpw/data.php#united.

58 SO LITTLE BULLION WAS AVAILABLE Hopton Haynes, *Brief Memoires Relating to the Silver and Gold Coins of England,* quoted in Li, *Great Recoinage,* p. 48.

58 "I STAYED SO LONG ABOUT IT" In fact, the cost of the Nine Years' War (also known as the War of the Grand Alliance) demonstrated that the military mobilizations both William and Louis had attempted exceeded the capacity of their states to sustain. For both the French and the British (as the state became after the act of union between England and Scotland in 1707) the armies of the Nine Years' War were the largest either nation fielded until the Napoleonic Wars. It was just too damn expensive. The problem with raising money from local bankers is discussed in Li, *Great Recoinage,* p. 58; and is documented in the letter from Richard Hill to Sir WilliamTrumbull, August 21, 1695, quoted in Childs, *The Nine Years' War and the British Army, 1688–1697* (Manchester: Manchester University Press 2013), p. 297.

59 ENGLAND NEEDED TO UNDERTAKE A COMPLETE RECOINAGE There was in fact a vigorous debate beyond the small group Lowndes consulted over the need for recoinage, with a significant party opposed to any remaking of the money during the war, given plausible fears of the disruption the calling-in and reminting process would involve. See Li, *Great Recoinage,* pp. 65–67.

59 NEWTON WROTE BACK TO LOWNDES "Isaac Newton Concerning the Amendment of English Coins," Goldsmiths Library, University of London, MS 62, reprinted in Li, *Great Recoinage,* p. 217.

60 "IN STATING THE VALUE OF GOLD AND SILVER" "Isaac Newton Concerning the Amendment," in Li, *Great Recoinage,* p. 218.

60 "TIS MERE OPINION" Isaac Newton, cited in John Craig, *Newton at the Mint* (Cambridge: University Press, 1946), p. 42.

60 CHANGING THE NUMBER ASSOCIATED WITH A COIN John Locke's

response to Lowndes, Goldsmiths Library, University of London, MS 62, quoted in Li, *Great Recoinage,* p. 227.

61 "SOME ARE OF THE OPINION" John Locke, quoted in Li, *Great Recoinage,* p. 102.

62 DEVALUATION WOULD "ONLY SERVE TO DEFRAUD THE KING" John Locke, quoted in Li, *Great Recoinage,* p. 104.

62 PARLIAMENT APPROVED THE RECOINAGE An Act for Remedying the Ill State of the Coin of the Kingdome, 1695–1696, 7 & 8 Will. 3, c. 1, in *Statutes of the Realm, vol. 7, 1695–1701,* ed. John Raithby (Great Britain Record Commission, 1820), pp. 1–4.

63 IT WAS A DISASTER On the lack of circulating coin, see Li, *Great Recoinage,* pp. 135–6; Edmund Bohun to John Cary, July 21, 1696, quoted in C. E. Challis, ed., *A New History of the Royal Mint* (Cambridge: Cambridge University Press, 1992), p. 387; John Evelyn, diary entry for June 11, 1696, quoted in Li, *Great Recoinage,* p. 135; and D. W. Jones, *War and Economy in the Age of William III and Marlborough* (Oxford: Basil Blackwell, 1988), p. 137.

63 IN THE NICK OF TIME, ENTER ISAAC NEWTON Hopton Haynes, quoted in Challis, *New History of the Royal Mint,* p. 394. The description of the work comes from the same passage. The figure of five hundred men working at the mint comes from Craig, *Newton at the Mint,* p. 14. See also Isaac Newton, "Observations Concerning the Mint," 1697, in Newton, *Correspondence,* vol. 4, document 579, p. 258, online at http://www.newtonproject.ox.ac.uk/view/texts/normalized/MINT00883.

64 TO "JUDGE OF THE WORKMEN'S DILIGENCE" Hopton Haynes, *Briefe Memoires,* cited in Westfall, *Never at Rest,* p. 561.

64 HE IDENTIFIED THE PERFECT PACE Challis, *New History of the Royal Mint,* p. 394. Challis cites Haynes for Newton's calculations. The figure of fourteen men working each press comes from Isaac Newton, "Observations Concerning the Mint."

64 £6,722,970 0S. 2D., TO BE EXACT Li, *Great Recoinage,* p. 140.

64 TO SWITCH FROM SILVER TO GOLD See Tamim A. Bayoumi, Barry J. Eichengreen, and Mark P. Taylor, eds., *Modern Perspectives on the*

Gold Standard (Cambridge: Cambridge University Press, 2005), p. 65; and T. S. Ashton, *Economic History of England: The Eighteenth Century* (London: Routledge, 1955 and 2006), p. 171. Newton's own report can be found in "Sir Isaac Newton's State of the Gold and Silver Coin," published in William Arthur Shaw, *Select Tracts and Documents Illustrative of English Monetary History 1626–1730: Comprising Works of Sir Robert Cotton, Henry Robinson, Sir Richard Temple* [. . .] (London: C. Wilson, 1896), pp. 189–95.

66 ABOUT FOURTEEN THOUSAND MEN UNDER ARMS John Brewer, *The Sinews of Power: War, Money and the English State, 1688–1783* (1988; Cambridge, MA: Harvard University Press, 1990), p. 8.

66 FOLLOWING THE GLORIOUS REVOLUTION John Brewer, *Sinews of Power,* p. 30.

66 THE PROCESS REQUIRED A SECOND ARMY OF CLERKS Geoffrey Parker, "The 'Military Revolution,' 1560–1660—a Myth?" *Journal of Modern History* 48, no. 2 (June 1976): 209.

66 ENGLAND HAD SPENT ABOUT £2 MILLION A YEAR P. G. M. Dickson, *Financial Revolution in England: A Study in the Development of Public Credit, 1688–1756* (Aldershot, UK: Gregg Revivals, 1993), p. 46.

66 JUST £3.64 MILLION PER YEAR John Brewer, *Sinews of Power,* p. 89.

66 OVER THE ENTIRE WAR Dickson, *Financial Revolution,* p. 10.

66 "THE WHOLE ART OF WAR" Charles Davenant, *An Essay upon Ways and Means of Supplying the War* (London: Jacob Tonson, 1695), quoted in Anne L. Murphy, *The Origins of English Financial Markets* (Cambridge: Cambridge University Press, 2009), p. 39.

67 THE INFAMOUS STOP OF THE EXCHEQUER William Cobbett, *Parliamentary History of England, from the Earliest Period to the Year 1803, vol. 4, Comprising the Period from the Restoration of Charles the Second in 1660 to the Revolution* (London: T. C. Hansard, 1808), p. 498.

68 ENGLISH SOVEREIGNS WERE REQUIRED Coronation Oath Act of 1688, 1688, 1 W. & M., c. 6., http://www.legislation.gov.uk/aep/WillandMar/1/6/section/III.

68 THEY WERE EXPLICITLY FORBIDDEN An Act Declaring the Rights and Liberties of the Subject and Settling the Succession of the Crown, 1689, 1 W. & M., c. 2, http://avalon.law.yale.edu/17th_century/england.asp.

68 THE SOLE SOURCE OF PUBLIC FUNDS As William Deringer points out in *Calculated Values* (Cambridge, MA: Harvard University Press, 2018), pp. 43–45, the monarchy did not lose all access to funds independent of Parliament's intentions with the settlement to the Glorious Revolution. He tells the story of William Jephson, secretary to the Treasury until his death in 1691, who handled payments for the "Secret Service," in which capacity he could disburse substantial sums of public funds—£100,000 a year—without reference to a legislated purpose or, crucially, any requirement to account for his expenditures. But even though it would take some time for Parliament to command full power of the purse, the ambition to acquire that authority was integral to the settlement that placed William and Mary on the throne. See Deringer, *Calculated Values,* pp. 51–57.

68 IT FELL TO THE HOUSE OF COMMONS This account of the background to the Million Act is derived from Dickson, *Financial Revolution,* pp. 50–52.

68 THE OTHER SCHEME Dickson, *Financial Revolution,* p. 51. Paterson's plan, as Dickson notes, seemed to suggest that the government create a form of fiat money—paper, backed by only one fifth of its value in coin—in the form of a loan at 6 percent that would also circulate as legal tender. But the proposal was so generally worded that it's not entirely clear precisely what Paterson had in mind, nor what the committee rejected.

69 PATERSON ACCEPTED THE CHALLENGE There is a hint of uncertainty about Paterson's role in the proposal, as the scheme was veiled by anonymity. But his name is the one most commonly associated with the Million Act.

69 PARLIAMENT SWIFTLY PASSED THE MEASURE Dickson, *Financial Revolution,* pp. 52–53.

70 NEALE'S TWIST Thomas Neale, *The Profitable Adventure to the Fortunate* (London: F. Collins, 1694), p. 1.

70 ONE RESPONSE WAS TO CREATE THE BANK OF ENGLAND Andreas Michael Andreades, *History of the Bank of England,* trans. Christabel Meredith (London: P.S. King & Son, 1909), pp. 73–74.

71 THE BANK OF ENGLAND, BY CONTRAST Andreades, *History of the Bank of England,* p. 82. The Bank of Sweden issued banknotes decoupled from its reserves in the 1660s. The experiment failed, and the bank collapsed within a few years. For more examples of early fractional reserve operations, see John Clapham, *The Bank of England,* vol. 1 (Cambridge: Cambridge University Press, 1970), pp. 20–25.

72 THE TREASURY MISSED SOME PAYMENTS Murphy, *Origins of English Financial Markets,* p. 57.

72 THE TREASURY WENT TO THE PUBLIC WITH ANOTHER LOTTERY OFFER Anne Murphy, "Lotteries in the 1690s: Investment or Gamble?" *Financial History Review* 12, no. 2 (October 2005): 227–46.

74 "SOME OF THE GREAT FOLLIES OF LIFE" Daniel Defoe, *The Anatomy of Exchange-Alley; or, A System of Stock-Jobbing* (London: E. Smith, 1719), p. 35.

CHAPTER 5

76 MAKING COFFEE IN THE TURKISH MANNER John Houghton, "A Discourse of Coffee," *Philosophical Transactions of the Royal Society* 21 (January 1, 1699): 312.

76 THE PARTNERS OPENED THEIR OPERATION IN 1652 Edwards's and Rosee's introduction of the coffeehouse to England draws on Markman Ellis, *The Coffee House: A Cultural History* (London: Weidenfeld & Nicholson, 2004), chapter 3. The anecdote about Edwards's coffee habit can be found in Houghton, "Discourse of Coffee," p. 312.

77 BUT THE OPENING OF THE FIRST COFFEEHOUSE There is some controversy about the priority claim. Several histories of coffee place the first public vendor in England in Oxford. I'm persuaded by

Ellis's argument that Rosee's has the better documented historical trail. See *Coffee House,* Kindle location 691 of 7491.

77 "BURNT BREAK AND PUDLE WATER," John Tatham, *Knavery in All Trades; or, The Coffee-house: a Comedy,* act III.

77 "GENTRY, TRADESMAN, ALL ARE WELCOME HITHER" *A Brief Description of the Excellent Vertues of That Sober and Wholesome Drink, Called Coffee* (London: Paul Greenwood, 1674).

78 "[THE] COFFEE-HOUSE MAKES ALL SORTS OF PEOPLE SOCIABLE" Houghton, "Discourse of Coffee," p. 317.

78 THE POLYMATHIC ROBERT HOOKE Robert Hooke, entries for January 17, 1680, and September 20, 1677, in Robert Hooke, *The Diary of Robert Hooke,* ed. Henry W. Robinson and Walter Adams (London: Taylor and Francis, 1935), pp. 436 and 314.

78 HOOKE THERE READ THE LETTER FROM ANTONI VAN LEEUWENHOEK Antoni van Leeuwenhoek, "Brief No. 99 [54], 9 May 1687, addressed to the Gentlemen of the Royal Society," in *The Collected Letters of Antoni van Leeuwenhoek,* part 4, a committee of Dutch scientists, ed. (Amsterdam: N.V. Swets & Zeitlinger, 1961), p. 222, http://www.dbnl.org/tekst/leeu027alle06_01/leeu027alle06_01_0011.php#b0099. Atypically, this letter was not published in the *Proceedings of the Royal Society.* The connection to Hooke lies with the fact that Hooke is recorded as commenting on the letter at the Society meeting on May 25, 1687—and, more broadly, in the common interest the two men had in both microscopy and coffee.

80 THE EAST INDIA COMPANY ITSELF Stefania Gialdroni, "Incorporation and Limited Liability in Seventeenth-Century England: The Case of the East India Company," in *The Company in Law and Practice,* ed. Dave De ruysscher, Albrecht Cordes, Serge Dauchy, and Heikki Pihlajamäki (The Hague: Brill/Nijhoff, 2017), p. 111. More generally, Gialdroni's paper provides context for the evolution of the joint-stock form through her examination of the East India Company case study.

80 IN 1660, THE ROYAL AFRICAN COMPANY Gold from the Royal African Company supplied the Royal Mint for over fifty years, to the point that the region from which the company extracted its supply,

Guinea, gave its name to the first gold coin to be struck on the machines at the Tower of London.

80 SHARES COULD BE TRADED ON EXCHANGE ALLEY Anne L. Murphy, *The Origins of English Financial Markets* (Cambridge: Cambridge University Press, 2009), pp. 15–16. For share transaction volume, see her table 1.1, p. 16.

81 EACH HUNDRED POUNDS INVESTED RETURNED £10,000 Murphy, *Origins of English Financial Markets,* pp. 10–11.

81 THEY RAN THE ENTIRE RANGE OF HUMAN IMAGINATION William Robert Scott, *The Constitution and Finance of English, Scottish and Irish Joint-Stock Companies to 1720* (Cambridge: Cambridge University Press, 1910), p. 111 et seq.

82 THE CONCENTRATION OF EXOTIC FIGURES Ned Ward, *The London Spy Compleat, in Eighteen-Parts* (London: J. How, 1703), pp. 66–69.

83 LICENSED BROKERS WERE FORMALLY EJECTED See Richard Dale, *The First Crash: Lessons from the South Sea Bubble* (Princeton, NJ: Princeton University Press, 2004), p. 31.

85 "BANKRUPTS AND BEGGARS" Daniel Defoe, *The Villainy of Stock-Jobbers Detected, and the Causes of the Late Run upon the Bank and Bankers Discovered and Considered* (London, 1701), p. 26.

85 NED WARD AGREED Ward, *London Spy,* p. 298, quoted in Murphy, *Origins of the English Financial Markets,* p. 161.

85 THOSE KNAVES CAME TO LIFE Thomas Shadwell, *The Volunteers; or, The Stock-Jobbers* (London: James Knapton, 1693), pp. 23–24 (act 2, scene 1).

86 FAKE NEWS WAS A DAILY FACT OF LIFE IN EXCHANGE ALLEY *The Case of the Coffee House Men* (undated), p. 7, quoted in Dale, *First Crash,* p. 10.

86 THE CASE OF ESTCOURT'S LEAD MINE COMPANY Anne L. Murphy, "Trading Options Before Black-Scholes: A Study of the Market in Late Seventeenth-Century London," *Economic History Review,* vol. 62, no. S1 (August 2009): 17–18.

86 "JOBBING THEIR STOCKS ABOUT" Defoe, *Villainy of Stock-Jobbers Detected,* p. 17.

87 "The only reasonable conclusion" Murphy, *Origins of English Financial Markets,* p. 183.

Chapter 6

89 Jonathan Castaing washed up in London Natasha Glaiser, "Calculating Credibility: Print Culture, Trust and Economic Figures in Early Eighteenth-Century England," *Economic History Review* 60, no. 4 (2007): 698.

96 The Treasury borrowed almost exactly one-third P. G. M. Dickson, *The Financial Revolution in England: A Study in the Development of Public Credit, 1688–1756* (Aldershot, UK: Gregg Revivals, 1993), p. 10, table 1.

96 John Pollexfen had become a wealthy man "John Pollexfen," in *The History of Parliament: the House of Commons 1660–1690,* ed. B. D. Henning (London: Secker & Warburg, 1983), http://www.historyofparliamentonline.org/volume/1660-1690/member/pollexfen-john-1638-1715.

96 Walbrook House John F. Bold and Edward Chaney, eds. *English Architecture: Public and Private* (London: Hambledon Continuum, 1993), p. 92.

97 "it may occasion that a Nation that relies much on Paper Credit" John Pollexfen, *A Discourse of Trade, Coyn, and Paper Credit, and of Ways and Means to Gain, and Retain Riches* (London: Brabazon Aylmer 1697), pp. 65–69.

98 "too much paper credit" Isaac Newton, "Untitled Holograph Draft Memorandum on John Pollexfen's *A Discourse on Trade, Coyn and Paper Credit,*" 1697, Mint 19/II, 608–11, National Archives, Kew, UK, http://www.newtonproject.ox.ac.uk/view/texts/normalized/MINT00261.

101 Others, knowledgeable and worldly, disagreed William Lowndes, secretary to the Treasury, was much more suspicious of credit than Newton was. See G. Findlay Shirras and J. H. Craig, "Sir Isaac Newton and the Currency," *Economic Journal* 55, no. 218/219 (June–September 1945): 231.

CHAPTER 7

102 IT PROVED JUST AS COSTLY Henry Roseveare, *The Financial Revolution: 1660–1760* (London and New York, 1991), p. 33, quoted in Richard Dale, *The First Crash: Lessons from the South Sea Bubble* (Princeton, NJ: Princeton University Press, 2004), p. 41.

102 THE HUGE (FOR ITS DAY), NINETY-THOUSAND-STRONG ARMY John Brewer, *The Sinews of Power: War, Money and the English State, 1688–1783* (1988; Cambridge, MA: Harvard University Press, 1990), page 30, table 2.1.

102 10 PERCENT OF ENGLAND'S INCOME See estimates for GDP to 1700 in Alexander Apostolides et al., "English Gross Domestic Product, 1300–1700: Some Preliminary Estimates," November 26, 2008, part of the project "Reconstructing the National Income of Britain and Holland, c.1270/1500 to 1850," ref. no. F/00215AR, https://warwick.ac.uk/fac/soc/economics/staff/sbroadberry/wp/pre1700v2.pdf. Calculating historical GDP is a tricky problem, and there is at least a fifty-year trail of scholarship on the question of England/Britain's national income. The totals generally fall between sixty and eighty million pounds per year.

103 MORE OBLIGATIONS FOLLOWED See P. G. M. Dickson, *The Financial Revolution in England: A Study in the Development of Public Credit, 1688–1756* (Aldershot, UK: Gregg Revivals, 1993), p. 80, table 7.

103 IT TOOK A HEROIC OFFER Dickson, *Financial Revolution*, p. 74.

104 BRIMSTONE SHOWERED DOWN Henry Sacheverell, *The Perils of False Brethren: Both in Church, and State: Set Forth in a Sermon Preach'd before the Right Honourable the Lord-Mayor* [. . .] (London: Henry Clements, 1709).

104 GODOLPHIN'S WHIGS BROUGHT HIM TO TRIAL Henry Sacheverell, *The Tryal of Dr. Henry Sacheverell, Before the House of Peers, for High Crimes and Misdemeanors* (London: Jacob Tonson, 1710), p. 455.

104 GODOLPHIN LOST That all such party labels were exceptionally fluid in those days can be seen from the fact that Godolphin was nominally a Tory, while Harley had gotten his start in politics as a Whig.

105 THE BANK OF ENGLAND HAD TRIED SOMETHING SIMILAR Stephen Quinn, "Money, Finance and Capital Markets" in *The Cambridge Economic History of Modern Britain*, vol. 1 (Cambridge: Cambridge University Press, 2003), Foderick Foud, Paul Johnson, eds., p. 168.

106 "SUFFICIENT FOR MUCH MORE BUSINESS" Daniel Defoe, *The Anatomy of Exchange-Alley; or, A System of Stock-Jobbing* (London: E. Smith, 1719).

106 A KIND OF FINANCIAL GRAVE ROBBING John Carswell, *The South Sea Bubble* (Phoenix Mill, UK: Alan Sutton, 1993), pp. 30–32.

107 A PERFECT AVATAR OF EXCHANGE ALLEY John G. Sperling, *The South Sea Company: An Historical Essay and Bibliographical Finding List* (Boston: Baker Library, Harvard Graduate School of Business Administration, 1962), p. 6.

107 "BURLY AND OVERBEARING" Carswell, *South Sea Bubble,* p. 19.

107 A DODGY LAND DEAL Dale, *First Crash,* p. 44, drawing on Carswell, *South Sea Bubble,* p. 35.

108 IN A MOVE THAT PLAUSIBLY ORIGINATED WITH BLUNT HIMSELF Carswell, *South Sea Bubble,* p. 35.

108 BY ONE ESTIMATE, BLUNT AND HIS ASSOCIATES MADE AT LEAST £25,000 For the £25,000 figure, see Carswell, *South Sea Bubble,* p. 35. Converting that number into a twenty-first-century equivalent is difficult. The Bank of England's inflation calculator puts it at just over £4 million. To get a finer-grained perspective on comparing prices across long periods of time and significant changes in consumption, see, e.g., E. H. Phelps Brown and Sheila V. Hopkins, "Seven Centuries of the Prices of Consumables, Compared with Builders' Wage-Rates," *Economica* (New Series) 23, no. 92 (November 1956): 296–314; and Robert Twigger, "Inflation: The Value of the Pound 1750–1996," House of Commons Library Research Paper 97/76, June 6, 1997.

108 BANK OF ENGLAND INSIDERS HAD DONE MUCH THE SAME Dale, *First Crash,* p. 44.

109 SWORD BLADE STOCK VALUES FELL Carswell, *South Sea Bubble,* p. 38.

109 THE FLOATING DEBT ALONE WAS STILL TERRIFYING Sperling, *South Sea Company*, p. 3.

110 IT LOOKED LESS AND LESS LIKELY THAT IT WOULD EVER BE PAID OFF Sperling, *South Sea Company*, p. 25.

110 QUEEN ANNE ALMOST BEGGED Carswell, *South Sea Bubble*, p. 48.

111 HARLEY TOLD THE HOUSE OF COMMONS Sperling, *South Sea Company*, p. 6.

PART 2

113 ALL *BRITAINS* REJOICE Arthur Maynwaring, "An Excellent New Song, Call'd Credit Restor'd," 1711, quoted in Carl Wennerlind, *Casualties of Credit: The English Financial Revolution, 1620–1720* (Cambridge, MA: Harvard University Press, 2011), p. 209.

CHAPTER 8

116 A RELIABLE MOUTHPIECE FOR GODOLPHIN For this quick step from Whig to Tory pamphleteering, see Daniel Defoe, *An Essay on the South-Sea Trade* (London: J. Baker, 1712), pp. 14–16.

117 "IT FALLS AMONG PEOPLE UNACQUAINTED WITH TRADE" Defoe *Essay on the South-Sea Trade*, pp. 33–34.

117 "NOT A SINGLE ONE HAD ANY EXPERIENCE" John Carswell, *The South Sea Bubble* (Phoenix Mill, UK: Alan Sutton, 1993), p. 58.

117 "THIS TRADE IS NOT ONLY PROBABLY TO BE GREAT" Defoe, *Essay on the South-Sea Trade*, p. 38.

118 TRADE WITH SPANISH POSSESSIONS REQUIRED SPAIN'S CONSENT John G. Sperling, *The South Sea Company: An Historical Essay and Bibliographical Finding List* (Boston: Baker Library, Harvard Graduate School of Business Administration, 1962), p. 18. This account of the notional South Sea expedition of 1712 comes from Sperling's account.

118 A CARGO OF 1,100 TONS Sperling, *The South Sea Company*, p. 18.

118 THE GOODS WITHIN THE HOLDS OF THE WAITING SHIPS HAD BEGUN TO ROT Richard Dale, *The First Crash: Lessons from the South Sea Bubble* (Princeton, NJ: Princeton University Press, 2004), p. 49.

119 ABOUT £100,000 IN PROFITS Dale, *First Crash,* p. 49. For more detail on the Company's early activities, see Carswell, *The South Sea Bubble,* pp. 47–67.

120 THE COMPANY COULDN'T EVEN GET PAID PROPERLY Dale, *First Crash,* p. 49.

121 SELLING STOLEN LIVES TOOK SKILL Carswell, *South Sea Bubble,* pp. 66–68.

121 THE FLOW OF GOODS AND BODIES For this overview of the Company's trading efforts I relied most heavily on Helen Paul's work and that of several historians whose primary interest was in understanding the financial side of the company. Since I wrote these sections, Harvard University professor Adrian Finucane has released a book that offers the first comprehensive academic history of the South Sea Company as a trade monopoly. It covers both the details of its business and the politics that surrounded it, from the Company's founding to the final collapse of its mercantile aspirations in 1748. For an account of the first hard lessons discussed here, see chapter 2, "Britain Hopes for the 'Riches of America,' 1713–1716." Adrian Finucane, *The Temptations of Trade: Britain, Spain, and the Struggle for Empire* (Philadelphia: University of Pennsylvania Press, 2016), pp. 21–52.

123 HER MAJESTY ADDED THAT LAPSE TO THE TALLY OF HARLEY'S SINS Brian W. Hill, *Robert Harley, Speaker, Secretary of State, and Premier Minister* (New Haven, CT: Yale University Press, 1988), p. 223. Jonathan Swift in his correspondence reported a similar series of complaints, omitting the charge that Harley "often came drunk." See Jonathan Swift and John Hawkesworth, *Letters, Written by Jonathan Swift: D.D., Dean of St Patrick's, Dublin. And Several of His Friends* [. . .], 6th ed., vol. 2 (London: T. Davies, 1767), p. 72.

123 SO, AT FOUR IN THE AFTERNOON AFTER ANNE'S DEATH Edward Gregg, *Queen Anne* (New Haven: Yale University Press, 2001), p. 397. For the tally of Catholics who were more closely related to Queen Anne than George was, see the family tree on pp. xviii and xix.

124 BRITAIN'S FINANCES REMAINED BOTH DIRE AND DESPERATELY CONFUSED Henry Roseveare, *The Financial Revolution: 1660–1760* (London: Taylor and Francis, 1991), pp. 52–53.

124 MORE THAN HALF A MILLION POUNDS STILL DUE J. Keith Horsefield, "The 'Stop of the Exchequer' Revisited," *Economic History Review* 35, no. 4 (November 1982): 513.

125 THE RESULTS WERE GRIM INDEED P. G. M. Dickson, *The Financial Revolution in England: A Study in the Development of Public Credit, 1688–1756* (Aldershot, UK: Gregg Revivals, 1993), p. 80, table 7.

125 GOLDSMITH-BANKERS CHARGED THEIR CUSTOMERS JUST 5 PERCENT, *Financial Revolution,* p. 53.

CHAPTER 9

127 IT WAS PERFECTLY LEGAL TO TAKE OUT INSURANCE ON A STRANGER Life Assurance Act 1774, 1774, 14 Geo. 3, c. 48, https://www.legislation.gov.uk/apgb/Geo3/14/48/1991-02-01.

127 SUCH WAGERS COULD TAKE MANY FORMS These examples all come from Geoffrey Clark, *Betting on Lives: The Culture of Life Insurance in England, 1695–1775* (Manchester, UK: Manchester University Press, 1999), pp. 44 and 49–50.

128 AN AMBIGUOUSLY MALE FRENCH DIPLOMAT Clark, *Betting on Lives,* pp. 45–48.

128 DOZENS OF BETS ON THE GREAT AND THE GOOD Clark, *Betting on Lives,* p. 52.

129 A KIND OF INDEX FOR TROUBLE Clark, *Betting on Lives,* p. 49.

129 A RESPECTABLE CAREER IN THE CHURCH William Coxe, *Memoirs of the Life and Administration of Sir Robert Walpole, Earl of Orford,* vol. 1 (London: T. Cadell, Jun., and W. Davies, 1798), p. 5.

130 "HIS EVENINGS PASSED IN FESTIVE SOCIETY" Both quotations in this paragraph are from Coxe, *Memoirs of the Life,* vol. 1, p. 5.

130 "A WOMAN OF EXQUISITE BEAUTY" Coxe, *Memoirs of the Life,* vol. 1, p. 5.

130 "HE WAS SHORT AND PLUMP" J. H. Plumb, *Sir Robert Walpole: The Making of a Statesman* (London: Cresset, 1956), p. 114.

131 AN "ENGINEER, SO SKILL'D IN THE FORTIFICATION OF *CHEESE CAKES*" Ned Ward, *The Secret History of Clubs: Particularly the Kit-Cat, Beef-Stake, Uertuocsos, Quacks, Knights of the Golden Fleece, Florists, Beaus, &c* [. . .]. (London: Booksellers, 1709), p. 361.

131 AMONG THEM ISAAC NEWTON'S NIECE, CATHERINE BARTON Richard Westfall, *Never at Rest* (Cambridge: Cambridge University Press, 1980), p. 595.

131 A WHIFF OF WALPOLE'S NEW WORLD See, for a rather elevated memoir of the Kit Kats, "The Kit-Cat Club," *Blackwood's Edinburgh Magazine* 11, no. 61 (1822): 201–6.

132 THE COST OF LIFE IN LONDON Plumb, *Sir Robert Walpole,* p. 91; and Edward Pearce, *The Great Man: Scoundrel, Genius and Britain's First Prime Minister* (London: Jonathan Cape, 2007), pp. 28–31.

132 HE DOVE EVER DEEPER IN DEBT Plumb, *Sir Robert Walpole,* pp. 108–10.

132 "SOME REPORTS SPREAD" Charles Taylor to Robert Walpole, August 7, 1706, quoted in Plumb, *Sir Robert Walpole,* p. 110.

133 WALPOLE FOUND HIS WAY ONTO A COMMITTEE Jeremy Black, *Robert Walpole and the Nature of Politics in Early Eighteenth Century Britain* (New York: St. Martin's Press, 1990), p. 4.

133 THE PERKS OF THAT OFFICE SERVED HIM WELL Plumb, *Sir Robert Walpole,* p. 121.

133 HIS RISE HALTED Stephen Taylor, "Robert Walpole," in *Oxford Dictionary of National Biography,* ed. David Cannadine, online ed. (Oxford: Oxford University Press, 2004–16).

134 WALPOLE GREW MUCH RICHER IN OFFICE See, e.g., Plumb, *Sir Robert Walpole,* pp. 180 and 203–6.

134 TITLED *ON THE JEWEL IN THE TOWER* Coxe, *Memoirs of the Life,* vol. 1, pp. 65–67.

135 THE MORE ACUTE TORIES Jonathan Swift, writing on December 18, 1711, quoted in "Walpole, Robert," in *Dictionary of National Biography* (1900), https://en.wikisource.org/wiki/Walpole,_Robert_(1676-1745)_(DNB00).

135 THE SEEMINGLY MINOR BUT EXCEPTIONALLY LUCRATIVE POST OF PAYMASTER GENERAL Plumb, *Sir Robert Walpole,* pp. 204–6.

135 HE OVERSAW A RIVER OF CASH Plumb, *Sir Robert Walpole,* pp. 207–9.

135 BRITAIN STILL OWED HUGE SUMS Henry Roseveare, *The Financial Revolution: 1660–1760* (London: Taylor and Francis, 1991), p. 53.

136 TO DEFEND "THE RELIGION, LAWS AND LIBERTIES OF *GREAT BRITAIN*" Robert Walpole, *Some Considerations Concerning the Publick Funds, the Publick Revenues and the Annual Supplies Granted by Parliament,* 2nd ed. (London: J. Roberts, 1735), p. 11.

136 THESE "IRREDEEMABLES" P. G. M. Dickson, *The Financial Revolution in England: A Study in the Development of Public Credit, 1688–1756* (Aldershot, UK: Gregg Revivals, 1993), p. 81.

136 WALPOLE WROTE, "IT CANNOT BE DOUBTED" Robert Walpole, *Some Considerations,* p. 13, quoted in Dickson, *Financial Revolution,* p. 84.

137 WALPOLE'S DEAL INCLUDED AN OFFER Helen Paul, "The 'South Sea Bubble,' 1720," *European History Online,* November 4, 2015, http://ieg-ego.eu/en/threads/european-media/european-media-events/helen-j-paul-the-south-sea-bubble-1720, section 2.

139 THE NATION'S DEBT HAD JUST BECOME CHEAPER For more on Walpole's 1717 efforts, see Dickson, *Financial Revolution,* pp. 82–87.

139 THE BIGGEST TASK WAS LEFT UNDONE This is P. G. M. Dickson's view (*Financial Revolution,* p. 85). He argues here with J. H. Plumb and prior scholars, who saw this as a minor alteration in the scheme Walpole had in mind in 1717.

140 AMBUSHED BY FELLOW WHIGS A blow-by-blow of this phase in the Stanhope-Sunderland v. Walpole-Townshend factional feud can be found in Plumb, *Sir Robert Walpole,* pp. 224–42. Taylor, "Robert Walpole."

CHAPTER 10

142 THE COMPANY'S FIRST PURPOSE-BUILT SHIP D. Templeman, *The Secret History of the Late Directors of the South-Sea Company* (London: Printed for the Author, 1735), p. 56.

142 RAYMOND MADE LANDFALL AT VERA CRUZ Geoffrey J. Walker,

Spanish Politics and Imperial Trade, 1700–1789 (London: Macmillan, 1979), p. 91.

145 "IT IS EASY TO PROPOSE SUFFICIENT FUNDS" Daniel Defoe, *An Essay upon Publick Credit* (London: Printed, and Sold by the Booksellers, 1710), pp. 27–28.

146 WHEN PARLIAMENT APPROVED THE LONGITUDE ACT *Acts of Parliament and awards: Anno Regni Annae, Reginae, Magnae Britannia, Francia, & Hibernia,* RGO 14/1 (London: John Baskett, Printer to the Queen's most Excellent Majesty, 1714), pp. 355–57.

146 BOTH NEWTON AND HALLEY ADVISED PARLIAMENT Peter Johnson, "The Board of Longitude 1714–1828," *Journal of the British Astronomical Association* 99, no. 2 (April 1989): 63. Newton's meticulous interest in the problem can be seen in the repeated drafts of a report to Parliament in holograph documents held in the Cambridge University Library, "Papers on Finding the Longitude at Sea," MS Add.3972.2:27–36, Cambridge University Library, Department of Manuscripts and University Archives, https://cudl.lib.cam.ac.uk/view/MS-ADD-03972/67.

146 "THIS OR THAT WHEEL IN THE GOVERNMENT" Defoe, *Essay upon Publick Credit,* p. 16.

147 DEFOE'S THEORY WASN'T AS PRECISE AS NEWTON'S Defoe, *Essay upon Publick Credit,* pp. 16–17.

148 THE IDEA WAS TO PERFORM A PRESENT-VALUE CALCULATION William Deringer, "Compound Interest Corrected: The Imaginative Mathematics of the Financial Future in Early Modern England," *Osiris* 33, no. 1 (2018): 116.

149 AS THE DEAL EMERGED P. G. M. Dickson, *The Financial Revolution in England: A Study in the Development of Public Credit, 1688–1756* (Aldershot, UK: Gregg Revivals, 1993), pp. 88–89.

151 THE "METHODS WHICH MADE OUR NATIONAL CREDIT RISE" Daniel Defoe, *The Chimera; or, The French Way of Paying National Debts Laid Open* (London: T. Warner, 1720), pp. 2–3.

151 THE SALE OF NEW SHARES TO THE PUBLIC These figures are drawn from Dickson, *Financial Revolution,* pp. 88–89.

152 "SIR, HERE IS A GREAT PIECE OF NEWS" Daniel Defoe, *The Anatomy*

of Exchange-Alley; or, A System of Stock-Jobbing (London: E. Smith, 1719), pp. 4–6.

154 "There is a gulf, where thousands fell" Jonathan Swift, *The Bubble* (Edinburgh: Benj. Took, 1721).

155 Could public credit survive Daniel Defoe, *The Villainy of Stock-Jobbers Detected, and the Causes of the Late Run upon the Bank and Bankers Discovered and Considered* (London, 1701), p. 5.

156 a gloriously uninteresting range All South Sea Company share prices in this and subsequent chapters come from the data set developed by Rik G. P. Frehen, William N. Goetzmann, and K. Geert Rouwenhorst, "New Evidence on the First Financial Bubble" (NBER Working Paper 15332, National Bureau of Economic Research, Cambridge, MA, September 2009). The data is available at http://som.yale.edu/faculty-research/our-centers-initiatives/international-center-finance/data/historical-southseasbubble.

157 the two sides were contemplating a stunningly ambitious second act See "The Secret History of the South Sea Scheme" (1721), in John Toland, *A Collection of Several Pieces of Mr. John Toland,* vol. 1 (London: J. Peele, 1726), p. 407. Toland, a philosopher and pamphleteer, did not write this piece; it was found unsigned among his papers and is most likely by one of the directors of the Company.

Chapter 11

158 "A thousand pointed rays" Alexander Pope to Edward Blount, June 2, 1725, in *The Works of Alexander Pope, Esq.*, vol. 5 (London: C. Bathurst etc., 1788), pp. 287–88.

159 the nonfictitious James Figg Figg's status as the first British champion is as much the result of those who publicized him—including Hogarth—as it is to any certain claim over any of the other bare-knuckle boxers of his day. For a characteristic account of his life and triumphs, see Henry Downes Miles, *Pugilistica: The History of British Boxing,* vol. 1 (Edinburgh: John Grant), 1906, pp. 8–12.

160 WALPOLE'S SUCCESSOR AS CHANCELLOR OF THE EXCHEQUER, JOHN AISLABIE For details on Aislabie's involvement, see P. G. M. Dickson, *The Financial Revolution in England: A Study in the Development of Public Credit, 1688–1756* (Aldershot, UK: Gregg Revivals, 1993), pp. 93–96. Dickson quotes Onslow on p. 95.

160 HE WAS OFTEN SEEN IN CONVERSATION WITH THE COMPANY'S CASHIER Robert Surman, the South Sea Company deputy cashier, to a Parliamentary committee of inquiry in 1721, quoted in Dickson, *Financial Revolution,* p. 96.

160 OBLIQUE INSIGHT INTO KNIGHT'S CHARACTER Stuart Handley, "Robert Knight," in *Oxford Dictionary of National Biography,* David Cannadine, general editor (Oxford: Oxford University Press, 2004–2016).

160 "RESTRAINED BY NO SCRUPLES OF CONSCIENCE" Quoted in Dickson, *Financial Revolution,* p. 95.

161 A POOL OF ABOUT £31 MILLION Dickson, *Financial Revolution,* p. 525.

162 WHICH MUST HAVE SURPRISED THE EAST INDIA DIRECTORS *St. James Weekly Journal,* January 2, 1720, quoted in Dickson, *Financial Revolution*, p. 96.

162 "THE BARGAIN WITH THE SOUTH SEA COMPANY WAS AGREED" Thomas Brodrick to his brother, January 24, 1720, quoted in "Brodrick, Thomas," in *The History of Parliament: The House of Commons 1715–1754,* ed. R. Sedgwick (London: The Stationery Office, 1970); online at The History of Parliament Trust, "Thomas Brodrick," Member Biographies, http://www.historyofparliamentonline.org/volume/1715-1754/member/brodrick-thomas-1654-1730; and discussed and quoted in Dickson, *Financial Revolution,* p. 96.

162 AISLABIE TOLD THE HOUSE OF COMMONS William Coxe, *Memoirs of the Life and Administration of Sir Robert Walpole,* vol. 2 (London: Longman, Hurst, Rees Orme, and Brown, 1816), p. 5.

162 CRAGGS STOOD NEXT Coxe, *Memoirs of the Life,* p. 5.

162 THOMAS BRODRICK, MEMBER FOR STOCKBRIDGE Stockbridge was

a famously rotten borough, featuring about one hundred voters whose favor was determined by the local bailiff.

162 "WE COULD NOT PROPERLY SPEAKING, CALL OURSELVES A NATION" Coxe, *Memoirs of the Life,* p. 6.

163 SUAVELY, SMOOTHLY, HE TOO PRAISED AISLABIE'S FISCAL PRUDENCE Coxe, *Memoirs of the Life,* p. 7.

164 WALPOLE'S STRATAGEM WAS WORKING EXACTLY AS INTENDED See Dickson, *Financial Revolution,* p. 521.

165 AS AISLABIE PUT IT John Aislabie's speech in Parliament on July 20, 1721, quoted in John Carswell, *The South Sea Bubble* (Phoenix Mill, UK: Alan Sutton, 1993), p. 112.

165 THE COMPANY'S NEGOTIATORS WERE ORDERED Coxe, *Memoirs of the Life,* p. 8.

165 A VICTORY FOR THE GOVERNMENT Brodrick to Lord Midleton, February 2, 1720, quoted in Dickson, *Financial Revolution,* p. 100.

165 THE SOUTH SEA ACT FILLED THIRTY-FIVE PRINTED PAGES Dickson, *Financial Revolution,* p. 525.

166 THE BANK OF ENGLAND'S LATEST OFFER Dickson, *Financial Revolution,* p. 521.

166 "AT SUCH PRICES . . . AS SHALL BE AGREED" South Sea Act, 6 Geo. 1, c. 4, quoted in Dickson, *Financial Revolution,* p. 104.

167 THIS WAS THE BEAUTY OF THE DEAL It should be noted that Richard Dale, among others, vehemently argues that the failure to agree on a conversion price in advance was not the original sin from which the South Sea disaster flowed. He writes that the Company could not treat its surplus shares as "profit," that existing shareholders would claim any surplus as a dividend; that the Company had no need of extra capital, as it wasn't an active trading company; and that it didn't need this mechanism to create new stock to sell, despite what he describes as "an anomalous clause" in its charter that constrained such new issues. Dale instead points to the huge payment the Company had committed itself to make to secure the rights to the deal.

These arguments are unconvincing: there is no evidence that the primary motivation of new South Sea shareholders anticipated dividends based on the gap between par and the market price of the stock as exchanged. It is clear that the Company's governing inner circle were perfectly prepared to treat surplus stock as wealth for the taking. The history of the bribery with which governmental support for the deal was nailed down turns on transactions that assume surplus stock is available to fund such expenses. And above all, the structure of the deal itself creates a difference of interest between existing company insiders (and prior shareholders) and those coming in later by exchanging other assets at a fixed valuation but a floating price.

I do agree with Dale that the high price of the deal, inflated by the competition with the bank, was a significant driver of Company actions and the need, ultimately, to pump up share prices. But there was more than one driver of that urge, and the creating of a huge pool of uncommitted shares was one of the most powerful.

167 "THE WHOLE SUCCESS OF THE SCHEME" Coxe, *Memoirs of the Life*, p. 9.

168 "IF THE PRICE OF THE STOCK HAD FALLEN" Archibald Hutcheson (uncredited), *The Several Reports of the Committee of Secrecy to the Honourable House of Commons, Relating to the Late South-Sea Directors* (London: A. Moore, 1721), p. 2.

169 "PERSONS WHOSE NAMES WERE NOT PROPER TO BE KNOWN" Hutcheson, *Several Reports of the Committee,* pp. 4–6.

170 NO PROOF OF WALPOLE'S PROBITY This isn't to say that Walpole was a model of probity, or even that he did not gain from privileged transactions. Walpole owned South Sea shares acquired in the decade before the bubble. Then, in February, he spent £8,770 for £9,000 in South Sea stock to cover what was in effect a forward sale from the month before. The slight discount is hard to explain, but this was clearly not one of Blunt and company's well-placed bribes. He had sold all of his holdings in the company by March 18, 1720, when the price stood at £194, or well before the precipitous rise to come. See J. H. Plumb, *Sir Robert Walpole: The Making of a Statesman* (London: Cresset, 1956), pp. 306–8.

170 THE BILL RECEIVED ITS FIRST READING The sequence of parlia-

mentary actions is drawn from Dickson, *Financial Revolution*, pp. 102–3, which differs slightly from Carswell's account in *The South Sea Bubble.*

170 RUNNERS WERE STAGED IN THE LOBBIES Carswell, *South Sea Bubble*, p. 122.

CHAPTER 12

173 "A MAN THAT IS OUT OF THE *STOCKS*" Daniel Defoe, *The Commentator*, no. 29, April 8, 1720 (London: J. Roberts, 1720).

175 THOSE BUYING INTO THE SCHEME Defoe, *Commentator.*

175 "SUCH TERMS AND CONDITIONS" P. G. M. Dickson, *The Financial Revolution in England: A Study in the Development of Public Credit, 1688–1756* (Aldershot, UK: Gregg Revivals, 1993), p. 131 and Table 16, p. 135.

176 THAT NUMBER TURNED OUT TO BE £375 Dickson, *Financial Revolution*, p. 135, table 16, and p. 139, table 17.

180 "THE PROFIT OF THE COMPANY" Dickson, *Financial Revolution*, p. 141.

180 "FOR THE COMPANY'S INTEREST" Quoted in Dickson, *Financial Revolution*, p. 142.

180 "A FINANCIAL PUMP" John Carswell, *The South Sea Bubble* (Phoenix Mill, UK: Alan Sutton, 1993), p. 135.

180 "THE HUMOUR OF THE TOWN" Defoe, *Commentator*, May 6, 1720.

181 FINANCIER AND COMPANY DIRECTOR SIR THEODORE JANSSEN Carswell, *South Sea Bubble*, p. 154.

181 "AS I PASS'D BY ONE OF THE PEWS" Defoe, *Commentator*, May 9, 1720.

182 "TO MAKE YOUR GRACE AS EASY AS I CAN" Sunderland to the Duchess of Marlborough, June 23, 1720, quoted in G. M. Townsend, *The Political Career of Charles Spencer, Third Earl of Sunderland 1695–1722* (Ph.D. diss., University of Edinburgh, 1984), p. 278.

183 "I GROW RICH SO FAST" James Windham to K. Windham, 1720, undated further, presumably spring, in *The Manuscripts of the Duke of Beaufort, K.G., the Earl of Donoughmore, and Others*, 12th report,

appendix, part 9 (London: Her Majesty's Stationery Office, 1891), p. 200.

183 MORE SIGNS OF THE TIMES *Applebee's Original Weekly Journal,* July 7, 1720, cited in Richard Dale, *The First Crash: Lessons from the South Sea Bubble* (Princeton, NJ: Princeton University Press, 2004), p. 109.

183 "OUR GREATEST LADIES" Edward Ward, "A South-Sea BALLAD, or, Merry Remarks upon Exchange Alley BUBBLES," broadside, in the Baker Business Library Kress Collection, online at https://iiif.lib.harvard.edu/manifests/view/ids:1245023.

184 THE COMPANY'S NEW RICH Dickson, *Financial Revolution,* p. 146.

185 ONE LONDONER WROTE THAT HE "HAD A FANCY TO GO AND TAKE A LOOK" Charles Wilson, *Anglo Dutch Commerce & Finance in the 18th Century* (Cambridge: Cambridge University Press, 1941), p. 122, quoted in Dale, *First Crash,* pp. 110–11.

185 "ANY IMPUDENT IMPOSTER" Adam Anderson, *An Historical and Chronological Deduction of the Origin of Commerce from the Earliest Accounts,* vol. 2 (London: J. Walter, 1764), p. 291.

186 HIS "EPISTLE TO BLUNT" Nicholas Amhurst, *Mr. Amhurst's Epistle to Sir John Blunt, Bart.* (London: R. Francklin, 1720), p. 9.

186 THE COMPANY OFFERED A NEW ISSUE Richard Westfall, *Never at Rest* (Cambridge: Cambridge University Press, 1980), p. 861.

CHAPTER 13

188 BORN IN 1660 OR THEREABOUTS This account of Hutcheson's early career is derived from William Deringer's *Calculated Values* (Cambridge, MA: Harvard University Press, 2018), p. 167.

189 HUTCHESON RAPIDLY BECAME ONE OF THE LEADING QUANTITATIVE DISPUTANTS A. A. Hanham, "Hutcheson, Archibald," in *Oxford Dictionary of National Biography,* ed. David Cannadine, online ed. (Oxford: Oxford University Press, 2004), https://doi.org/10.1093/ref:odnb/53923.

189 PAGE AFTER PAGE OF FIGURES See, for example, Hutcheson's addendum "Abstracts of the Aforgoing States," in *Some Calculations*

and Remarks Relating to the Present State of Public Debts and Funds (London: no publisher listed, 1718), pp. 30–31, as well as the tables that accompany each section of the pamphlet except the preface and conclusion.

191 AT FIRST, HUTCHESON DID NOT OPPOSE THE SCHEME Archibald Hutcheson, *Some Calculations Relating to the Proposals Made by the South-Sea Company, and the Bank of England, to the House of Commons* [...]., 2nd ed. (London: J. Morphew, 1720), published in Hutcheson's *A Collection of Calculations and Remarks Relating to the South Sea Scheme & Stock* (London: no publisher listed, 1720), p. 7.

192 WHAT ALL THAT MIGHT BE WORTH See Deringer, *Calculated Values,* pp. 195–98.

194 THE GENUINE FLASH OF BRILLIANCE See Deringer, *Calculated Values,* pp. 202–8.

195 HUTCHESON MADE HIS DISTINCTIVE CONTRIBUTION Deringer emphasizes here Hutcheson's technical wizardry in finding a simple way to quantify and calculate expectations for a company whose stock was both a financial instrument—basically, a bond with a return defined by the interest payment the government agreed to pay for all the debt swapped for shares—and an ongoing mercantile enterprise, trading goods and slaves across the Atlantic. That's a key point; Hutcheson was able to achieve the scientific revolutionary goal of subsuming a variety of phenomena under a single mathematical treatment. But the speculative reach of Hutcheson's model was at least equally important: he made it possible to examine a range of futures, and thus deepen an understanding of how matters would have to work in order for an investment or an enterprise to make sense, and in some ways, the whole notion of financial analysis flows from this kind of work.

196 EVEN SO, THE NUMBERS DIDN'T WORK Hutcheson, *Collection of Calculations,* p. 11.

197 WHAT REMAINED UNKNOWN Hutcheson, *Collection of Calculations,* p. 15.

198 "IF THE COMPUTATIONS I HAVE MADE, BE RIGHT" Hutcheson, *Collection of Calculations,* pp. 15–16.

199 HUTCHESON USED THAT CULTURAL AUTHORITY FOR HIS OWN ENDS Some historians have downplayed what others, including me, see as Hutcheson's significance during the bubble. See, e.g., Helen J. Paul, *The South Sea Bubble: An Economic History of Its Origins and Consequences* (London: Routledge, 2011), pp. 75–77. The response, I think, is (a) he was right about the South Sea Company's prospects and he framed his argument in a mathematically coherent way, and (b) one can be an early-modern thinker, not a modern one, but still anticipate in important ways lines of reasoning that would be brought to their full development later.

200 AN EXERCISE IN ABSURDITY Hutcheson, *Some Calculations Relating to the Proposals,* p. 7.

CHAPTER 14

202 THE NAME BY WHICH IT BECAME FAMOUS Pat Rogers, "South Sea Bubble Myths," *Times Literary Supplement,* April 9, 2014.

202 "ABOUT 10 SHARES [BOUGHT]" William Chetwood, *The Stock-Jobbers; or, The Humours of Exchange-Alley* (London: J. Roberts, 1720), pp. 3–4.

203 "BUBLED OUT OF ¼ OF THE INTRINSICK VALUE" Transmitted by Archibald Hutcheson as "Some Memorandums Relating to Exchange, by an Eminent Merchant," possibly to John Dalrymple, Earl of Stair and British ambassador to France, November 16, 1718, quoted in William Deringer, *Calculated Values* (Cambridge, MA: Harvard University Press, 2018), p. 201.

203 THOSE WHO "MAKE THE *EXCHANGE* A GAMING TABLE" Daniel Defoe, *The Free-Holders Plea Against Stock-Jobbing Elections of Parliament Men* (London: 1701), pp. 21–22.

204 "*RESOLVING TO BE RICH*" Daniel Defoe, *The Anatomy of Exchange-Alley; or, A System of Stock-Jobbing* (London: E. Smith, 1719), p. 38.

205 THOMAS GUY WAS NO MAN'S FOOL Roger Jones, "Roger Mead, Thomas Guy, the South Sea Bubble and the Founding of Guy's Hospital," *Journal of the Royal Society of Medicine* 103, no. 3 (March 1, 2010): 87–92.

205 GUY WAS A CHARITABLE MAN Jones, "Roger Mead, Thomas Guy."

205 HE DID WELL WITH HIS FIRST TRANSACTIONS John Carswell, *The South Sea Bubble* (Phoenix Mill, UK: Alan Sutton, 1993), p. 26.

205 HE OWNED 54,040 SHARES Carswell, *South Sea Bubble,* p. 136.

206 JUST UNDER THEIR PAR VALUE OF £100 Andrew Odlyzko, "Newton's Financial Misadventures in the South Sea Bubble," *Notes and Records of the Royal Society* 73, no. 1 (February 2019): 40.

206 HE BEGAN TO SELL Odlyzko, "Newton's Financial Misadventures," p. 40. Guy's sales can be seen in figure 1 on p. 34. The final sale is noted on p. 40.

206 A TOTAL OF £250,000 Jones, "Roger Mead, Thomas Guy."

206 "THE LARGEST HONEST FORTUNE" Carswell, *South Sea Bubble,* p. 137.

207 HE DID NOT CHASE THE STOCK Guy did re-enter the market for South Sea stock over the summer, but Odlyzko argues that this was most likely a maneuver required to cover a forward sale (what we would now call a short sale), in which he was required to deliver stock he did not still own. He lost a small amount of money on this move, but it did not change his overall result. See Odlyzko, "Newton's Financial Misadventures," p. 41.

208 "NOT WORTH SO MUCH AS 2 OR 300L" Mary Cowper, *Diary of Mary, Countess Cowper* (London: John Murray, 1864), p. 184.

208 SARAH CHURCHILL TOOK THE TEMPERATURE OF THE MARKET Carswell, *South Sea Bubble,* pp. 154–55, and J. H. Plumb, *Sir Robert Walpole: The Making of a Statesman* (London: Cresset, 1956), p. 307.

208 HE SOLD ABOUT HALF Plumb, *Sir Robert Walpole,* pp. 306–8.

208 SIR ISAAC NEWTON DID BETTER Odlyzko, "Newton's Financial Misadventures," p. 50.

210 EVERY SOUTH SEA STOCK TRANSACTION THE BANK MADE IN 1720 The details of Hoare's trading come from Peter Temin and Hans-Joachim Voth, "Riding the South Sea Bubble" (Massachusetts Institute of Technology Department of Economics Working Paper Series, no. 04-02, December 21, 2003), especially pp. 13–25. For a similar discussion drawing on the same data in a broader context,

Peter Temin and Hans-Joachim Voth, *Prometheus Shackled: Goldsmith Banks and England's Financial Revolution After 1700* (Oxford: Oxford University Press, 2013), chapter 5. I follow these two authors in their interpretation of Hoare's results: these were, as they argue from several lines of evidence not included here, the product of a well-thought-out trading strategy that turned on empirical reasoning about both market behavior and the prospects of the South Sea Company.

210 £140,029 CHANGING HANDS Temin and Voth, *Prometheus Shackled,* p. 111.

210 "THE BANKERS EARNED AS MUCH IN 1720–21" Temin and Voth, *Prometheus Shackled,* p. 120.

211 LENDING BARELY MORE THAN £40 ON £100 OF STOCK Temin and Voth, *Prometheus Shackled,* p. 118.

211 HE WAS LUCKY UP TO A POINT Leo Damrosch, *Jonathan Swift: His Life and His World* (New Haven, CT: Yale University Press, 2013), p. 339.

212 CONCEIVE THE WORKS OF MIDNIGHT HAGS Jonathan Swift, "The Run upon the Bankers," lines 29–32. *Dean Swift's Works,* vol. 7 (London: Luke Hansard, 1801), p. 177, online at https://en.wikisource.org/wiki/The_Works_of_the_Rev._Jonathan_Swift/Volume_7/The_Run_upon_the_Bankers.

CHAPTER 15

213 "THERE IS NO GAIN UNTIL THE STOCK IS SOLD" Alexander Pope to John Caryll, April or May 1720, in *The Works of Alexander Pope,* new ed., vol. 6 (London: John Murray, 1871), pp. 271–72.

213 COME, FILL THE SOUTH SEA GOBLET FULL Alexander Pope, "An Inscription upon a Punch-Bowl," in *The Complete Poetical Works* (1902), online at http://www.bartleby.com/203/67.html.

214 "THERE HAPPENED SUCH SUDDEN FLUCTUATIONS" Adam Anderson, *An Historical and Chronological Deduction of the Origin of Commerce, from the Earliest Accounts,* vol. 3 (London: J. Walter, 1787), p. 97.

215 "RAIS'D UP ON HOPE'S ASPIRING PLUMES" Jonathan Swift, *The Bubble* (Edinburgh: Benj. Took, 1721).

215 DUTCH SPECULATORS For foreign investment in the bubble, see P. G. M. Dickson, *The Financial Revolution in England: A Study in the Development of Public Credit, 1688–1756* (Aldershot, UK: Gregg Revivals, 1993), pp. 140–41.

215 PACKED IN HOGSHEADS Hogsheads were standardized barrels, but the standard was different depending on the commodity carried. The most common volume was in the range of sixty (US) or fifty (Imperial) gallons. The term was also used for a tobacco barrel of a standard, much larger size. Whatever had originally been carried in the barrels the Dutch used to deliver their wealth to Exchange Alley, a single hogshead could hold a lot of cash.

217 IT BEGAN TO LEND REAL CASH The terms and amounts of these loans are drawn from Dickson, *Financial Revolution,* pp. 141–44.

218 THE TOTAL IN PLAY WAS STAGGERING Dickson, *Financial Revolution,* p. 142.

218 ALLEY MEN SPECULATED ABOUT THE PRICE Anderson, *Historical and Chronological Deduction,* p. 96.

219 THE ENTIRE ISSUE—£5 MILLION—SOLD OUT Dickson, *Financial Revolution,* p. 127, citing *Applebee's Original Weekly Journal,* June 18, 1720.

219 AT LEAST HALF OF BOTH HOUSES OF PARLIAMENT SIGNED UP Dickson, *Financial Revolution,* p. 127.

219 153 MEMBERS ON THE LIST John Carswell, *The South Sea Bubble* (Phoenix Mill, UK: Alan Sutton, 1993), pp. 160–61; and Dickson, *Financial Revolution,* p. 126. Dickson says Craggs tapped 153 members of parliament; Carswell 154. I follow Dickson as the more authoritative (and recent) researcher.

219 THOSE WHO BOUGHT IN JUNE Dickson, *Financial Revolution,* p. 125, table 13.

219 WHAT WE NOW CALL AN OPTION Over the last decade and a half, economic historians have waged an active debate over this payment structure: Was the South Sea installment plan an early form

of option trading, a payment that permits an investor to decide whether or not to buy shares at a later date? Or did signing up for the money subscription bind one to a contract, an obligation to pay when the balance came due? As events played out, the option interpretation seems to be closest to right, as investors could and did walk away from their deals. But from the Company's side in April and on the later subscriptions this was a distinction with very little difference. It is impossible to reach a wholly benign reading of the money subscriptions: the goal was to prop up the market for South Sea stock, and the increasingly forgiving payment plans did just that. See Anderson, *Historical and Chronological Deduction,* p. 97, for what is close to an eyewitness account of the Company's behavior. For the debate between economists, see Richard Dale, *The First Crash: Lessons from the South Sea Bubble* (Princeton, NJ: Princeton University Press, 2004), in which he treats subscription sales as a binding agreement, and Gary Shea, "Understanding Financial Derivatives During the South Sea Bubble: The Case of the South Sea Subscription Shares," *Oxford Economic Papers* 59 (2007): 73–104, especially Shea's conclusions on pp. 100–104.

220 ANOTHER PHILOSOPHER'S SUCCESSFUL OPTION TRADE Aristotle, *Politics,* book 1, section 1259a.

220 OPTIONS WERE IN COMMON USE Anne L. Murphy, *The Origins of English Financial Markets* (Cambridge: Cambridge University Press, 2009), pp. 24–30.

220 A VARIETY OF PRIVATE CALL CONTRACTS See Dickson, *Financial Revolution,* pp. 497–505, and Gary S. Shea, "Sir George Caswall vs. the Duke of Portland: Financial Contracts and Litigation in the Wake of the South Sea Bubble" (2007), in Jeremy Atack and Larry Neal, *The Origin and Development of Financial Markets and Institutions* (Cambridge: Cambridge University Press, 2009), pp. 130–31.

221 WHEN HE SOLD OUT HIS COMPANY HOLDINGS Andrew Odlyzko, "Newton's Financial Misadventures in the South Sea Bubble," *Notes and Records of the Royal Society* 73, no. 1 (February 2018): 13.

222 Such contracts were common See, e.g., Dickson, *Financial Revolution,* pp. 499–500.

222 he could not resist the lure of playing for the highest stakes See Shea, "Sir George Caswall," especially pp. 15–24.

224 derivatives are not inevitably dangerous For a first overview of derivatives, see, e.g., Paul Wilmott, Susan Howson, Sam Howison, *The Mathematics of Financial Derivatives: A Student Introduction* (Cambridge: Cambridge University Press, 1995).

225 Eager projectors floated twenty-seven companies William Robert Scott, *Constitution and Finance of English, Scottish and Irish Joint Stock Companies to 1720,* vol. 3 (Cambridge: Cambridge University Press, 1901), pp. 449–57.

226 "melting down saw dust and chips" Anderson, *Historical and Chronological Deduction,* vol. 3, p. 103.

226 "for raising the sum of Six Millions sterling" Quoted in John Carswell, *South Sea Bubble,* p. 156.

226 a 1764 survey of the wreckage See Dale, *First Crash,* p. 107. He got his numbers from Anderson, *Historical and Chronological Deduction.*

227 "the traffic [of bubbles]" Quoted in Ron Harris, "The Bubble Act: Its Passage and Its Effects on Business Organization," *Journal of Economic History* 54, no. 3 (September 1994): 615.

227 Parliament passed a bill known as the Bubble Act For details on the legislative background to the Bubble Act, see Dickson, *Financial Revolution,* pp. 147–49.

227 shares changed hands at £1,050 Harris, "Bubble Act," p. 613. See his footnote 11 for a discussion of the attempts to reconstruct intraday peaks, as distinct from the printed reports of closings like Castaing's and others.

228 Exchange Alley priced the South Sea Company Anderson, *Historical and Chronological Deduction,* p. 97.

228 How goes the Stock Quoted in Arthur H. Cole, "The Bancroft Collection," *Bulletin of the Business Historical Society* 9, no. 6 (December 1935): 94.

CHAPTER 16

229 "THE DEMON MAKES HIS FULL DESCENT" Alexander Pope, "Of the Use of Riches" (Epistle to Bathurst), in *The Complete Poetical Works of Alexander Pope,* ed. Henry Boynton (Boston: Houghton, Mifflin, 1903), lines 370–74.

229 "THESE NEW SUBSCRIBERS" Archibald Hutcheson, *An Estimate of the Value of South-Sea Stock with Some Remarks Relating Thereto* (London: no publisher listed, 1720), pp. 3–4.

229 THE OPTIMISTS STILL SMILED James Windham to Ashe Windham, July 12, 1720, in *The Manuscripts of the Duke of Beaufort, K.G., the Earl of Donoughmore, and Others,* 12th report, appendix, part 9 (London: Her Majesty's Stationery Office, 1891), p. 200.

230 HE REFUSED ALL ADVICE TO CASH OUT Lewis Melville, *The Life and Letters of John Gay (1685–1732), Author of "The Beggar's Opera"* (London: Daniel O'Connor, 1921), chapter 6, https://openlibrary.org/books/OL13493940M/Life_and_letters_of_John_Gay_(1685-1732).

231 "WHAT MAGICK MAKES OUR MONEY RISE" Jonathan Swift, "The South Sea Project," 1721, lines 2–3.

231 THE LAST MAJOR PILE OF OFFICIAL DEBT IN PRIVATE HANDS P. G. M. Dickson, *The Financial Revolution in England: A Study in the Development of Public Credit, 1688–1756* (Aldershot, UK: Gregg Revivals, 1993), p. 93, table 9.

232 "THE PEOPLE CAME THITHER" *Weekly Journal, or, Saturday's Post,* July 16, 1720, quoted in Dickson, *Financial Revolution,* p. 133.

232 "WHAT A RAGE PREVAILS HERE" James Craggs to Earl Stanhope, July 15, 1720, in William Coxe, *Memoirs of the Life and Administration of Sir Robert Walpole, Earl of Orford,* vol. 2 (London: T. Cadell, Jun., and W. Davies, 1798), p. 189.

232 "WE DESIRE YOUR FAVOR" John Blunt, *A True State of the South-Sea Scheme* (London: J. Peele, 1722), p. 32.

233 ISAAC NEWTON, WHO TRADED ANNUITIES Isaac Newton to John Francis Fauquier, July 27, 1720, in *The Correspondence of Isaac New-*

ton, vol. 7, 1718–1727, ed. A. Rupert Hall and Laura Tilling (Cambridge: Cambridge University Press, 1959), p. 96; and Andrew Odlyzko, "Newton's Financial Misadventures in the South Sea Bubble," *Notes and Records of the Royal Society* 73, no. 1 (February 2018): 26.

233 THE DIRECTORS AUTHORIZED STILL MORE LOANS Dickson, *Financial Revolution,* p. 143.

233 THE COMPANY IGNORED ONE OF THE FEW EXISTING MARKET REGULATIONS John Carswell, *The South Sea Bubble* (Phoenix Mill, UK: Alan Sutton, 1993), p. 169.

233 "GREAT SUMS AT FORTY AND FIFTY *PER CENT*" James Milner, *Three Letters, Relating to the South-Sea Company and the Bank* (London: J. Roberts, 1720), p. 34.

234 "THIS SUBSCRIPTION . . . WAS COMPLETED IN THREE HOURS' TIME" Adam Anderson, *An Historical and Chronological Deduction of the Origin of Commerce,* vol. 3 (London: J. Walter, 1787), p. 113.

235 "THIS PROJECT MUST BURST" Sarah, Duchess of Marlborough, quoted in J. H. Plumb, *Sir Robert Walpole: The Making of a Statesman* (London: Cresset, 1956), p. 301.

235 "THEIR STOCK WOULD RECEIVE THE LEAST ALTERATION" Daniel Defoe, *The Commentator,* no. 65, August 15, 1720 (London: J. Roberts, 1720).

235 "BUY OF SOUTH SEA STOCK AT THE PRESENT PRICE" Alexander Pope to Lady Mary Wortley Montagu, August 22, 1720, Record ID 301686, Morgan Library & Museum, https://www.themorgan.org/blog/some-terrible-investment-advice-alexander-pope-buy-south-sea-stock.

236 A NEW ANALYSIS OF HIS ACCOUNTS Andrew Odlyzko, "Newton's Financial Misadventures in the South Sea Bubble," *Notes and Records of the Royal Society* 73, no. 1 (February 2018): 24–25.

236 HIS SUMMER-LONG BUYING SPREE Odlyzko, "Newton's Financial Misadventures," p. 22.

238 "I AM AT LAST SETTLED" Carswell, *South Sea Bubble,* p. 139.

238 BLUNT'S ACCOUNTS John Blunt, *The Particular and Inventory of Sir*

John Blunt, Bart. One of the late Directors of the South-Sea Company [. . .] (London: Jacob Tonson, 1721) p. 12.

238 "CONSIDERABLE ESTATES IN NORFOLK" Lewis Melville, *The South Sea Bubble* (London: Daniel O'Connor, 1921), p. 112; and Dickson, *Financial Revolution,* p. 147.

238 BLUNT TOSSED JUST £500 INTO THE KITTY See the not entirely unbiased "The Secret History of the South Sea Scheme" (1721), in John Toland, *A Collection of Several Pieces of Mr. John Toland,* vol. 1 (London: J. Peele, 1726), p. 434.

238 "THE CASHIERS LENT UPWARDS OF THREE MILLIONS IN ONE DAY" "Secret History of the South Sea Scheme," pp. 429–30.

239 "THE MORE CONFUSION THE BETTER" "Secret History of the South Sea Scheme," p. 429.

239 "ONE MILLION OR TWO" "Secret History of the South Sea Scheme," p. 430.

PART 3

241 YE WISE PHILOSOPHERS *The Works of Jonathan Swift,* vol. 1 (London: Henry Washbourne, 1841), p. 620.

CHAPTER 17

243 "FOR A FEW DAYS, INDEED" Adam Anderson, *An Historical and Chronological Deduction of the Origin of Commerce,* vol. 3 (London: J. Walter, 1764), p. 102.

243 THE PRINCE OF WALES John Carswell, *The South Sea Bubble* (Phoenix Mill, UK: Alan Sutton, 1993), p. 168.

244 THE "PRINCE OF WALES' BUBBLE" James Craggs the younger to Earl Stanhope, July 12, 1720, in William Coxe, *Memoirs of the Life and Administration of Sir Robert Walpole,* vol. 2 (London: T. Cadell, Jun., and W. Davies, 1798), p. 188.

244 "THE SOUTH SEA JUNTO" Anderson, *Historical and Chronological Deduction,* p. 113.

244 TREASURY OFFICIALS CONTACTED THE ATTORNEY GENERAL P. G. M. Dickson, *The Financial Revolution in England: A Study in*

the Development of Public Credit, 1688–1756 (Aldershot, UK: Gregg Revivals, 1993), p. 148.

244 THE YORK BUILDINGS COMPANY WENT FIRST Anderson, *Historical and Chronological Deduction,* p. 113.

245 "THE GROWING EVIL OF PROJECTING" Daniel Defoe, *The Commentator,* no. 68, August 26, 1720 (London: J. Roberts, 1720).

246 ACCELERANTS IN THE COLLAPSE See, e.g., Dickson, *Financial Revolution,* pp. 152–53.

246 "FATAL WRITS OF *SCIRE FACIAS*" Anderson, *Historical and Chronological Deduction,* pp. 113–14.

248 "PRAY, WHAT WOULD BE THE PRICE OF STOCKS" Defoe, *Commentator,* no. 66, August 15, 1720 (London: J. Roberts, 1720).

249 "THE COMMON PEOPLE" Daniel Defoe, *A Journal of the Plague Year* (Eugene: University of Oregon Press, Renascence Editions, 2008), p. 18.

249 "IT WAS NOT IN THE POWER OF THE MAGISTRATES" Defoe, *A Journal of the Plague Year,* p. 100.

249 "PEOPLE BEGAN TO GIVE UP THEMSELVES TO THEIR FEARS" Defoe, *A Journal of the Plague Year,* pp. 102–3.

CHAPTER 18

252 A BEAU From "An Essay in Defense of the Female Sex" (1696), quoted in Antoin E. Murphy, *John Law: Economic Theorist and Policy-Maker* (Oxford: Clarendon, 1997), p. 20.

252 "HANDSOME, TALL, WITH A GOOD ADDRESS" W. Gray, *The Memoirs, Life and Character of the Great Mr. Law and His Brother in Paris* (London: Sam Briscoe, 1721), p. 4.

252 THE WRONG PERSON'S LOVER W. Gray and John Evelyn, writing in the *London Journal,* quoted in Murphy, *John Law,* p. 21. Murphy discusses the duel and the suspicions around it beginning on p. 21.

252 JAILED, TRIED, AND CONDEMNED TO DEATH Murphy, *John Law,* p. 29.

253 "ROMANCES MUST BE EMBELLISHED" Gray, *Memoirs, Life and Character,* p. 9.

253 WALKED LAW OUT OF THE PRISON See Murphy, *John Law*, p. 33.

253 "SUPERIOR AND VERY UNCOMMON SKILL" Gray, *Memoirs, Life and Character*, p. 5.

254 "THE FIRST . . . THEORETICAL FRAMEWORK" François Velde, "Book Review: *John Law: Economic Theorist and Policy-Maker.* By Antoin E. Murphy," *Journal of Political Economy* 107, no. 1 (February 1999): 202.

254 THE VARIOUS THINGS MONEY DOES John Law, *Money and Trade Considered: with a Proposal for Supplying the Nation with Money*, quoted in Murphy, *John Law*, p. 54.

255 A MUCH MORE MODERN VIEW See Murphy, *John Law*, chapters 7 and 8.

255 "THE STOCKS OF THE EAST INDIA COMPANIES" John Law, *John Law's Essay on a Land Bank*, quoted in Murphy, *John Law*, p. 61.

256 FRANCE'S TRADITIONAL METHODS OF RAISING REVENUE For a description of Louis XIV's various revenue sources, see François R. Velde, "French Public Finance between 1683 and 1726" (preliminary and rough draft of paper prepared for the 14th International Economic History Congress, Helsinki, 2006, session 112), http://www.helsinki.fi/iehc2006/papers3/Velde.pdf, pp. 5–10.

256 FRANCE OWED 3.5 BILLION LIVRES Philip T. Hoffman, Gilles Postel-Vinay, and Jean-Laurent Rosenthal, *Priceless Markets: The Political Economy of Credit in Paris, 1660–1870* (Chicago: University of Chicago Press, 2000), p. 70. Currency conversion based on the amount of gold that could be bought in each currency.

257 LAW'S BANQUE GÉNÉRALE Hoffman, Postel-Vinay, and Rosenthal, *Priceless Markets*, p. 73.

258 THOSE WHO SWAPPED THEIR HOLDINGS These figures on the *Banque générale* business come from Murphy, *John Law*, pp. 158–61.

258 LAW GOT THE GO-AHEAD Murphy, *John Law*, p. 172. Currency conversion based on the amount of gold that could be bought in each currency.

259 A GLUTTONOUS EXPANSION See Antoin Murphy's discussion of Law's acquisition spree beginning on p. 110 of *John Law*.

260 MISSISSIPPI COMPANY SHARES MOVED François Velde, "John Law's System," *American Economic Review* 97, no. 2 (May 2007): 277.

261 LAW HASTILY RESTORED THE PEG François Velde, "Government Equity and Money: John Law's System in 1720 France," Working Paper, p. 30, https://www.heraldica.org/econ/law.pdf.

261 LAW AGAIN TACKLED THE PRICING FOR MISSISSIPPI COMPANY SHARES See Murphy, *John Law*, pp. 244–45, for these figures and a detailed account of the steps Law proposed; see also François Velde, "Government Equity and Money: John Law's System in 1720 France" (Federal Reserve Bank of Chicago Working Paper no. 2003-31, December 2003), pp. 29–31, https://www.heraldica.org/econ/law.pdf.

262 THE PARIS STOCK MARKET RESPONDED Murphy, *John Law*, p. 257.

262 THE FINAL BILL WAS STAGGERING Murphy, *John Law*, p. 307.

262 LAW FLED PARIS This gloss on Law's escape is derived from Murphy, *John Law*, pp. 309–11.

CHAPTER 19

264 "LET THE PRINCE LYE WITH HIS WIFE" Mary Cowper, *Diary of Mary, Countess Cowper* (London: John Murray, 1864), p. 134.

264 HE FLEXED HIS MUSCLES William Coxe, *Memoirs of the Life and Administration of Sir Robert Walpole, Earl of Orford*, vol. 1 (London: T. Cadell, Jun., and W. Davies, 1798), pp. 116–25.

265 WALPOLE LET IT BE KNOWN J. H. Plumb, *Sir Robert Walpole: The Making of a Statesman* (London: Cresset, 1956), pp. 291–92.

265 THE ROYAL FAMILY'S PUBLIC DETENTE Coxe, *Memoirs of the Life*, pp. 132–33.

265 "TO PROCURE *WALPOLE* AND *TOWNSHEND*" Cowper, *Diary*, p. 135.

266 WALPOLE'S STOCK EXCHANGE ADVENTURES Plumb, *Sir Robert Walpole*, p. 291.

266 HE BOUGHT BANK OF ENGLAND AND THE ROYAL AFRICAN COMPANY STOCK Plumb, *Sir Robert Walpole*, p. 309.

266 HE WAS SO SURE OF THE COMPANY'S PROSPECTS Cowper, *Diary*, p. 158.

267 A GOOD CHOICE OF AGENT For Walpole's summer investment decisions, see Plumb, *Sir Robert Walpole,* pp. 315–19.

267 THE COMPANY'S FIRST ATTEMPTS John Carswell, *The South Sea Bubble* (Phoenix Mill, UK: Alan Sutton, 1993), pp. 152–68.

268 WALPOLE ATTENDED FOR THE MINISTRY Carswell, *South Sea Bubble,* p. 183.

268 A CONTRACT IN NAME ONLY P. G. M. Dickson, *The Financial Revolution in England: A Study in the Development of Public Credit, 1688–1756* (Aldershot, UK: Gregg Revivals, 1993), pp. 166–67.

269 "THE UNIVERSAL DELUGE OF THE SOUTH SEA" Alexander Pope to the Bishop of Rochester, September 23, 1720, in *The Works of Alexander Pope, with Notes and Illustrations by Himself and Others,* vol. 7 (London: Longman, Brown, 1847), pp. 184–85.

270 "THE MOST UNLUCKY IN THE WORLD" Lady Mary Cowper to Lady Mary Wortley Montague, March 21, 1721, in Lady Mary Wortley Montague, *The Complete Letters of Lady Mary Wortley Montague,* vol. 2, ed. Robert Halsband (Oxford: Oxford University Press, 1966), pp. 3–4.

270 "POOR JIMMY'S AFFAIRS" William Windham to Ashe Windham, September 27, 1720, and November 26, 1720, in *The Manuscripts of the Duke of Beaufort, K.G., the Earl of Donoughmore, and Others,* 12th report, appendix, part 9 (London: Her Majesty's Stationery Office, 1891), p. 201.

270 "A VERY SILLY FOOL" James Windham to Ashe Windham, January 5, 1720/21, in *Manuscripts of the Duke,* pp. 201–2.

270 "AN UNDONE MAN" James Windham to Ashe Windham, January 3, 1721, in *Manuscripts of the Duke,* pp. 201–2.

271 A FARMER RETURNING HOME John Saunders in *Applebee's Original Weekly Journal,* October 15, 1720, p. 4.

271 "SOME HONEST SORROWFUL PERSONS" *Applebee's Original Weekly Journal,* October 1, 1720, p. 1.

271 THIRTY THOUSAND PEOPLE AND INSTITUTIONS Dickson, *Financial Revolution,* pp. 161–62.

271 THOSE BUYERS INCLUDED Dickson, *Financial Revolution,* p. 108.

271 ON RECORD AT SOUTH SEA HOUSE Adam Anderson, *An Historical and Chronological Deduction of the Origin of Commerce,* vol. 3 (London: J. Walter, 1764), p. 123.

272 WALPOLE HIMSELF FACED A SIMILAR RISK Plumb, *Sir Robert Walpole,* pp. 318–19.

272 "A GREAT MANY GOLDSMITHS ARE ALREADY GONE" Thomas Brodrick to Alan Brodrick, 1st Viscount Midleton, September 27, 1720, in Coxe, *Memoirs of the Life,* p. 191.

272 A "GREAT AND GENERAL CALAMITY" Archibald Hutcheson, *Some Paragraphs of Mr. Hutcheson's Treatises on the South-Sea Subject* (London, 1723), p. 8.

CHAPTER 20

277 "PLEASED WITH THE THOUGHT" Extracts of letters from Mr. Jacombe to Robert Walpole, in William Coxe, *Memoirs of the Life and Administration of Sir Robert Walpole, Earl of Orford,* vol. 2 (London: T. Cadell, Jun., and W. Davies, 1798), p. 193.

278 HE HAD THOUGHT ABOUT JACOMBE'S NOTION See P. G. M. Dickson, *The Financial Revolution in England: A Study in the Development of Public Credit, 1688–1756* (Aldershot, UK: Gregg Revivals, 1993), p. 159.

278 "IT WAS WITH GREAT RELUCTANCE" Robert Walpole to King George I, "Some Thought and Considerations Concerning the Present Posture of the South Sea Stock, Humbly Laid Before His Majesty," in Coxe, *Memoirs of the Life,* p. 197.

278 THE PLAN HAD THREE PARTS Walpole to the King, "Some Thought and Considerations," p. 198.

280 THE FORLORN HOPE OF SOUTH SEA INVESTORS Dickson, *Financial Revolution,* p. 171.

280 THE CENTERPIECE OF THE SETTLEMENT Dickson, *Financial Revolution,* p. 176.

280 WALPOLE REFUSED TO ENTERTAIN ONE OBVIOUS MOVE Dickson, *Financial Revolution,* p. 170.

281 "THE PLUNDERERS OF THEIR COUNTRY" Archibald Hutcheson,

Some Paragraphs of Mr. Hutcheson's Treatises on the South-Sea Subject (London, 1723), pp. 8–9.

282 "The South-Sea act was made" William Cobbett, *The Parliamentary History of England, from the Earliest Period to the Year 1803, vol. 7, 1714–1722* (London: T. C. Hansard, 1811), p. 690.

284 This Committee of Secrecy The term "secrecy" here meant that the committee was restricted to those selected for it; other MPs could not simply choose to attend its meetings.

284 a punishing schedule William Robert Scott, *The Constitution and Finance of English, Scottish and Irish Joint Stock Companies to 1720,* vol. 3 (Cambridge: Cambridge University Press, 1910), p. 334.

284 The South Sea Company's cashier, Robert Knight Abel Boyer, *The Political State of Great Britain,* vol. 21 (London: printed for the author, 1721), pp. 87–90.

285 "it would open such a Scene" "16 February 1721 (Anno 7° Georgii Regis 1720)," *Journal of the House of Commons, vol. 19, 1718–1721* (London: reprinted by order of the House of Commons, 1805), p. 436.

285 he "did not think it proper" "16 February 1721 (Anno 7° Georgii Regis 1720)," *Journal of the House of Commons,* p. 432.

285 "the weight of Inquiry" Boyer, *Political State of Great Britain,* p. 74.

285 "the risk of losing his own great share" Historical Manuscripts Commission, *The Manuscripts of the Earl of Buckinghamshire, the Earl of Lindsey, the Earl of Onslow, Lord Emly, Theodore J. Hare, Esq., and James Round, Esq., M.P.,* 14th report, appendix, part 9 (London: Her Majesty's Stationery Office, 1895), p. 507.

285 He had told Janssen "16 February 1721 (Anno 7° Georgii Regis 1720)," *Journal of the House of Commons,* pp. 426–27.

286 Knight's letter to his colleagues Boyer, *Political State of Great Britain,* p. 75.

286 it was common knowledge Historical Manuscripts Commission, p. 507.

Chapter 21

290 Sir John Blunt was many things I follow John Carswell's account of Blunt before the Lords from his *The South Sea Bubble* (Phoenix Mill, UK: Alan Sutton, 1993), p. 233.

291 that "favorite minister, by name Sejanus" "Wharton, Philip, Duke of Wharton (1698–1731)," *Oxford Dictionary of National Biography,* vol. 60 (Oxford: Oxford University Press, 1895–1900), p. 411.

292 James Craggs the Younger Carswell, *South Sea Bubble,* p. 234.

292 "Statesman, yet Friend to Truth!" Alexander Pope, "Epitaphs on James Craggs, Esq.," in *The Complete Poetical Works of Alexander Pope,* ed. Henry Boynton (Boston: Houghton, Mifflin, 1903), https://www.bartleby.com/203/124.html.

292 "having used violence to himself" Arthur Onslow, "A Manuscript Belonging to the Earl of Onslow," in *Report of the Historical Manuscript Commission,* vol. 14, part 9, p. 511, https://babel.hathitrust.org/cgi/pt?num=511&u=1&seq=514&view=image&size=100&id=coo.31924090788427.

292 the fate of Henry Bentinck, Duke of Portland Gary S. Shea, "Sir George Caswall vs. the Duke of Portland: Financial Contracts and Litigaton in the Wake of the South Sea Bubble" (2007), in Jeremy Atack and Larry Neal, *The Origin and Development of Financial Markets and Institutions* (Cambridge: Cambridge University Press, 2009), p. 29.

293 Blunt and a small inner group of directors Archibald Hutcheson (uncredited), *The Several Reports of the Committee of Secrecy to the Honourable House of Commons, Relating to the Late South-Sea Directors* (London: A. Moore, 1721), p. 2.

293 Blunt admitted Hutcheson, *Several Reports,* p. 3.

293 the example of Postmaster Craggs Hutcheson, *Several Reports,* p. 4.

294 In repeated rounds of questioning Hutcheson, *Several Reports,* p. 9 and p. 5.

294 The amount of money involved was staggering See Peter M.

Garber, *Famous First Bubbles: The Fundamentals of Early Manias* (Cambridge, MA: MIT Press, 2000), pp. 111–12.

294 "THE SAID MR. SURMAN" Hutcheson, *Several Reports*, p. 46.

295 THE WEALTH OF THE DIRECTORS J. H. Plumb, *Sir Robert Walpole: The Making of a Statesman* (London: Cresset, 1956), p. 341.

295 "MERCY MAY BE CRUELTY" John Trenchard and Thomas Gordon, "The Fatal Effects of the South-Sea Scheme, and the Necessity of Punishing the Directors" (November 12, 1720), in *Cato's Letters; or, Essays on Liberty,* vol. 1 (London: W. Wilkins, 1737), pp. 7–8.

296 "CORRUPT, INFAMOUS, AND DANGEROUS PRACTICES" Quoted in John Carswell, *The South Sea Bubble* (Phoenix Mill, UK: Alan Sutton, 1993), p. 241.

296 ON THE FACTS, STANHOPE WAS GUILTY See William Coxe, *Memoirs of the Life and Administration of Sir Robert Walpole, Earl of Orford,* vol. 1 (London: T. Cadell, Jun., and W. Davies, 1798), pp. 151–52.

297 "PUTT THE TOWNE IN A FLAME" Sir Thomas Brodrick to Lord Midleton, March 7, 1721, in Coxe, *Memoirs of the Life,* p. 209.

297 "MR. WALPOLE'S CORNER" Sir Thomas Brodrick to Lord Midleton, March 9, 1720–21, in Coxe, *Memoirs of the Life,* p. 210.

297 CELEBRATORY BONFIRES Sir Thomas Brodrick to Lord Midleton, March 11, 1720–21, in Coxe, *Memoirs of the Life,* p. 212.

297 "SUCH TRIFLING STUFF" Sir Thomas Brodrick to Lord Midleton, March 16, 1720–21, in Coxe, *Memoirs of the Life,* p. 214.

297 ONE OF WALPOLE'S ALLIES TOLD THE HOUSE Sir Thomas Brodrick to Lord Midleton, March 16, 1721.

299 "THE HARSH CAUTERY OF REALITY" P. G. M. Dickson, *The Financial Revolution in England: A Study in the Development of Public Credit, 1688–1756* (Aldershot, UK: Gregg Revivals, 1993), p. 176.

299 THE LOSS OF ABOUT HALF THEIR CAPITAL See Dickson, *Financial Revolution,* pp. 181–87, for a detailed accounting.

300 THE COMPANY'S DIRECTORS HAD TO SUFFER I draw all the directors' fates from John Carswell's account; for more details, see *South Sea Bubble,* pp. 257–59.

CHAPTER 22

303 FRANCE'S NEW FINANCIAL ADMINISTRATORS François R. Velde, "French Public Finance Between 1683 and 1726," preliminary and rough draft, July 20, 2006, http://www.helsinki.fi/iehc2006/papers3/Velde.pdf, pp. 22–26. See also François R. Velde, "What We Learn from a Sovereign Debt Restructuring in France in 1721," *Economic Perspectives* 40, no. 5 (2016), pp. 1–17. Velde, "French Public Finance," p. 32.

304 THE "GENEVAN HEADS" See Rebecca L. Spang's discussion of the Geneva annuity syndicates in *Stuff and Money in the Time of the French Revolution* (Cambridge, MA: Harvard University Press, 2015), chapter 1.

305 THE FRENCH GOVERNMENT COULDN'T RAISE MONEY EASILY Larry Neal, "How It All Began: The Monetary and Financial Architecture of Europe During the First Global Capital Markets, 1648–1815," *Financial History of Review* 7 (October 2000): 133.

305 FRANCE WAS AT LEAST TWICE AS RICH AS ITS ADVERSARY Thomas J. Sargent and François R. Velde, "Macroeconomic Features of the French Revolution," *Journal of Political Economy* 103, no. 3 (June 1995): 489.

305 NOR WAS FRANCE ITSELF REMOTELY A FAILED STATE Sargent and Velde, "Macroeconomic Features," p. 480.

305 FRANCE ITSELF REMAINED RICH See Phillip T. Hoffman, Gilles Postel-Vinay, and Jean-Laurent Rosenthal, *Priceless Markets: The Political Economy of Credit in Paris, 1660–1870* (Chicago: University of Chicago Press, 2000), p. 101, fig. 5.2. For a sense of Parisian elite consumption, see *Paris: Life & Luxury in the Eighteenth Century,* the catalog to the J. Paul Getty Museum exhibit of the same title, on display April 26–August 7, 2011.

306 "I UNEQUIVOCALLY DENY" Robert Walpole in the House of Commons, February 13, 1741, quoted in William Coxe, *Memoirs of the Life and Administration of Sir Robert Walpole,* vol. 1 (London: T. Cadell, Jun., and W. Davies, 1798), p. 668.

307 THE COMPANY'S CAPITAL P. G. M. Dickson, *The Financial Revolu-*

tion in England: A Study in the Development of Public Credit, 1688–1756 (Aldershot, UK: Gregg Revivals, 1993), pp. 179–80.

308 BREAK UP THE MONSTER Dickson, *Financial Revolution,* p. 181.

309 A NET PROFIT Donald L. Cherry, "The South Sea Company, 1711–1855," *Dalhousie Review* 13, no. 1 (1934): p. 62.

309 BUYING AND SELLING HUMAN BEINGS For data on the South Sea Company slave trade, see Helen Paul, *The South Sea Bubble: An Economic History of Its Origins and Consequences* (London: Routledge, 2011), pp. 59–65. Mortality estimates can be found on p. 65.

310 THE "SINKING FUND" See Dickson, *Financial Revolution,* pp. 85–87 and 205–11, for an account of the establishment and use of the Sinking Fund.

310 UNDER WALPOLE'S DIRECTION Dickson, *Financial Revolution,* p. 210, table 24.

311 A VIRTUOUS CYCLE Dickson, *Financial Revolution,* p. 210.

311 BRITAIN HAD CROSSED A THRESHOLD My thanks to William Deringer for conversations that clarified the distinction between the claims for the Sinking Fund and its actual, more complicated use.

311 HORRIFIED JOY The story of the ordeal of the *Rebecca* and its captain comes from a report published in the *Universal Spectator and Weekly Journal* on June 19, 1731, pp. 2–3. The report there is attributed to Captain Jenkins's testimony before the Duke of Newcastle, secretary of state of the southern parts.

314 THE EXTRAORDINARY COSTS OF SUCH A GLOBAL STRUGGLE Sargent and Velde, "Macroeconomic Features," pp. 479 and 481, figures 2 and 3.

315 "THEY ARE RINGING THEIR BELLS" Pat Rogers, *A Political Biography of Alexander Pope* (London: Routledge, 2015), p. 212.

316 AN INCREASINGLY SKITTISH MARKET FOR NEW DEBT The details on borrowing and rates, including the rate reduction and consolidation that followed the war, come from the work of Dickson, *Financial Revolution,* pp. 218–43.

317 CONSOLIDATED BONDS—CONSOLS FOR SHORT For an account of the twists and turns required to create the consols out of the vari-

ous instruments in play, see Dickson, *Financial Revolution*, pp. 228–243.

Chapter 23

321 THE SCALE OF BRITISH BORROWING B. R. Mitchell, *British Historical Statistics* (Cambridge University Press, 1988), pp. 392 (revenue) and 396 (expenditure).

321 THE PRICE FOR CONSOLS Michael D. Bordo and Eugene N. White, "A Tale of Two Currencies: British and French Finance During the Napoleonic Wars," *Journal of Economic History* 51, no. 2 (June 1991): 306, fig. 3.

321 THE FIRST INCOME TAX Customs and excise still accounted for the largest fraction of official revenue, but this was still a significant influx of new money.

322 INTEREST ON CONSOLS DROPPED Bordo and White, "Tale of Two Currencies," p. 306, fig. 3.

323 "FOREIGNERS HAD BEEN HEARD TO SAY" Daniel Defoe, *The Chimera; or, The French Way of Paying National Debts Laid Open* (London: T. Warner, 1720), pp. 2–3.

325 THE MOST OBVIOUS SOURCE OF CAPITAL See Richard Sylla, "Comparing the UK and US Financial Systems, 1790–1830," in Jeremy Atack and Larry Neal, *The Origins and Development of Financial Markets and Institutions* (Cambridge: Cambridge University Press, 2009), pp. 226–28.

325 BRITAIN'S BANKING SYSTEM WAS SIMILARLY HOBBLED Sylla, "Comparing the UK and US," p. 221.

326 MANY ORGANIZED AS CORPORATIONS For a discussion of these differences, see Sylla, "Comparing the UK and US," pp. 222–24.

326 3,500 SUCH NEW BUSINESSES Sylla, "Comparing the UK and US," p. 229.

326 US GDP Louis Johnston and Samuel H. Williamson, "What Was the U.S. GDP Then?" MeasuringWorth.com, 2020, https://www.measuringworth.com/datasets/usgdp/.

326 BRITAIN AT THE SAME MOMENT Christopher Chantrill, "UK

Gross Domestic Product GDP History," UKPublicSpending .co.uk, n.d., https://www.ukpublicspending.co.uk/spending_chart_1692_1790UKm_17c1li001mcn__UK_Gross_Domestic_Product_GDP_History?show=n.

327 CREDIT AVAILABLE TO THE PRIVATE CUSTOMERS See especially the discussion in chapter 7 of Peter Temin and Hans-Joachim Voth, *Prometheus Shackled: Goldsmith Banks and England's Financial Revolution After 1700* (Oxford: Oxford University Press, 2013), pp. 148–75.

328 "THE AMERICANS ARRIVED BUT AS YESTERDAY" Alexis de Tocqueville, *Democracy in America,* trans. Harry Reeve (1835–40; repr., State College: Pennsylvania State University, 2002), pp. 624–26.

330 A MARKET IN FUTURES For an account of the development of the Chicago system of grain collection, storage, classification, and futures trading, see William Cronon, *Nature's Metropolis* (New York: W. W. Norton, 1991), chapter 3, "Pricing the Future: Grain," especially the sections "The Golden Stream" and "Futures."

331 GREAT BRITAIN SHED MANY OF ITS RESTRICTIVE FINANCIAL RULES Richard Sylla, "Comparing the UK and US," p. 237.

332 ON THE FUTURE VALUE OF THE SONG "REBEL REBEL" See, among many news accounts, Tom Espiner, "'Bowie Bonds'—the Singer's Financial Innovation," *BBC News,* January 11, 2016, https://www.bbc.com/news/business-35280945.

EPILOGUE

339 THE LAWYERS STARTED TO PREPARE THE PAPERWORK For a good synopsis of the sequence of events in the financial crisis, see Mauro F. Guillén, "The Global Economic & Financial Crisis, a Timeline," released by the Lauder Institute at the University of Pennsylvania, n.d., https://lauder.wharton.upenn.edu/wp-content/uploads/2015/06/Chronology_Economic_Financial_Crisis.pdf.

339 THE CRASH THAT FOLLOWED Alexandra Twin, "Stocks Get Pummeled," *CNN Money,* September 21, 2008, https://money.cnn.com/2008/09/15/markets/markets_newyork2/.

340 SUCH WIDESPREAD MISERY COST LIVES See Thor Norström and Hans Grönqvist, "The Great Recession, Unemployment and Suicide," *Journal of Epidemiology & Community Health* 69 (2015): 110–16. See also Marinea Karanikolos et al., "Financial Crisis, Austerity, and Health in Europe," *Lancet* 381, no. 9874 (April 13–19, 2013): 1323–31.

Bibliography

Adams, Gavin John. *Letters to John Law*. Newton Page, 2012.

Anderson, Adam. *An Historical and Chronological Deduction of the Origin of Commerce from the Earliest Accounts.* 4 vols. London: J. Walter, 1787.

Andreades, Andreas Michael. *History of the Bank of England.* Translated by Christabel Meredith. London: P.S. King & Son, 1909.

Ashton, T. S. *Economic History of England: The Eighteenth Century.* London: Routledge, 1955 and 2006.

Atack, Jeremy, and Larry Neal. *The Origin and Development of Financial Markets and Institutions.* Cambridge: Cambridge University Press, 2009.

Aubrey, John. *"Brief Lives," Chiefly of Contemporaries, Set Down by John Aubrey, Between the Years 1669 & 1696.* 2 volumes. Edited by Andrew Clark. Oxford: Clarendon Press, 1898.

———. *My Own Life*. Edited by Ruth Scurr. New York: New York Review of Books, 2016.

Backscheider, Paula. *Daniel Defoe: His Life.* Baltimore: Johns Hopkins University Press, 1989.

Bayoumi, Tamim A., Barry J. Eichengreen, and Mark P. Taylor, eds. *Modern Perspectives on the Gold Standard.* Cambridge: Cambridge University Press, 2005.

Bell, John. *London's Remembrancer; or, A True Accompt of Every Particular Weeks Christnings and Mortality in All the Years of Pestilence Within the Cognizance of the Bills of Mortality, Being XVII Years.* London: E. Cotes, 1665.

Bellhouse, David. *Leases for Lives: Life Contingent Contracts and the Emergence of Actuarial Science in Eighteenth-Century England.* Cambridge: Cambridge University Press, 2017.

———. "A New Look at Halley's Life Table." *Journal of the Royal Statistical Society* 174, no. 3 (July 2011): 823–32.

Bennett, J. A. "Hooke and Wren and the System of the World: Some Points Towards an Historical Account." *British Journal for the History of Science* 8, no. 1 (March 1975): 32–61.

Biener, Zvi, and Eric Schliesser, eds. *Newton and Empiricism.* Oxford: Oxford University Press, 2014.

Black, Jeremy. *Robert Walpole and the Nature of Politics in Early Eighteenth Century Britain.* New York: St. Martin's Press, 1990.

Blunt, Sir John. *The Particular and Inventory of Sir John Blunt, Bart, One of the Late Directors of the South-Sea Company, Together with the Abstract of the Same.* 1721. Gale ECCO Print Editions, 2010.

———. *A True State of the South Sea Scheme.* London: J. Peele, 1722.

Bold, John F., and Edward Chaney, eds. *English Architecture: Public and Private.* London: Hambledon Continuum, 1993.

Bordo, Michael D., and Eugene N. White. "A Tale of Two Currencies: British and French Finance During the Napoleonic Wars." *Journal of Economic History* 51, no. 2 (June 1991): 303–16.

Bottigheimer, Karl S. "English Money and Irish Land: The 'Adventurers' in the Cromwellian Settlement of Ireland." *Journal of British Studies* 7, no. 1 (November 1967): 12–27.

Bremer-David, Charissa, ed. *Paris: Life & Luxury in the Eighteenth Century.* Los Angeles: Getty Publications, 2011.

Brewer, John. *The Sinews of Power: War, Money and the English State, 1688–1783.* 1988. Cambridge, MA: Harvard University Press, 1990.

Brewster, David. *Memoirs of the Life, Writings, and Discoveries of Sir Isaac Newton.* Vol. 1. Edinburgh: Thomas Constable, 1855.

A Brief Description of the Excellent Vertues of That Sober and Wholesome Drink, Called Coffee, and Its Incomparable Effects in Preventing or Curing Most Diseases Incident to Humane Bodies. London, 1674.

Brown, Ernest Henry Phelps, and Sheila V. Hopkins. "Seven Centuries of the Prices of Consumables, Compared with Builders' Wage-Rates." *Economics* (New Series) 23, no. 92 (November 1956): 296–314.

Buchwald, Jed Z., and Mordecai Feingold. *Newton and the Origin of Civilization.* Princeton, NJ: Princeton University Press, 2013.

Carlos, Anne M., and Larry Neal. "The Microfoundations of the Early London Capital Market: Bank of England Shareholders During and After the South Sea Bubble, 1720–25." *Economic History Review* 59, no. 3 (2006): 498–538.

Carswell, John. *The South Sea Bubble.* Phoenix Mill, UK: Alan Sutton, 1993.

Cawston, George, and A. H. Keane. *The Early Chartered Companies: A.D. 1296–1878.* London: Edward Arnold, 1896.

Challis, C. E., ed. *A New History of the Royal Mint.* Cambridge: Cambridge University Press, 1992.

Chandrasekhar, S. *Newton's Principia for the Common Reader.* Oxford: Clarendon Press, 1995.

Cherry, Donald L. "The South Sea Company, 1711–1855." *Dalhousie Review* 13, no. 1 (1934): 61–68.

Chetwood, William. *The South Sea; or, The Biters Bit.* London: J. Roberts, 1720.

———. *The Stock-Jobbers; or, The Humours of Exchange-Alley.* London: J. Roberts, 1720.

Ciecka, James. "Edmond Halley's Life Table and Its Uses." *Journal of Legal Economics* 15, no. 1 (2008): 65–74.

Clapham, Sir John. *The Bank of England.* 2 volumes. Cambridge: Cambridge University Press, 1970.

Clark, Geoffrey. *Betting on Lives: The Culture of Life Insurance in England, 1695–1775.* Manchester, UK: Manchester University Press, 1999.

Clark, Geoffrey, et al. *The Appeal of Insurance.* Toronto: University of Toronto Press, 2010.

Clark, Gregory. "Debt, Deficits and Crowding Out: England, 1727–1840," *European Review of Economic History* 5, no. 3 (2001): 403–36.

———. "The Political Foundations of Modern Economic Growth: England, 1540–1800." *Journal of Interdisciplinary History* 26, no. 4 (Spring 1996): 563–88.

Clowes, Gregory. "'The Devil's Interlude' in the South Sea Bubble." *Ex Historia* (2007), https://humanities.exeter.ac.uk/media/universityofexeter/collegeofhumanities/history/exhistoria/volume6/Devil_and_South_Sea_Bubble.pdf.

Cobbett, William. *The Parliamentary History of England, from the Earliest Period to the Year 1803. Vol. 6, 1702–1714,* and *Vol. 7, 1714–1722.* London: T. C. Hansard, 1810 and 1811.

Cohen, I. Bernard, and George E. Smith, eds. *The Cambridge Companion to Newton.* Cambridge: Cambridge University Press, 2002.

Cole, Arthur H. "The Bancroft Collection." *Bulletin of the Business Historical Society* 9, no. 6 (December 1935): 93–96.

Cook, Alan. *Edmond Halley: Charting the Heavens and the Seas.* Oxford: Oxford University Press, 1998.

Cowper, Mary. *Diary of Mary, Countess Cowper.* London: John Murray, 1864.

Coxe, William. *Memoirs of the Life and Administration of Sir Robert Walpole, Earl of Orford.* 3 volumes. London: Longman, T. Cadell, Jun., and W. Davies, 1798. New edition (4 volumes) London: Longman, Hurst, Rees Orme, and Brown, 1816.

Cronon, William. *Nature's Metropolis.* New York: W. W. Norton, 1991.

Dale, Richard. *The First Crash: Lessons from the South Sea Bubble.* Princeton, NJ: Princeton University Press, 2004.

Damrosch, Leo. *Jonathan Swift: His Life and His World.* New Haven, CT: Yale University Press, 2013.

Daston, Lorraine. *Classical Probability in the Enlightenment.* Princeton, NJ: Princeton University Press, 1988.

Davis, Andrew McFarland. "A Search for the Beginnings of Stock Speculation." *Transactions of the Colonial Society of Massachusetts.* Vol. 10. Cambridge, MA: John Wilson and Sons, 1907.

Dear, Peter. *Discipline and Experience: The Mathematical Way in the Scientific Revolution.* Chicago: University of Chicago Press, 1995.

Defoe, Daniel. *The Anatomy of Exchange-Alley; or, A System of Stock-Jobbing.* London: E. Smith, 1719. http://quod.lib.umich.edu/e/ecco/004843169.0001.000/1:2?rgn=div1;view=fulltext.

———. *A Brief Case of the Distillers: And of the Distilling Trade in England, etc.* London: Tho. Warner, 1726. http://quod.lib.umich.edu/e/ecco/004834050.0001.000/1:3?rgn=div1;view=toc.

———. *A Brief State of the Inland or Home Trade, of England: etc.* London: Tho. Warner, 1730. http://quod.lib.umich.edu/e/ecco/004834053.0001.000/1:2?rgn=div1;view=toc.

———. *The Chimera; or, The French Way of Paying National Debts Laid Open.* London: T. Warner, 1720.

———. *The Commentator.* London: J. Roberts, January 1–September 16, 1720. https://babel.hathitrust.org/cgi/pt?id=coo.31924007283595;view=1up;seq=21.

———. *The Director.* London: W. Boreham, October 5, 1720–January 16, 1721. https://babel.hathitrust.org/cgi/pt?id=coo.31924092977762;view=1up;seq=39;size=400.

———. *A Journal of the Plague Year* (originally published 1722). Eugene: University of Oregon Press (Renascence Editions), 2008. https://pdfs.semanticscholar.org/d783/6e710cf3ea0a80e2ade0f9ecbe58d1bc0a94.pdf.

———. *The Villainy of Stock-Jobbers Detected, and the Causes of the Late Run upon the Bank and Bankers Discovered and Considered.* London, 1701.

Delbourgo, James. "Sir Hans Sloane's Milk Chocolate and the Whole History of the Cacao." *Social Text 106* 29, no. 1 (Spring 2011): 71–101.

Deringer, William. *Calculated Values: Finance, Politics, and the Quantitative Age, 1668–1776.* Cambridge, MA: Harvard University Press, 2018.

———. "Compound Interest Corrected: The Imaginative Mathematics of the Financial Future in Early Modern England." *Osiris* 33, no. 1 (2018): 109–29.

———. "Finding the Money: Public Accounting, Political Arithmetic, and Probability in the 1690s." *Journal of British Studies* 52 (2013): 638–68.

———. "Pricing the Future in the Seventeenth Century: Calculating Technologies in Competition." *Technology and Culture* 58, no. 2 (April 2017): 506–28.

Dickson, P. G. M. *The Financial Revolution in England: A Study in the Development of Public Credit, 1688–1756.* Aldershot, UK: Gregg Revivals, 1993.

Dobbs, Betty Jo Teeter. *The Janus Faces of Genius.* Cambridge: Cambridge University Press, 1991.

———. "Newton's Alchemy and His Theory of Matter." *Isis* 73, no. 4 (December 1982): 511–28.

Ellis, Markman. *The Coffee House: A Cultural History.* London: Weidenfeld & Nicholson, 2004.

Engell, James. "Wealth and Words: Pope's 'Epistle to Bathurst.'" *Modern Philology* 85, no. 4 (1988): 433–46.

Erskine-Hill, Howard. "Blunt, Sir John." In *Dictionary of National Biography,* September 23, 2004. Oxford: Oxford University Press.

Finucane, Adrian. *The Temptations of Trade: Britain, Spain, and the Struggle for Empire.* Philadelphia: University of Pennsylvania Press, 2016.

Flood, Donal. "William Petty and 'The Double Bottom.'" *Dublin Historical Record* 30, no. 3 (June 1977): 96–110.

Francis, John. *History of the Bank of England: Its Times and Traditions.* 3rd ed. London: Willoughby, 1848.

Frehen, Rik G. P., William N. Goetzmann, and K. Geert Rouwenhorst. "New Evidence on the First Financial Bubble." NBER Working Paper no. w15332, National Bureau of Economic Research, September 2009. https://ssrn.com/abstract=1472270.

Garber, Peter M. *Famous First Bubbles: The Fundamentals of Early Manias.* Cambridge, MA: MIT Press, 2000.

Gialdroni, Stefania. "Incorporation and Limited Liability in Seventeenth-Century England: The Case of the East India Company." In *The Company in Law and Practice,* edited by Dave De ruysscher, Albrecht Cordes, Serge Dauchy, and Heikki Pihlajamäki. The Hague: Brill/Nijhoff, 2017.

Gibbon, Edward. *Miscellaneous Works of Edward Gibbon, Esquire: With Memoirs of His Life and Writings.* London: A. Strahan, and T. Cadell Jun., and W. Davies, 1796.

Giusti, Giovanni, Charles Noussair, and Hans-Joachim Voth. "Recreating the South Sea Bubble: Insights from an Experiment in Financial History." Working Paper no. 146, University of Zurich, Department of Economics, March 2014. http://www.zora.uzh.ch/id/eprint/94471/1/econwp146.pdf.

Glaiser, Natasha. "Calculating Credibility: Print Culture, Trust and Economic Figures in Early Eighteenth-Century England." *Economic History Review* 60, no. 4 (2007): 685–711.

Gleick, James. *Isaac Newton.* New York: Random House, 2003.

Goss, David. "The Ongoing Binomial Revolution." arXiv.org, May 18, 2011. arXiv:1105.3513v1.

Graunt, John. *Natural and Political Observations, Mentioned in a Following Index, and Made upon the Bills of Mortality.* 2nd ed. London: Tho. Roycroft, for John Martin, James Allestry, and Tho. Dicas, 1662.

Gray, Mr. *The Memoirs Life and Character of the Great Mr. Law and His Brother at Paris: Down to This Present Year 1721, with an Accurate and Particular Account of the Establishment of the Missisippi Company in France, the Rise and Fall of It's Stock, and All the Subtle Artifices Used to Support the National Credit of That Kingdom, by the Pernicious Project of Paper-Credit.* 2nd ed. London: Sam Briscoe, 1721.

Guillén, Mauro F. "The Global Economic & Financial Crisis: a Timeline." Lauder Institute at the University of Pennsylvania, n.d. https://lauder.wharton.upenn.edu/wp-content/uploads/2015/06/Chronology_Economic_Financial_Crisis.pdf.

Hacking, Ian. *The Emergence of Probability.* Cambridge: Cambridge University Press, 2006.

Hall, A. Rupert. *Isaac Newton: Adventurer in Thought.* Cambridge: Cambridge University Press, 1996.

Halley, Edmond. "Some Considerations About the Cause of the Universal Deluge, Laid Before the Royal Society, on the 12th of December 1694." *Philosophical Transactions* 33 (January 1, 1724): 118–23.

Hannam, A. A. "Hutcheson, Archibald." *Oxford Dictionary of National Biography,* September 23, 2004. https://doi.org/10.1093/ref:odnb/53923.

Harris, Ellen T. "Courting Gentility: Handel at the Bank of England." *Music and Letters* 91, no. 3 (August 2010): 357–75.

———. "Handel the Investor." *Music and Letters* 85, no. 4 (November 2004): 521–75.

Harris, Ron. "The Bubble Act: Its Passage and Its Effects on Business Organization." *Journal of Economic History* 54, no. 3 (September 1994): 610–27.

Hattox, Ralph. *Coffee and Coffeehouses: The Origins of a Social Beverage in the Medieval Near East.* Seattle: University of Washington Press, 1985.

Head, Richard. *Canting Academy; or The Devil's Cabinet Opened.* London: F. Leach for Mat. Drew, 1674. https://quod.lib.umich.edu/e/eebo/A43142.0001.001/1:5.39?rgn=div2;view=toc.

Hill, Brian W. *Robert Harley, Speaker, Secretary of State, and Premier Minister.* New Haven, CT: Yale University Press, 1988.

———. *Sir Robert Walpole: "Sole and Prime Minister."* London: Hamish Hamilton, 1989.

Historical Manuscripts Commission. *The Manuscripts of His Grace the Duke of Portland.* Vol. 7. London: Her Majesty's Stationery Office, 1901.

———. *The Manuscripts of the Duke of Beaufort, K.G., the Earl of Donoughmore, and Others.* 12th report, part 9. London: Her Majesty's Stationery Office, 1891.

———. *The Manuscripts of the Earl of Buckinghamshire, the Earl of Lindsey, the Earl of Onslow, Lord Emly, Theodore J. Hare, Esq., and James Round, Esq., M.P.* 14th report, appendix, part 9. London: Her Majesty's Stationery Office, 1895.

An Historical Treatise Concerning Jews and Judaism, in England. London, 1720.

The History of the Lives and Actions of Jonathan Wild, Thief Taker; Joseph Blake Alias Buskin, Foot-pad; and John Sheppard, Housebreaker. London: Edward Midwinter, 1725.

Hoffman, Philip T., Gilles Postel-Vinay, and Jean-Laurent Rosenthal. *Priceless Markets: The Political Economy of Credit in Paris, 1660–1870,* Chicago: University of Chicago Press, 2000.

Hooke, Robert. *The Diary of Robert Hooke.* Henry W. Robinson and Walter Adams, eds. London: Taylor and Francis, 1935.

Hoppit, Julian. "The Myths of the South Sea Bubble." *Transactions of the Royal Historical Society* 12 (2002): 141–64.

Horsefield, J. Keith. *British Monetary Experiments: 1650–1710.* Cambridge, MA: Harvard University Press, 1960.

———. "The 'Stop of the Exchequer' Revisited." *Economic History Review* 35, no. 4 (November 1982): 511–28.

Houghton, John. "A Discourse of Coffee." *Philosophical Transactions of the Royal Society* 21 (January 1, 1699): 311–17.

Houghton, Walter E., Jr. "The History of Trades: Its Relation to Seventeenth-Century Thought: As Seen in Bacon, Petty, Evelyn, and Boyle." *Journal of the History of Ideas* 2, no. 1 (January 1941): 33–60.

Hutcheson, Archibald. *A Collection of Calculations and Remarks Relating to the South Sea Scheme & Stock.* London: no publisher listed, 1720.

———. *Proposal for the Payment of the Publick Debts.* London: no publisher listed, 1715.

——— (uncredited). *The Several Reports of the Committee of Secrecy to the Honourable House of Commons, Relating to the Late South-Sea Directors.* London: A. Moore, 1721.

———. *Some Calculations and Remarks Relating to the Present State of the Publick Debts and Funds.* London: no publisher listed, 1718.

———. *Some Calculations Relating to the Proposals Made by the South-Sea Company, and the Bank of England, to the House of Commons.* London: J. Morphew, 1720.

———. *Some Paragraphs of Mr. Hutcheson's Treatises on the South-Sea Subject.* London, 1723.

———. *Some Seasonable Considerations for Those Who Are Desirous, by Subscription or Purchase, to Become Proprietors of South-Sea Stock: With Remarks on the Surprizing Method of Valuing South-Sea Stock.* London: J. Morphew, 1720.

Jackson, Robert V. "Government Expenditure and British Economic Growth in the Eighteenth Century: Some Problems of Measurement." *Economic History Review* 43, no. 2 (May 1990): 217–35.

Johanisson, Karin. "Society in Numbers: The Debate over Quantification in 18th Century Political Economy." In *The Quantifying Spirit in the 18th Century,* edited by Tore Frängsmyr, J. L. Heilbron, and Robin E. Rider. Berkeley: University of California Press, 1990.

Johnson, Peter. "The Board of Longitude 1714–1828." *Journal of the British Astronomical Association* 99, no. 2 (1989): 63–69.

Jones, D. W. *War and Economy in the Age of William III and Marlborough.* Oxford: Basil Blackwell, 1988.

Karanikolos, Marinea, et al. "Financial Crisis, Austerity, and Health in Europe." *Lancet* 381, no. 9874 (April 13–19, 2013): 1323–31.

Krüger, Loren, Lorraine Daston, and Michael Heidelberger. *The Probabilistic Revolution.* Vol. 1, *Ideas in History.* Cambridge: MIT Press, 1987.

Lang, Paul Henry. *George Frideric Handel.* Mineola, NY: Dover, 1966.

Laurence, Anne, Joseph Maltby, and Janette Rutherford, eds. *Women and Their Money 1700–1950.* London and New York: Routledge, 2009.

Li, Ming-hsun. *The Great Recoinage of 1696–9.* London: Weidenfeld & Nicholson, 1963.

Lord, Evelyn. *The Great Plague: A People's History.* New Haven, CT: Yale University Press, 2014.

Lowndes, William. *A Report Containing an Essay for the Amendment of the Silver Coins.* London: C. Bill, 1695.

Lyons, H. G. *The Royal Society, 1660–1940: A History of Its Administration Under Its Charters.* New York: Greenwood Press, 1968.

Mackay, Charles. *Extraordinary Popular Delusions and the Madness of Crowds.* New York: Crown, 1980. (Volume 1 originally published in London in 1841.)

Macky, John. *A Journey Through England in Familiar Letters from a Gentleman Here to His Friend Abroad.* 2nd ed. London: J. Hooke, 1722.

Mason, A. E. W. *The Royal Exchange.* London: Royal Exchange, 1920.

McCormick, Ted. *William Petty and the Ambitions of Political Arithmetic.* Oxford: Oxford University Press, 2009.

McCusker, John. *Essays in the Economic History of the Atlantic World.* London: Routledge, 2005.

McKendrick, Neil, John Brewer, and J. H. Plumb. *The Birth of a Consumer Society: The Commercialization of Eighteenth-Century England.* Bloomington: Indiana University Press, 1982.

McKie, D., and G. R. de Beer. "Newton's Apple." *Notes and Records of the Royal Society of London* 9, no. 1 (October 1951): 46–54.

McLeod, Christine. "The 1690s Patents Boom: Invention or Stock-Jobbing?" *Economic History Review* 39, no. 4 (1986): 549–71.

McVeigh, John, ed. *Political and Economic Writings of Daniel Defoe.* Vol. 6. London: Routledge, 2000.

Melville, Lewis. *The Life and Letters of John Gay (1685–1732), Author of "The Beggar's Opera."* London: Daniel O'Connor, 1921. https://openlibrary.org/books/OL13493940M/Life_and_letters_of_John_Gay_(1685-1732).

———. *The South Sea Bubble.* London: Daniel O'Connor, 1921.

Merrett, Robert James. *Daniel Defoe, Contrarian.* Toronto: University of Toronto Press, 2013.

Michie, Ranald. *The London Stock Exchange: A History.* Oxford: Oxford University Press, 2001.

Milner, James. *Three Letters, Relating to the South-Sea Company and the Bank.* London: J. Roberts, 1720. http://find.galegroup.com.libproxy.mit.edu/ecco/retrieve.do?scale=0.50&sort=Author&docLevel=FASCIMILE&prodId=ECCO&tabID=T001&resultListType=RESULT_LIST&qrySerId=Locale%28en%2C%2C%29%3AFQE%3D%28BN%2CNone%2C7%29T050279%24&retrieveFormat=MULTIPAGE_DOCUMENT&inPS=true&userGroupName=camb27002&docId=CW3307369868&relevancePageBatch=CW107369834&forRelevantNavigation=true&pageNumber=-1&contentSet=&workId=0533201900&callistoContentSet=ECSS¤tPosition=1&showLOI=&quickSearchTerm=&stwFuzzy=.

Mitchell, B. R. *British Historical Statistics.* Cambridge: Cambridge University Press, 1988.

Mokyr, Joel. *The Enlightened Economy.* New Haven, CT: Yale University Press, 2009.

Montague, Lady Mary Worley. *The Complete Letters of Lady Mary Worley Montague.* Edited by Robert Halsband. Oxford: Oxford University Press, 1966.

Moote, A. Lloyd, and Dorothy C. Moote. *The Great Plague: The Story of London's Most Deadly Year.* Baltimore: Johns Hopkins University Press, 2004.

Morgan, William Thomas. "The Origins of the South Sea Company." *Political Science Quarterly* 44, no. 1 (March 1929): 16–38.

Mortimer, Thomas. *Every Man His Own Broker; or, A Guide to Exchange-Alley.* 7th ed. London: S. Hooper, 1769. https://archive.org/details/everymanhisownbr00mort.

Murphy, Anne L. "Lotteries in the 1690s: Investment or Gamble?" *Financial History Review* 12, no. 2 (October 2005): 227–46.

———. *The Origins of English Financial Markets.* Cambridge: Cambridge University Press, 2009.

———. "Trading Options Before Black-Scholes: A Study of the Market in Late Seventeenth-Century London." *Economic History Review* 62, no. S1 (August 2009): 8–30.

Murphy, Antoin E. *The Genesis of Macroeconomics.* Oxford: Oxford University Press, 2009.

———. *John Law: Economic Theorist and Policy-Maker.* Oxford: Clarendon, 1997.

Neal, Larry. "How It All Began: The Monetary and Financial Architecture of Europe During the First Global Capital Markets, 1648–1815." *Financial History Review* 7 (October 2000): 117–40.

———. "The Microstructure of First Emerging Markets in Europe in the 18th Century." Paper for Manufacturing Markets: Legal, Political and Economic Dynamics, Villa Finaly, Florence, June 11–13, 2009. https://www.economix.fr/uploads/source/doc/colloques/2009_Florence/Larry-Neal.pdf.

———. *The Rise of Financial Capitalism.* Cambridge: Cambridge University Press, 1990.

Neale, Thomas. *The Profitable Adventure to the Fortunate.* London: F. Collins, 1694.

Newton, Isaac. *The Correspondence of Isaac Newton.* Various editors, 7 volumes. Cambridge: Cambridge University Press, 1959.

———. *The Principia.* 1687. Translated by I. Bernard Cohen and Ann Whitman. Berkeley: University of California Press, 1999.

———. "Sir Isaac Newton's State of the Gold and Silver Coin." In William Arthur Shaw, *Select Tracts and Documents Illustrative of English Monetary History 1626–1730: Comprising Works of Sir Robert Cotton, Henry Robinson, Sir Richard Temple* [. . .], 189–95. London: C. Wilson, 1896.

———. "Untitled Holograph Draft Memorandum on John Pollexfen's *A Discourse on Trade, Coyn and Paper Credit.*" 1697. Mint 19/II. 608–11. National Archives, Kew, UK. http://www.newtonproject.ox.ac.uk/view/texts/normalized/MINT00261.

Nicholson, Colin. *Writing and the Rise of Finance: Capital Satires of the Early Eighteenth Century.* Cambridge: Cambridge University Press, 2004.

Norström, Thör, and Hans Grönqvist. "The Great Recession, Unemployment and Suicide." *Journal of Epidemiology & Community Health* 69 (2015): 110–16.

North, Douglass C., and Barry R. Weingast. "Constitutions and Commitment: The Evolution of Institutions Governing Public Choice in Seventeenth-Century England." *Journal of Economic History* 49, no. 4 (December 1989): 803–32.

Novak, Maximillian. *Daniel Defoe: Master of Fictions.* Oxford: Oxford University Press, 2001.

O'Brien, John J. "Samuel Hartlib's Influence on Robert Boyle's Scientific Development." *Annals of Science* 21, no. 1 (1965): 1–14.

Odlyzko, Andrew. "Newton's Financial Misadventures in the South Sea Bubble." *Notes and Records of the Royal Society* 73, no. 1 (February 2019): 29–59.

Ogburn, Miles. *Spaces of Modernity: London's Geographies 1680–1780.* New York: Guildford, 1998.

Parker, Geoffrey. "The 'Military Revolution,' 1560–1660—a Myth?" *Journal of Modern History* 48, no. 2 (June 1976): 195–214.

Paterson, William. *A Brief Account of the Intended Bank of England.* London: Randal Taylor, 1694. http://quod.lib.umich.edu/e/eebo/A56581.0001.001/1:2?rgn=div1;view=fulltext.

Paul, Helen J. *The South Sea Bubble: An Economic History of Its Origins and Consequences.* London: Routledge, 2011.

———. "The 'South Sea Bubble,' 1720." *European History Online* (November 4, 2015). http://ieg-ego.eu/en/threads/european-media/european-media-events/helen-j-paul-the-south-sea-bubble-1720.

Pearce, Edward. *The Great Man: Scoundrel, Genius and Britain's First Prime Minister.* London: Jonathan Cape, 2007.

Pepys, Samuel. *Diary.* http://www.pepysdiary.com/.

Petty, William. *The Advice of W. P. to Mr. Samuel Hartlib, for the Advancement of Some Particular Parts of Learning.* London: no publisher listed, 1647. http://quod.lib.umich.edu/e/eebo/A54605.0001.001/1:1?rgn=div1;view=fulltext.

———. *The History of the Survey of Ireland, Commonly Called the Down Survey.* Edited by Thomas Aiskey Larcom. Dublin: Irish Archaeological Survey, 1851.

———. *Political Arithmetick; or, A Discourse Concerning, the Extent and Value of Lands* [. . .]. 3rd ed. London: Robert Clavel and Hen. Mortlock, 1690.

———. *Reflections upon Some Persons and Things in Ireland.* London: John Martin, James Allstraye and Thomas Dicas, 1660.

———. *A Treatise of Taxes & Contributions* [. . .]. London: N. Brooke, 1662.

Plumb, J. H. *Sir Robert Walpole: The Making of a Statesman.* London: Cresset, 1956.

Pollexfen, John. *A Discourse of Trade, Coyn, and Paper Credit, and of Ways and Means to Gain, and Retain Riches.* London: Brabazon Aylmer, 1697.

Pope, Alexander. "Of the Use of Riches" (Epistle to Bathurst). *The Complete Poetical Works of Alexander Pope.* Edited by Henry Boynton. Boston: Houghton, Mifflin, 1903.

———. *The Works of Alexander Pope, New Edition.* Vol. 6. London: John Murray, 1871.

———. *The Works of Alexander Pope, with Notes and Illustrations by Himself and Others.* 8 volumes. London: Longman, Brown, 1847.

Porter, Stephen. *The Great Plague.* Phoenix Mill, UK: Sutton, 1999.

Quinn, Stephen. "The Glorious Revolution's Effect on English Private Finance: A Microhistory 1680–1705." *Journal of Economic History* 61, no. 3 (September 2001): 593–615.

———. "Money, finance and capital markets" in *The Cambridge Economic History of Modern Britain,* Vol. 1 (Cambridge: Cambridge University Press, 2003), Foderick Foud, Paul Johnson, editors, pp. 147–74.

Richetti, John, ed. *The Cambridge Companion to Daniel Defoe.* Cambridge: Cambridge University Press, 2008.

Richetti, John. *The Life of Daniel Defoe.* Oxford: Blackwell, 2005.

Rogers, Pat. *A Political Biography of Alexander Pope.* London: Routledge, 2015.

———. "South Sea Bubble Myths." *Times Literary Supplement,* April 9, 2014.

Roncaglia, Alessandro. *The Wealth of Ideas: A History of Economic Thought.* Cambridge and New York: Cambridge University Press, 2005.

Roseveare, Henry. *The Financial Revolution, 1660–1760.* London and New York: Taylor and Francis, 1991.

Rusnock, Andrea A. *Vital Accounts: Quantifying Health and Population in Eighteenth-Century England and France.* Cambridge: Cambridge University Press, 2002.

Russell, G. A. *The "Arabick" Interest of the Natural Philosophers in Seventeenth-Century England.* Leiden, Netherlands: E. J. Brill, 1994.

Sacheverell, Henry. *The Perils of False Brethren: Both in Church, and State: Set Forth in a Sermon Preach'd before the Right Honourable the Lord-Mayor* [. . .]. London: Henry Clements, 1709.

———. *The Tryal of Dr. Henry Sacheverell, Before the House of Peers, for High Crimes and Misdemeanors.* London: Jacob Tonson, 1710.

Sargent, Thomas J., and François R. Velde. *The Big Problem of Small Change.* Princeton, NJ: Princeton University Press, 2002.

Schaffer, Simon. "Defoe, Natural Philosophy and the Worlds of Credit." In *Nature Transfigured,* edited by John Christie and Sally Shuttleworth. Manchester, UK: Manchester University Press, 1989.

Schubert, Eric S. "Innovations, Debts, and Bubbles: International Integration of Financial Markets in Western Europe, 1688–1720." *Journal of Economic History* 48, no. 2 (June 1988): 299–306.

Scott, William Robert. *The Constitution and Finance of English, Scottish and Irish Joint-Stock Companies to 1720.* 3 vols. Cambridge: Cambridge University Press, 1911.

Shadwell, Thomas. *The Volunteers; or, The Stock-Jobbers.* London: James Knapton, 1693.

Shea, Gary S. "Financial Market Analysis Can Go Mad (in the Search for Irrational Behavior During the South Sea Bubble.)" *Economic History Review* 60, no. 4 (2007): 742–65.

———. "Rational Pricing of Options During the South Sea Bubble: Valuing the 22 August 1720 Options." CRIEFF Discussion Paper no. 0410, Centre for Research into Industry, Enterprise, Finance and the Firm, University of St. Andrews, St. Andrews, Scotland, 2004.

———. "Sir George Caswall vs. the Duke of Portland: Financial Contracts and Litigation in the Wake of the South Sea Bubble." Centre for Dynamic Macroeconomic Analysis Working Paper no. 06/05, Centre for Dynamic Macroeconomic Analysis, University of St. Andrews, St. Andrews, Scotland, April 2007.

———. "Understanding Financial Derivatives During the South Sea Bubble: The Case of the South Sea Subscription Shares." *Oxford Economic Papers* 59 (2007): 73–104.

Sherman, Sandra. *Finance and Fictionality in the Early Eighteenth Century.* Cambridge: Cambridge University Press, 1996.

Shirras, G. Findlay, and J. H. Craig. "Sir Isaac Newton and the Currency." *Economic Journal* 55, no. 218/219 (June–September 1945): 217–41.

Shovlin, John. "Jealousy of Credit: John Law's 'System' and the Geopolitics of Financial Revolution." *Journal of Modern History* 88 no. 2 (June 2016): 275–305.

Simmel, Georg. *The Philosophy of Money.* Oxford: Routledge and Kegan Paul, 2011.

Sinclair, John. *The History of the Public Revenue of the British Empire.* 3rd ed. London: A. Strahan, 1803.

Smith, Adam. *An Inquiry into the Nature and Causes of the Wealth of Nations.* 1776. Electronic ed.: Amsterdam: MetaLibri, 2007.

Smythe, William J. *Map-making, Landscapes and Memory: A Geography of Colonial and Early Modern Ireland c. 1530–1750.* Notre Dame, IN: University of Notre Dame Press, 2006.

Spang, Rebecca L. *Stuff and Money in the Time of the French Revolution.* Cambridge, MA: Harvard University Press, 2015.

Speck, W. A. "Robert Harley, First Earl of Oxford and Mortimer (1661–1724)." In *Oxford Dictionary of National Biography,* edited by David Cannadine. Online edition. Oxford: Oxford University Press, 2004–16.

Sperling, John G. *The South Sea Company: An Historical Essay and Bibliographical Finding List.* Boston: Baker Library, Harvard Graduate School of Business Administration, 1962.

Spiegel, Henry William. *The Growth of Economic Thought.* 3rd ed. Durham, NC: Duke University Press, 1991.

Stigler, Stephen M. "Isaac Newton as Probabilist." *Statistical Science* 21, no. 3. (2006): 400–403.

Stone, Lawrence. *An Imperial State at War: Britain from 1689–1815.* London: Routledge, 1993.

Stone, Richard. "The Accounts of Society." Nobel Prize Lecture, December 8, 1984. https://www.nobelprize.org/uploads/2018/06/stone-lecture.pdf.

Stringham, Edward Peter. *Private Governance: Creating Order in Economic and Social Life.* Oxford: Oxford University Press, 2015.

Stukeley, William. *Memoirs of Sir Isaac Newton's Life.* 1752. MS/142, Royal Society Library, London. http://www.newtonproject.sussex.ac.uk/view/texts/normalized/OTHE00001.

Sussman, Nathan, and Yishay Yafeh. "Institutional Reforms, Financial Development and Sovereign Debt: Britain 1690–1790." *Journal of Economic History* 66, no. 4 (December 2006): 906–35.

Swift, Jonathan. *The Bubble.* Farmington Hills, MI: Gale ECCO, 2010.

Swift, Jonathan, and John Hawkesworth. *Letters, Written by Jonathan Swift: D.D., Dean of St Patrick's, Dublin. And Several of His Friends* [. . .]. 6th ed. Vol. 2. London: T. Davies, 1767.

Taylor, Stephen. "Robert Walpole." In *Oxford Dictionary of National Biography,* edited by David Cannadine. Online edition. Oxford: Oxford University Press, 2004–16.

Temin, Peter, and Hans-Joachim Voth. *Prometheus Shackled: Goldsmith Banks and England's Financial Revolution After 1700.* Oxford: Oxford University Press, 2013.

———. "Riding the South Sea Bubble." Massachusetts Institute of Technology Department of Economics Working Paper no. 04-02. Massachusetts Institute of Technology, Cambridge, MA, 2003.

Templeton, Daniel. *The Secret History of the Late Directors of the South-Sea Company*. London: self-published, 1735.

Thiers, Adolphe, and Francis Skinner Fiske. *The Mississippi Bubble: A Memoir of John Law*. New York: W. A. Townshend, 1859.

Tocqueville, Alexis de. *Democracy in America*. 1835–40. Translated by Harry Reeve. (State College: Pennsylvania State University, 2002).

Townsend, G. M. *The Political Career of Charles Spencer, Third Earl of Sunderland 1695–1722*. Ph.D. dissertation, University of Edinburgh, 1984.

Trenchard, John, and Thomas Gordon. *Cato's Letters; or, Essays on Liberty*. 4 volumes. 4th ed. London: W. Wilkins, 1737. https://books.google.com/books?id=QYgDAAAAQAAJ&printsec=frontcover&source=gbs_ge_summary_r&cad=0#v=onepage&q&f=false.

Twigger, Robert. "Inflation: The Value of the Pound 1750–1996." House of Commons Library Research Paper 97/76, June 6, 1997.

Velde, François R. "French Public Finance Between 1683 and 1726." Preliminary and rough draft dated July 20, 2006. Paper prepared for the 14th International Economic History Congress, Helsinki, 2006, session 112. http://www.helsinki.fi/iehc2006/papers3/Velde.pdf.

———. "Government Equity and Money: John Law's System in 1720 France." Federal Reserve Bank of Chicago Working Paper no. 2003-31, December 2003. https://www.heraldica.org/econ/law.pdf.

———. "John Law's System." *American Economic Review* 97, no. 2 (May 2007): 276–79.

———. "What We Learn from a Sovereign Debt Restructuring in France in 1721." *Economic Perspectives* 40, no. 5 (2016).

Walker, Geoffrey J. *Spanish Politics and Imperial Trade, 1700–1789*. London: Macmillan, 1979.

Walpole, Robert. *Some Considerations Concerning the Publick Funds, the Publick Revenues and the Annual Supplies Granted by Parliament.* 2nd ed. London: J. Roberts, 1735.

Walsh, Patrick. "Irish Money on the London Market: Ireland, the Anglo-Irish, and the South Sea Bubble of 1720." *Eighteenth Century Life* 39, no. 1 (January 2015): 131–54.

Ward, Ned. *The London Spy Compleat, in Eighteen-Parts.* London: J. How, 1703. http://grubstreetproject.net/works/T119938?func=title&display=text.

———. *The Secret History of Clubs: Particularly the Kit-Cat, Beef-Stake, Uertuocsos, Quacks, Knights of the Golden Fleece, Florists, Beaus, &c* [. . .]. London: Booksellers, 1709.

Weir, David. "Tontines, Public Finance, and Revolution in France and England, 1688–1789." *Journal of Economic History* 49, no. 1 (March 1989): 95–124.

Wennerlind, Carl. *Casualties of Credit: The English Financial Revolution, 1620–1720.* Cambridge, MA: Harvard University Press, 2011.

Westfall, Richard. *Never at Rest.* Cambridge: Cambridge University Press, 1980.

Wilczek, Frank. "Whence the Force of F = ma?" *Physics Today* 57, no. 10 (October 2004): 11. https://physicstoday.scitation.org/doi/10.1063/1.1825251.

Wilkins, John. *Of the Principles and Duties of Natural Religion.* London: J. Walthoe, 1734.

Winch, Donald, and Patrick K. O'Brien. *The Political Economy of British Historical Experience, 1688–1914.* Oxford: Oxford University Press, 2002.

Wright, Thomas. *England Under the House of Hanover: Its History and Condition During the Reigns of the Three Georges, Illustrated from the Caricatures and Satires of the Day.* London: R. Bentley, 1848.

Image Credits

Page	Credit
xi	Wikimedia Commons
10	Courtesy of Cambridge University Library
12	Wikimedia Commons
17	Reproduced by kind permission of the Syndics of the Cambridge University Library
26	Wikimedia Commons
40	Wikimedia Commons
48	Wikimedia Commons
49	Wikimedia Commons)
71	© Bank of England
84	© The Trustees of the British Museum
87	© The Trustees of the British Museum
116	Wikimedia Commons
121	Wikimedia Commons
135	Wikimedia Commons
142	Wikimedia Commons
159	Wikimedia Commons
182	Wikimedia Commons
184	Wikimedia Commons
201	Bancroft Collection, Kress Collection, Baker Library, Harvard Business School

206 Courtesy of the Wellcome Collection under a Creative Commons BY license
226 Bancroft Collection, Kress Collection, Baker Library, Harvard Business School
230 © The Trustees of the British Museum
257 From *Groote tafereel der dwaasheid,* 1720; courtesy of the Harvard Business School Baker Library
262 From *Groote tafereel der dwaasheid,* 1720; Kress Collection, Baker Library, Harvard Business School
269 Wikimedia Commons
286 Bancroft Collection, Kress Collection, Baker Library, Harvard Business School
296 © The Trustees of the British Museum
313 © The Trustees of the British Museum
320 Wikimedia Commons
330 Wikimedia Commons

Index

About the Author

Thomas Levenson is a professor of science writing at MIT. He is the author of several books, including *The Hunt for Vulcan, Einstein in Berlin,* and *Newton and the Counterfeiter*. He has also made ten feature-length documentaries (including a two-hour *Nova* episode on Albert Einstein), for which he has won numerous awards.

About the Type

This book was set in Caslon, a typeface first designed in 1722 by William Caslon (1692–1766). Its widespread use by most English printers in the early eighteenth century soon supplanted the Dutch typefaces that had formerly prevailed. The roman is considered a "workhorse" typeface due to its pleasant, open appearance, while the italic is exceedingly decorative.